PROSPERITY WITHOUT POLLUTION:

The Prevention Strategy for Industry and Consumers

PROSPERITY WITHOUT POLLUTION:

The Prevention Strategy for Industry and Consumers

Joel S. Hirschhorn
Kirsten U. Oldenburg

VNR Van Nostrand Reinhold
New York

Library of Congress Catalog Card Number 90-46375
ISBN 0-442-00225-4

Manufactured in the United States of America

Published by Van Nostrand Reinhold
115 Fifth Avenue
New York, New York 10003

Chapman and Hall
2-6 Boundary Row
London, SE1 8HN

Thomas Nelson Australia
102 Dodds Street
South Melbourne 3205
Victoria, Australia

Nelson Canada
1120 Birchmount Road
Scarborough, Ontario M1K 5G4, Canada

16 15 14 13 12 11 10 9 8 7 6 5 4 3 2 1

Library of Congress Cataloging-in-Publication Data

Hirschhorn, Joel S.
    Prosperity without pollution : the prevention strategy for
industry and consumers / by Joel S. Hirschhorn, Kirsten U.
Oldenburg.
        p. cm.
    Includes bibliographical references and index.
    ISBN 0-442-00225-4
    1. Pollution—Environmental aspects. 2. Environmental protection.
3. Environmental policy. I. Oldenburg, Kirsten U. II. Title.
TD174.H57   1990
363.7′057—dc20   90-46375
                                                            CIP

7-4933

CONTENTS

PREFACE

We have written this book to fill the vacuum of environmental strategy and public understanding of environmental alternatives. Achieving effective, unified action at the local, national, and global levels to *prevent* further planetary destruction requires public demand for a *preventive* environmental strategy.

In the past, narrow, shortsighted interests and objectives gripped environmental activities in a vise of conflict and confrontation. Many environmental battles were won, but their long-term benefits were illusory at best. Tensions between environmental progress and economic concerns fostered compromise and negotiation. Compromise and negotiation sound good and reasonable, but they have diluted efforts and have often resulted in avoiding doing what is environmentally right and necessary technically, socially, and morally. Confronted by less than 100 percent scientific certainty and by industries that fear loss of business, protecting the economic wealth of the few often wins out over protecting everyone's natural planetary riches. Nations have unwisely searched for a middle ground between ecology and economy, as if they are comparable. Mistakenly, people may think that some sacrifice of natural assets is needed to build economic wealth. But using up or destroying irreplaceable planetary assets ultimately must limit economic growth.

No nation has explicitly adopted a pollution prevention strategy. None has required that economic growth be based on the use of environmentally clean technology and clean products so that irreplaceable natural assets are not squandered. The result is predictable. Economic growth has been based on environmentally unfriendly technologies and products. An environmental debt has been created, and cleanup costs continue to escalate. But not all planetary damage can be corrected. If you don't prevent pollution, you get it. The planet is injured permanently, slowly but surely, leaving little time to stop the damage or to clean it up.

What of the future? Conventional wisdom says that the 1990s will be the decade of the environment. Environmental problems have attracted historically high levels of government attention, industry concern, public involvement, and news media coverage, especially television. Terms like *global warming, ozone depletion, toxic waste, acid rain,* and *municipal solid waste* are part of everyday language. The "green" movement that started in Europe has invaded the United States and other countries. To say *green* these days designates one as savvy about the environment.

The daily load of junk mail at home is likely to contain solicitations from environmental organizations, whose representatives are also knocking at the door with increasing frequency. Politicians emphasize environmental issues, and nearly everyone has heard that the United States has an environmental president. Has anyone ever met a person who is *not* an admitted environmentalist? Communities oppose the siting of new waste incinerators and landfills and struggle to obtain effective cleanup of toxic waste sites. Grass-roots environmental activism and the fight for environmental justice are defining a major social movement.

Moreover, professionals, such as engineers, corporate managers, and lawyers, are bombarded by notices of conferences and courses to cope with the barrage of perplexing new environmental requirements and developments. Every business decision may have an environmental dimension. And help is everywhere. The environmental services and equipment industry has exploded in size and profits. Most people may fear pollution, but many others are becoming rich because of it. Some of the world's largest corporations have chosen pollution and toxic waste as new business opportunities.

Threatened manufacturers have quickly learned how to take advantage of the new environmental consciousness of consumers, but not always by offering authentic improvements. As usual, environmental progress is pitted against economic interests, so that steps forward are often offset by backward actions. Well-intentioned consumers may be steered to the wrong purchases. Contributing to the uneven progress are debates by scientists and economists. Are we certain about the problem? Will the costs for protection harm the economy? Technologically optimistic scenarios are constructed to allay public fears grounded in common sense and experience. Government consistently calls for more studies, as much to delay taking significant action as to resolve scientific uncertainty. Why, the public asks, do we keep spending so much to study the problem instead of to fix it?

Books for the general public proliferated with the arrival of the 20th anniversary of Earth Day. Their common aim seems to be to reduce everyday actions of benefit to the environment to utterly simplistic terms. More often than not, terms like *green* and *environmentally friendly* are used, as if anyone really knows the exact meaning of such buzz words. The attention to the problem is commendable, but the simplistic actions suggested for consumers are unlikely to produce genuine solutions on a scale that really matters. Token actions by individuals, governments, and industries may make people feel good, but they will not protect the planet effectively. Many actions are not what they are purported to be and will not eliminate environmental problems. Recycling, which has become so popular, is in this category.

The environment is "in." The environment has people's attention. But is this enough?

Hardly. Amid all this talk and activity, there is an alarming lack of depth of thought. Environmental literacy barely exists. One characteristic of the whole history of environmentalism during the past 20 years is this: people and organizations nearly always seem to be *reacting* to nightmarish events and startling new information. Attention to environmental issues has lurched between peaks and valleys as one new environmental crisis came into focus and then receded while the world waited for the next fit of fear. Now the buildup of unsolved environmental problems and the heightened level of collective paranoia about global environmental destruction may have made the environment a permanent fixture of fear. But concern and fear do not necessarily reflect understanding, and without understanding, real solutions may never catch up with old and new environmental problems. The question that needs to be asked is: Does all the current environmental talk and activity reflect a coherent strategy by governments, industries, and environmental organizations? No.

Our thesis is simple. All environmental problems have a common bond that can serve as the basis for a comprehensive strategy that is immediately implementable. The common bond is that the best solution is to practice prevention instead of reaction. But the pollution prevention strategy has received little attention. The concept is simple: if environmental threats result from the creation of pollutants and wastes that enter the environment, then preventing pollution means eliminating the sources of the problem. The universal mandate must be: reduce or eliminate the production of *all* wastes and pollutants from every place where they are created in industry, commercial establishments, homes, and institutions. But that is not what has been going on for the past 20 years.

This book is about *all* environmental problems, and it can help *everyone* in their roles as workers, consumers, investors, and concerned citizens. People in industry need more insight into waste reduction and about the role of consumers. Consumers need more information about industrial opportunities and practices. This is more of a self-help than a how-to book. To truly solve environmental problems, it is necessary for everyone to change behavior. Doing so requires a heartfelt understanding of the origins of environmental problems and the best way to solve them. Better understanding can then motivate individuals to change their behavior in the home, marketplace, and workplace. If enough persons change, then broad-scale social and political change is more likely to occur. The detailed information in this book can help anyone start a new lifestyle and a new workstyle based on the principles of pollution prevention and proven successful practices in every type of human endeavor. Feelings of being a pollution victim or pawn can be replaced by strength and control by practicing pollution prevention.

Planet Earth sustains and houses every one of us. Like personal health care, preventive planet care is the responsibility of each of us. However, capability is needed to implement responsible actions. This book can supply that capability.

PROSPERITY WITHOUT POLLUTION:

The Prevention Strategy for Industry and Consumers

POLLUTION PREVENTION PAYS FOR EVERYONE

Do you *personally,* as a worker and consumer want to produce pollutants, toxic waste, and garbage?

Do you *personally* want to spend your money on products which, in whole or in part, quickly become garbage?

Do you *personally* want to depend for your safety, health, and peace of mind on government environmental protection and waste management programs and to pay more taxes for them?

Answering "no" to one or more of these questions may seem like a reflex action because of moral values, common sense, or knowledge about environmental issues. But moving beyond idealistic intellectual positions to everyday actions is considerably more difficult. Implementing the values behind these questions requires serious introspective behavioral change. It is enlightening to advise others to think globally and act locally. But how are people to act?

Is it enough for people to march in protest, to watch television documentaries, to join environmental organizations, to donate money for good causes, and to vote for politicians claiming to be environmentalists and backing environmental causes? Or is it necessary for people to accept responsibility, use their latent power, and change their daily behavior as consumers, workers, and even investors? The deteriorating physical state of planet Earth and the failure of existing institutions, technologies, policies, and political "leaders" to anticipate and solve environmental problems point to the need for individuals to take more responsibility.

The need is to enhance everyone's motivation to implement true environmental values centered on the concept of prevention because everyone will benefit from doing so. The idea of prevention should and can be used as the unifying principle for a comprehensive environmental protection strategy. Even animals know better than to live in their own wastes, to not foul their own nests. Humans can no longer count on moving away from their waste or on the "out of sight, out of mind" mentality of traditional waste management. Planet Earth is finite. All wastes and pollutants may affect human life adversely as wastes and pollutants accumulate, migrate, and degrade our ecology.

Pollution prevention means eliminating or reducing wastes and pollutants where they first are (or would otherwise be) produced so that there is less to manage and less entering air, water, and soil. Moreover, this must happen for *all* wastes and pollutants without trying to prove beyond a shadow of a doubt that a particular waste or pollutant is a big enough or certain enough problem or threat.

Ben Franklin was correct when he said that an ounce of prevention is worth a pound of cure. This is a simple statement about the importance of prevention as a general principle. Spending $1 to prevent a problem may indeed give the benefits of spending $16 to react to the problem. In the language of economic analysis, prevention offers a benefit:cost ratio of 16:1, which is impressive. As nearly everyone has discovered, however, practicing prevention on a daily basis requires considerable personal commitment and discipline. For example, whether it be preventive automobile or house maintenance or health care, nearly everyone has trouble acting preventively all the time. People forget. People think that the worst will not happen.

DON'T COUNT ON GOVERNMENT

Planet Earth houses us and determines our health. Preventive care is, therefore, necessary and prudent. In theory, we might count on government to do what we individuals may not always do. But it has also become clear that governments have trouble acting preventively. Collapsing bridges and other failures of our infrastructure ultimately cost much more to replace than the cost of preventive maintenance, such as routine painting of steel bridges to prevent rusting. Local, state, and federal government agencies do not spend enough on preventive maintenance.

Even outer space has filled up with garbage because of all the ventures there. The government calls this garbage *orbital debris*. About 95 percent of the 500 million pounds of material in outer space is garbage. Space debris threatens future activities there because even small pieces are like bullets to satellites and space stations. There are more than 48,000 pieces 1 centimeter or more in size moving at 17,000 miles per hour. A paint fleck much smaller than 1 centimeter gouged a 0.25-inch pit in a window of the space shuttle Challenger in 1983. In addition to trashing outer space, the United States and other nations have been dumping trash and hazardous waste in Antarctica. Land, ice, and water are contaminated. In January 1990, the National Science Foundation announced a $30 million plan to start cleaning up U.S. outposts in Antarctica. These problems could have been prevented.

No matter where we look, it is obvious that counting on government to practice preventive planet care to avoid costly or irreversible damage to the health of the planet, to life-sustaining resources, and to human life itself is, therefore, unjustified and unwise. Losing a bridge, a highway, an underground water main, a satellite, or a jumbo jet aircraft is one thing; losing planet Earth is quite another matter. The margin for error has disappeared. The margin of safety has been eaten up by a lack of pollution prevention in the past. It is irrational to ignore pollution prevention for the future.

The difficulty government has in practicing and promoting prevention has been identified for four critical and interrelated areas of spending in the United States: health care, environmental protection, agricultural use of pesticides, and energy production. In each case, independent studies concluded that preventive actions could cut problems massively but that too little was being done to promote prevention.

J. Michael McGinnis, director of the Office of Disease Prevention and Health Promotion in the Department of Health and Human Services, said (1988), "More than 95 percent of the half-trillion dollars spent for medical care in the United States each year goes to treat rather than prevent disease. A better balance is clearly in order." This advocate of preventive health practices estimated that about 50 percent of premature deaths could be prevented by applying *current* knowledge.

The congressional Office of Technology Assessment (OTA), in a study on industrial waste reduction (1986), concluded: "Although there are many environmental and economic benefits to waste reduction, over 99 percent of Federal and State environmental spending is devoted to controlling pollution after waste is generated. Less than 1 percent is spent to reduce the generation of waste." OTA estimated that 50 percent of all industrial waste outputs could be prevented with *existing* technology. With successful research and development, another 25 percent reduction of industrial waste is possible. In other words, there is a comparable degree of prevention for industrial wastes and human illness.

OTA also said in 1983 that it costs 10 to 100 times more money to clean up toxic waste contamination than it would have cost to prevent the original releases into the environment. For example, a $280 million cleanup of polychlorinated biphenyls (PCBs) in a 40-mile stretch of Hudson River sediments will remove 250,000 pounds of the toxic chemical. That equals over $1,000 per pound of chemical that was dumped in the river between 1946 and 1975 by General Electric as waste products from its manufacturing plants. Surely prevention would have been a lot cheaper.

As with diseases and industrial hazardous waste, a 1987 study of pesticide use in agriculture by the Worldwatch Institute (Postel 1987) concluded, "With technologies and methods now available, pesticide use could probably be halved." Reduced pesticide use means less contamination of underground water supplies and foods. Many Superfund sites must be cleaned up because of pesticide contamination of soil, such as small airfields used by crop-dusting airplanes.

The fourth example of predicted preventive success is for alternative energy sources to replace the burning of fossil fuels, which is the major contributor to global warming. Michael Brower (1990) concluded that is is possible "to reduce world consumption of fossil fuels by 50 percent or more over the next several decades." Using renewable energy sources such as solar energy, wind, and biomass is true pollution prevention.

There is so much evidence that encourages us to be optimistic about prevention, and yet there is so much trouble in achieving success. Common sense says that everyone would rather not get a disease then get it and then have to seek surgical or medical help. So too with pollution effects on health and the environment. Human experience and social history, however, confirm that it is not enough to know intellectually that preventive behavior will result in large benefits. Something must turn wisdom into action. This book aims to provide the tools to help build and sustain personal behavioral change, consistent with pollution prevention, in all phases of life. The personal practice of pollution prevention requires new knowledge and new skills. Learning to enjoy life by consuming and wasting less and learning to define success at work by reducing waste and pollutant production is a demanding effort. Individual actions may provide individual satisfaction, but they may not be sufficient for a positive planetary effect. Many people will not practice pollution prevention out of ignorance, laziness, false optimism about other solutions, or a belief that other people will not do it. Preventive planet care will be assured only when copious numbers of consumers and workers act preventively, showing its feasibility and benefits and causing other people, industry, and government to respond vigorously. Making government policies and programs reflect new public values and objectives is, therefore, necessary, but it will not be easy. Later in this book, we will specify many personal actions and government policies to implement preventive planet care.

Difficult as any preventive human behavior is, there is cause for optimism. Consider that so many people have changed the way they think about their health care. Over time, people in industrialized, affluent societies came to depend more and more on solutions from physicians, hospitals, pharmaceuticals, and medical technology. But in recent years the limits to modern medicine became more obvious, and the costs of reacting to disease and illness became a heavy burden for individuals, companies, and government. The result was an upsurge in preventive medicine and health care. Individuals began to take more responsibility for their own health. It became practical and cost effective to change habits to stay healthy. For example, people pay more attention to diet, exercise, and the avoidance of addictive drugs and alcohol.

The oat bran phenomenon of the late 1980s vividly demonstrated how consumers react with speed and gusto to information and opportunities in the marketplace. People like simple solutions. In 1988, the public desire to reduce blood cholesterol caused the sale of oat cereals to jump more than 900 percent. Oat bran became ubiquitous; there are even potato chips with oat bran. Behind this and myriad other personal actions are the personal and social benefits of reducing the need for institutional health solutions, such as physicians, hospitals, and medicines. Preventive health care requires personal commitment and discipline, along with options in the marketplace and supportive government policies and programs. Preventive planet care will require even more commitment and discipline because the benefits are likely to seem more distant and uncertain to most people.

PERSONAL CHOICE

Even if you answered "yes" or "I don't care" to the three questions asked at the beginning of this chapter, you still have reason to read on because you might change your views. It is not just a matter of what is right or wrong; it is also a question of newly perceived personal opportunities, solutions, and benefits. Many environmental advocates try to reach people by stressing their obligation to treat future generations fairly, but our point of view is that concerns about fairness for the future—important though they are—will not have broad impact on most people's current behavior. When confronted with the statement that it is necessary to change personal behavior, desires, and consumption patterns to achieve some future good, it is human nature to question the need for change. In the environmental area, many people believe that forecasts of future doom are speculative fantasy. For every scientist and report that predicts doom, another one says that it is not necessarily so. Moreover, given a choice between perceived sacrifice and optimism about future technological solutions, many people will choose to believe that technology will come to the rescue or that only industrial polluters have to act.

Arguably, the most important lesson from the past few decades and from mountains of technical reports is that severe and perhaps catastrophic environmental damage has already occurred or been set in motion. This is absolutely the case because of destruction of the protective ozone layer, which has resulted from past uses of chemicals. Contaminated air and water have already harmed many people. The general perception that more environmental damage could occur for most of the planet's inhabitants within a few years or a decade or two is backed up by a lot of solid scientific evidence—even without 100 percent certainty. As with the destruction of the ozone layer, so much damage has already been set in motion that, regardless of what we now do, additional planetary deterioration is unavoidable. The only uncertainty is what level of damage to health and the environment will result. The choice facing each of us is whether to stem the tide of deterioration as soon as possible. Individuals will change industrial processes, practices, and products.

True, the full scope of future pollution damages is uncertain. But wouldn't it be better to be safe than sorry? Wouldn't it be prudent to assume the worst and take preventive actions in order to have some insurance to protect what we value? In other words, to achieve real solutions *in time,* it is necessary to change personal behavior and learn new preventive survival skills. People must focus on the immediate future—*their future*—to assess the risks and benefits in order to motivate themselves. The world's children need to learn preventive planet care by watching family members and other adults practicing it because it will take a long time to integrate preventive skills into school curricula. The planet cannot wait for a few generations until children grow up with better preventive skills. It is not a matter of being pessimistic about the future, but of being optimistic about humanity's ingenuity and rationality to act preventively.

TO THE ENVIRONMENTAL EDGE OF HISTORY

Millions of people are concerned about pollution and waste, garbage, or trash, but there are over 5 *billion* people living on the planet. Many people see severe environmental problems, but many others do not. Many people hasten the problems and slow the solutions. An important aspect of the environmental push on people is that threats seem to come from all directions, literally from everywhere and everything. Environmental problems range from the upper atmosphere, where the protective ozone layer is being weakened enough to cause people to worry about being able to step out into the sunshine without getting cancer, to deep underground water supplies which are being contaminated with enough toxic chemicals to make people worry about every glass of water they drink. In addition to worrying about bad air and water, consumers have learned to fear chemically contaminated food, radiation of all kinds, and a host of consumer products as well.

And then there is the news that if global warming continues and the Antarctic ice melts, the oceans will rise 230 feet and wipe out most of the world's major population centers. In January 1990, the Meteorological Office of the United Kingdom, the world's oldest keeper of climatic data, reported that the decade of the 1980s was the hottest in over 100 years. Six of the 10 warmest years on record occurred in the 1980s. Leading scientists said that this data confirmed the likelihood that global warming has been set in motion, although complete scientific certainty will not be proven for perhaps another two decades. It is not surprising that some scientists resist attributing the warming to emission of greenhouse gaseous pollutants such as carbon dioxide.

The business community takes advantage of the uncertainty that always exists. In January 1990 *Forbes* magazine editorialized that "the evidence supporting this [global warming] catastrophe theory is flimsy to nonexistent." But in February 1990, almost half of all the Nobel prize winners living in the United States and members of the National Academy of Sciences formally appealed to President George Bush to take the threat of global warming seriously and to begin immediately to reduce the use of artificial gases that cause the greenhouse effect. The Bush administration favored large-scale research. The demand for certainty is routinely used to block preventive environmental problem solving. Public experience has been that when scientific certainty is finally achieved, the original fear is almost always confirmed. But after the uncertainty battle is won, as it usually is, the next struggle is to institute the rapid use of preventive measures against the resistance of those who profit from current polluting activities.

Less debatable is the fact that pictures of medical waste and other garbage on beaches have caused people to cancel their vacations and ruin businesses. In 2 hours one summer a beach cleanup crew in New Jersey collected 700 plastic tampon applicators. Trains filled with sludge and ships filled with toxic waste trying to find a place to off load their dreaded loads seem commonplace. Someone always seems to be trying to move toxic waste or ordinary garbage somewhere else—to another state in the United States or to another country. One of the more interesting developments has been the stubborn refusal of people in rich and poor nations alike to accept other people's wastes. People are too fed up, frustrated, and afraid to accept other people's garbage. They have enough of their own. But often there is little that an American community can do about the use of their area as a waste disposal site.

Toxic terror and pollution paranoia aside, waste (like illness) seems natural both biologically and socially. If waste is inevitable, then all forms of waste have to go somewhere. Such an uncritical view ignores the environmental concern: What happens to waste, and what all the impacts on public health and the environment? *Pollution* refers to adverse effects of waste that enters the environment. But the concept of pollution has become more complex. This book's perspective is that *all* waste outputs and by-products are, either in the short or the long term, a form of pollution, although the adverse effects may vary enormously. Negative effects include acute or immediate health effects like neurological disorders; chronic or long-term health effects like cancer; damage to and contamination of animals, insects, or plants; chemical contamination of water and air; and unsightly littering, loss of property values, and the inconvenience and cost of using bottled drinking water. Many wastes and pollutants will have multiple adverse effects. There may also be synergistic interactions and cumulative exposures from different wastes and pollutants in the environment which worsen these ill effects.

One of the greatest complications is that people or sensitive parts of the physical environment may be exposed to relatively large numbers of harmful substances. Looked at individually, such exposures to specific environmental pollutants may not seem harmful, on the basis of usually very limited scientific data, because of the relatively low concentrations of the pollutants in air, water, or food, for example. But an individual person or plant or animal may be exposed to a specific chemical coming from many diverse sources in the environment. A toxic chemical may be in drinking water, in food, in household and workplace air, and in consumer products. Rarely do government agencies and others take into account *cumulative* exposures of people and the environment to harmful chemicals. Thousands of toxic pollutants are not regulated at all. Further, for many harmful chemicals and types of radiation there is no "safe" level of exposure; this means that even very small exposures can result eventually in serious and perhaps fatal health effects. A newly recognized phenomenon is that more and more people are becoming "chemically sensitive." These people have a lower threshold to chemicals than others. Chemically sensitive people have health effects which are often dismissed as being psychosomatic or undiagnosed allergies, but which may result from some unexplained dysfunction of the immune, nervous, or hormonal system that destroys the ability to adapt to chemical exposures.

Because it is better to be safe than sorry, more and more people have come to understand that it is wise to pluck a conclusion from the endless debates about the health effects of pollutants. That conclusion is this: There *is* a life-or-death, physical dimension to environmental problems because of myriad health effects (many more than cancer) and many effects on our ecosystem. The collective American experience includes an endless stream of revelations that chemicals previously believed to be harmless are now known—by the experts—to be harmful. This revelation also involves much more than having new information on health effects; time and time again, the public has heard that government agencies and companies have not disclosed the truth about harmful chemicals or products. The decline in public trust in government to protect people is evidenced just as much by the continuing increase in consumption of bottled drinking water in the United States as by public criticism of government programs and policies.

In the 1980s, the use of bottled drinking water increased by 500 percent in many parts of the United States—a faster growth than for any other beverage. Americans are spending $2.2 billion a year on bottled water and near $300 million for household water purification. Such spending reflects the fact that 80 percent of Americans surveyed indicate that they fear contaminated tap water. The latest craze is mining Alaskan glacial icebergs, which are said to be ultrapure and free of pollutants because they are thousands of years old. Alaskan glacial water is very expensive. One reason for the consumption trend has been the significant publicity about lead and pesticide contamination of U.S. drinking water. The discovery in early 1990 that Perrier mineral water imported from France was, in some instances, contaminated with four times as much benzene as allowed under U.S. drinking water standard startled the public. But that event may do more to promote government regulation of bottled water than to curb the American appetite for it.

It is estimated that more than 17 percent of Americans—that is, more than 40 million people—have household water with a lead concentration above 20 parts per billion, enough to pose a serious health threat to children. But the safe level is now recognized to be 10 parts per billion or less. Studies have demonstrated that exposure of young children to lead is correlated with high school dropout rates, reading disabilities, absenteeism, and lower vocabulary scores, for example. Lead poisoning vividly illustrates the need to consider more than cancer when evaluating environmental health risks, which the government still does not do routinely for other kinds of exposure. The federal government estimates that 16 percent of all U.S. children have high enough lead levels to cause neurological damage. Preventive approaches to keep lead out of the environment include prohibiting the use of lead in plumbing, water coolers, and paints and replacing old plumbing with lead-free materials. In other words, if technology is to come to the rescue, in spite of the health damages and destruction of the planet that have already occurred or been set in motion, it must be in the form of prevention.

ENVIRONMENTAL THREATS ARE UNIQUE
IN QUANTITY AND CHARACTER

Pope John Paul II has given an historic and apocalyptic appraisal of the global environmental situation. The Pope considers environmental problems a "moral crisis" and views the need to solve them as comparable to calls for social and economic justice. Interestingly, American environmental activists have been speaking of the need to achieve environmental justice. The Pope (1990) said:

> The ecological crisis has assumed such proportions as to be everyone's responsibility. Modern society will find no solution to the ecological problem unless it reviews its own lifestyle. In many areas of the world today, society is given to instant gratification and consumerism while remaining indifferent to the damage these cause.

This is a cogent religious indictment of America's throwaway society. How would a visitor from another planet see us? Perhaps in the throes of our own disposal.

Of the many recent significant changes in the world we work and consume in, one of the most basic is the emergence of *artificial* environmental threats that seem more dangerous than natural threats to peace, health, and welfare. These environmental threats seem, to a large number of people, increasingly complex, uncontrolled, and life-threatening. Psychologically, the attack on personal peace of mind has traditionally come from fears about natural catastrophes, world wars, and immediate health and economic problems. Today, however, personal and global environmental catastrophe has become a prominent, *plausible* threat at both local and global levels, even though scientists and politicians argue about whether the "proof" is in about environmental catastrophe in general and about nearly every specific environmental risk.

Most fears are, in truth, justified. Every year, thousands of people in the United States and elsewhere discover that they are living near or on top of a toxic waste dump; that their drinking water is contaminated by toxic chemicals; or that years ago they were exposed to a chemical that once was thought harmless but now is known to be toxic, like asbestos, PCBs, various pesticides, and industrial solvents. The case of asbestos is especially disheartening. It is taking the federal government several decades to ban asbestos use comprehensively. Even though commercial use of asbestos decreased from 617,000 tons in 1979 to 93,500 tons in 1988, this life-threatening substance is still used in countless applications. Similar stories come from people already affected by environmental problems and risks, such as those living near thousands of chemical plants, government facilities, nuclear reactors, and hundreds of toxic waste sites like Love Canal in New York, the Stringfellow Acid Pits in California, and the Westinghouse PCB sites in Indiana. Environmental victims feel like victims of violent crime. They also feel abandoned by society and its institutions, especially the government agencies that are supposed to protect them. Through the power of television, real and perceived environmental threats are quickly shared globally. The result is that many other people fear becoming environmental victims or being informed that they have been unknowing ones for years.

Some people may believe otherwise, but many people all over the planet see environmental threats as deadly real. More and more people sense that they are being pushed closer to the edge of history, and they are rightfully afraid of falling off. Kai Erikson (1990), a professor of sociology, has provided powerful insights into the new nature of environmental concerns which have instilled a terror of historically new proportions:

Natural disasters are almost always experienced as acts of God or caprices of nature. They happen *to* us. They *visit* us, as if from afar. Technological disasters, however, being of human manufacture, are at least in principle preventable, so there is always a story to tell about them, always a moral to draw, always a share of blame to assign. They provoke outrage rather than acceptance or resignation. They move people to a feeling that this thing should not have happened, that someone is at fault, and that the victims deserve not only compassion and compensation but also something akin to what lawyers call punitive damages.

Erikson goes on to identify a unique characteristic of toxic emergencies: "their capacity to induce a lasting sense of dread is a unique—and legitimate—property." Moreover, the new age of toxic problems is not bounded with easily perceived beginnings and endings. Although some environmental problems have clear beginnings,

others begin years before anyone senses that anything is wrong, as was the case at Love Canal. But toxic accidents never end. Invisible contaminants remain a part of the surroundings—absorbed into the grain of the landscape, the tissues of the body, and, worst of all, into the genetic material of the survivors. An "all clear" is never sounded. The book of accounts is never closed.

Environmental victims surely appreciate Erikson's description of pollution and toxicology:

Toxic poisons provoke a special dread because they contaminate, because they are undetectable and uncanny and so can deceive the body's alarm systems, and because they can become absorbed into the very tissues of the body and crouch there for years, even generations, before doing their deadly work.

The artificial character of environmental problems is of essential importance:

The knowledge that these catastrophes are the work of other humans only intensifies the dread they inspire. They are brought on not by declared enemies but by compatriots, even neighbors, many of whom respond by denying the importance of such crises or ignoring them. . . . The point is not that a particular region is now spoiled but that the whole world has been revealed as a place of danger and numbing uncertainty.

Over the years, as people have heard about the new, insidious forms of pollution and have built up painful feelings of dread, fear, and anxiety, almost always industry has been seen as the sole culprit. Even the above descriptions assume that the threats are external to the victims. This perception is wrong. Even though industrial and many government facilities are major waste generators and polluters, a lot of pollution also comes directly from the activities of individuals in their role as consumers. There is automobile pollution. There is power plant pollution that results from energy use by individuals. There are wastes from the consumption of nearly all goods and services, including transportation.

Ordinary household garbage can be—and often has been—the cause of pollution, because ordinary municipal solid waste typically contains hazardous chemicals from a multitude of products or produces them during waste management. When garbage is disposed of in landfills or incinerators, some of those hazardous chemicals contaminate the groundwater under the landfills or the air receiving incinerator emissions. On the other hand, industry has often tried to shift public attention away from itself to individuals, in ways that lack credibility. For example, public outcries about dioxin emissions from waste incinerators have been confronted by industry statements that all fires, such as those from barbecue pits and home fireplaces, produce dioxins. Not so. A recent study by General Electric scientists, for example, found that 3,000-year-old mummies had no dioxins, even though in life those people had been exposed routinely to wood fires.

And so, we ask this question: In this new world of rising environmental passions and uniquely persistent fears, do individuals feel responsible for pollutants and wastes or in control of them? For the most part, the answer seems to be "no." People feel victimized by forces that are so ubiquitous that it doesn't seem to matter very much which ones are specified. Nor is there much confidence in existing institutions, whether they be in government or the private sector. We believe that individuals will increasingly have to pay attention to the waste produced at work, in the home, and in industry *because of their purchases;* otherwise, they will suffer more risks, fears, and costs. Moreover, people face personal harm and cannot feel safe because certain problems are distant or in the future. People will feel more victimized, terrorized, and paranoid unless they take control. *The relief of personal anxiety may be a more effective incentive for behavioral change than the desire to do good environmentally or to save money.*

The "edge of history" is more than oratory. It is the worst-case scenario of a time when, regardless of how much we spend, there is no good cure, no possible complete cleanup, no technological salvation. It is when too many environmental problems happen at about the same time, when our planet and our ability to solve the problems are both overloaded. The environmental short circuit is no more or less unthinkable than the nuclear holocaust the world has contemplated for decades. The joke is that affluent nations have been raising their standard of living while lowering their quality of life. And it is funny because it is plausible. But, of course, many people see things differently; they reject the plausibility of this worst-case scenario, as well as the view that consumption- and waste-intensive modern life is a decline in quality. Preventive planet care is about being safe instead of sorry. Rejecting the prevention strategy means accepting an uncertain risk. Escalating uncertain risks increase personal anxiety. This may explain why a survey of people in 15 prosperous and developing nations accounting for half of the world's population revealed a universal preference for a *lower* standard of living with a cleaner environment than for a higher standard of living with a dirtier environment.

ECONOMIC THREATS ARE ALSO FRIGHTENING

Economically, environmental problems pose a threat to affluent societies because of higher costs for pollution control, waste management, and cleanup. The economic threat for developing nations is that attention to environmental problems, largely because of threats to industrialized nations, may frustrate the economic growth of these poorer nations. As the affluent industrialized world confronts ugly new environmental threats tied to its escalating conspicuous consumption, poor developing nations may be asked to curb their appetites. Ironically, those appetites have been whetted by years of watching and envying the hedonistic consumption of the United States. Television has made that consumption visible and alluring to the majority of the world's population, and that majority is relatively poor. The affluent industrialized nations have only about one-quarter of the world's population, but they consume about 80 percent of the world's goods. What if the rest of the world catches up? Listen to government officials speak about economic growth and prosperity, and what you hear is their hopes about consumers increasing their consumption of goods.

The World Commission on Environment and Development estimated that a fivefold to tenfold increase in economic activity would be required over the next 50 years to meet the needs and demands of the growing world population. If such an increase is based on today's technologies and products, the planet will be buried in its own wastes. It has been predicted that by the year 2025, only 35 years from now, the planet's population will rise to 10 billion, about twice the current level. It has also been estimated (Frosch and Gallopoulos 1989) that "at current U.S. rates 10 billion people would generate 400 billion tons of solid waste every year—enough to bury greater Los Angeles 100 meters deep."

And let there be no doubt about the hedonistic nature of contemporary American consumption. In the 1980s, the expanding American economy went into massive debt and was unable to spend more money on capital investment for the future or to save more money. But personal consumption surged. Americans, who total 250 million people, bought nearly 100 million new cars, 144 million television sets, and 72 million VCRs. These are so called durable, long-lasting goods. However, after 10 years of use, nearly all those automobiles, television sets, and VCRs will become garbage—probably over 100 million tons—to be disposed of in the United States. Worldwide, as the population grew from 2.5 billion in 1950 to 5 billion in 1988, the number of motor vehicles rose from 50 million to 350 million. The continued dependence on automobiles for transportation must be addressed within the context of pollution prevention.

What about nondurable goods? About 250 million Americans consume 1 *billion* foil-lined boxes of fruit juice, complete with plastic shrink wrapping and a plastic-encased plastic straw on the side, every year. Disposable, single-use, throwaway products are increasingly popular: 25 *billion* Styrofoam cups, 1.6 *billion* disposable pens, 2 *billion* disposable razors, and 16 *billion* disposable diapers are being used and discarded annually in the United States. These numbers mean that, on average, 100 Styrofoam cups, 6 disposable pens, and 8 disposable razors are purchased for *every* man, woman, and child in the United States every year. Could American society function without single-use, disposable products? Could *you* live well without them? These are just examples of America's fast trash consumption, a lifestyle just the opposite of the philosophy "waste not, want not." Verily, the American lifestyle—which is increasingly the global lifestyle—is "waste more, want more." Will consumers be able to resist Sony's disposable paper peel-off watches made in China and costing less then $4 each? Do Americans really want bright white tissue paper? Could they learn to accept unbleached tissue paper? Would it help to inform them that the bleaching process to make tissue paper so white results in 965 pounds of environmental waste for every 2,000 pounds of tissue compared to 367 pounds of waste for unbleached tissue?

Although many think of the 1980s as a period of deindustrialization in the United States and the rise of the service economy, factory outputs increased by 30 percent to help make the products craved by Americans. Foreign industries have bested us in a multitude of expensive products, such as electronic consumer goods and automobiles, but American industries have been able to hold on to many fast trash products, like paper towels, Styrofoam food containers, and disposable pens and markers. The decline of American industries that did occur during the 1980s may have resulted not merely from insufficient capital investment, research and development, and technological innovation but also from a shift from making competitive, high-quality, high-technology products to producing garbage-intensive products. Conversely, foreign nations profit from exporting products, and Americans end up with the garbage.

Reflecting on the 1980s, President Bush's budget director, Richard G. Darman, said in the summer of 1989: "we consume today as if there were no tomorrow. Collectively, we are engaged in a massive backward Robin

Hood transaction: robbing the future to give to the present.'' What Darman probably had in mind was only the repayment of the enormous U.S. budget deficit, but the statement also rings true for the inevitable *environmental deficit* to come due. The environmental deficit covers the future costs of pollution control and waste management, as well as the costs of cleaning up contamination of the environment and of caring for environmental victims. Michael Silverstein (1989) has considered the full range of future environmental costs and concluded that ''America's environmental debt is larger than its national debt.'' In terms of economic theory, the environmental deficit is an inevitable consequence of the externalization of the true, long-term environmental costs of goods and services. In other words, manufacturers, in paying the costs of production, and consumers, in paying for goods, do not pay originally for the full costs of producing pollutants and wastes. Environmental resources, such as air, water, and land, have traditionally been used as free common resources: free to use, overuse, abuse, and destroy.

A key characteristic of externalizing environmental costs is that they are transferred to the general society, and often beyond, to other nations. Externalization of environmental costs cuts the link between specific benefits, from the original production and use of products, and ultimate damages and costs. Those who enjoy do not pay the complete costs. Those who do not enjoy help to pay the environmental costs. Because of the shift of manufacturing to developing nations, consumers in the United States and other affluent nations externalize environmental costs to people in other nations, where the goods were manufactured or the necessary raw materials produced. Future U.S. citizens pay other costs because of postconsumer wastes and pollutants which reside in the United States, and people in other nations pay because of long-range environmental effects like acid rain.

In the past, society found it easy to dismiss inevitable environmental costs because those costs were unknown, or either seemed low or were uncertain and difficult to estimate. The current new age of environmentalism has changed this situation. The economic costs of waste and pollution have escalated, and estimates of future costs can no longer be viewed as inconsequential. For example, in 1988 the United States spent about $100 billion to deal with environmental management and problems, according to federal government statistics. That figure represents about 2 percent of the U.S. Gross National Product (GNP), which is a lot. Environmental spending has been increasing steadily over the past 20 years as the regulatory system has expanded. In 1985, it was $70 billion. Indeed, the United States spends a higher fraction of its GNP on the environment than any other industrialized nation—without, necessarily achieving better improvements in environmental quality.

In 1980, for example, the United States spent nearly four times more of its GNP than Japan and France and nearly three times more than West Germany. Those differences have probably persisted because U.S. regulations and spending have maintained their historic increases. There is some possibility, however, that the 1992 formation of what may become the United States of Europe may increase environmental spending in many of these nations. The reason is that more stringent pollution control regulations will be applied overall, and considerable money will also be spent on major environmental improvements in Eastern Europe with the help of Western European nations. In early 1990, the West German government estimated that it would take $200 billion over 20 years to lift East European industry to Western environmental standards, not counting the enormous sums of money needed to clean up contaminated land, water, and buildings or to address existing damages to human health, which are considerable.

The future promises much more environmental spending as America's environmental debt comes due. According to the OTA (1989), cleaning up toxic waste sites could cost $500 billion over the next 50 years, and cleaning up nuclear waste sites could cost hundreds of billions of dollars more. Such cleanup costs equal more than 2 percent of the total current value of the physical assets in the United States (about $30 trillion). These are direct cleanup costs; they do not include addressing health effects, for example. And many subtle but significant costs are not accounted for in the federal statistics on environmental spending. A good example is spending on lawyers and litigation by industrial and insurance companies, probably about $1 billion annually and rising. The future costs of dealing with massive damage from acid rain, global warming, and upper atmosphere ozone depletion could make toxic waste site cleanup costs seem minuscule.

Acid rain effects seem remote to most people but, despite continued claims of unproven effects, they are real, and literally everyone contributes to acid rain because electric utilities and transportation account for so much of the problem. For example, of the 23 million tons of sulfur dioxide emissions each year, 70 percent is from utilities—and about half of that comes from 24 electric utilities in the lower Midwest and the South. Of the 20.5 million tons of nitrogen oxides emissions, 32 percent is from utilities and 43 percent from transportation. Of the 22 million tons of volatile organic compounds, transportation accounts for 40 percent. (Canada produces 13 percent of the total U.S. emissions.) For most people, the effects of acid rain seem remote because they consist mostly of damage to natural resources, not human health. For example, maple tree blight has been linked to acid rain, and since 1981 maple syrup production has decreased by 45 percent in the northeastern United States and Canada. The destruction of irreplaceable forests is well known. But there is also troubling speculation that acid deposition and subsequent leaching of toxic metals (from soils and from water pipes, for example) into drinking water may contribute to a number of health problems, including lead poisoning of children and Alzheimer's disease in older people. The acid rain problem can be solved technically, but it will mean an enormous change in energy production and transportation, the costs of which will be passed on to consumers. Such costs and changes are resisted. Instead of embracing preventive measures that act on the source of acid rain pollutant production, there are calls for "measured responses that avoid the costly mistakes that come from panicky overreaction" (Singer 1990). As in so many environmental areas, experts use scientific uncertainty to resist acting preventively.

We conclude that there is more to fear than health effects and global environmental catastrophe. When the environmental deficit we have been creating for decades comes due, the bill will be extraordinarily expensive. For the United States, the environmental deficit will come due at the same time that people will be paying off the budget deficit. In others words, *future* workers, consumers, and taxpayers—we in our older years or our children—will pay twice for today's hedonistic consumption: for the goods currently purchased and for the environmental mess created by the production and consumption of those goods. Thinking abstractly about the future makes it easy to conclude that *future consumption* may be curtailed out of economic necessity. However, that future is moving toward *us*. The increasing economic, health, and environmental costs are, in fact, increasing incrementally. It is not merely that *they* in the future must pay for our excesses, but that we too are and will be paying for our own excesses as we are pushed closer to the edge of history.

THE GREEN SHIFT
TO INDIVIDUAL RESPONSIBILITY

Many people are calling the new attention to environmental problems throughout the world the *green movement* or *revolution*. The theme of this book, and a way to understand the significance of the green movement, is that individuals have a crucial and historically new environmental role to play. To minimize local and global environmental risks, individuals can and—if their values dictate—must prevent the production of pollutants, toxic waste, and ordinary household garbage, which they do control, as much as possible. To do so means shifting from being a victim or cause of problems to being part of their solution. The cost of accepting personal environmental responsibility is work, and the benefit is control of the planet to keep it livable. Naturally, government policies and programs must also change to reflect the environmental values held by people. But first, a large number of people will probably have to express themselves in the marketplace in order to drive governments to enact the policies, laws, and programs to cause the bulk of the population to change their consumption and behavior, as well as to overcome the resistance of large industries.

Because the new global green movement emphasizes pollution prevention, it is different fundamentally from the environmentalism of the past 20 years, which depended on pollution control and waste management for solutions. Traditional environmental protection has focused on the *effects* of pollutants and wastes, but pollution prevention focuses on their *production,* the use of products, processes, and practices. Traditional pollution control and waste management is often referred to as the *end-of-pipe* approach because pollutants and wastes are dealt with *after* they have been created in industrial facilities, homes, offices, and nearly everywhere where human activities occur.

Pollution prevention, like all forms of prevention, requires considerable individual responsibility, commitment, and discipline. In contrast, our traditional environmental system has focused on institutions, such as government regulations and agencies to control pollution, courts to resolve disputes and force changes, environmental organizations to identify problems and encourage solutions, and companies to invent and sell pollution control equipment and operate waste management facilities. The shift of responsibility from institutions to individuals is a key feature of the international green movement and the grass-roots activism that has been shaping new environmental approaches in the United States.

A good example of the latter is the 1986 federal statute called the Community Right-to-Know Act. It requires companies which release significant amounts of toxic chemicals into the environment—even legally permitted amounts—to make data on some of them directly available to the public. The Toxic Release Inventory database has eliminated a traditional role of government and made industry directly accountable to communities. The lack of public confidence in government and awareness of a regulatory system that sanctioned pollution created this new legislative approach. The theory of environmental activists was that raw data on the enormous amounts of chemicals being released into the environment would brighten the public and put new pressures on industry. A special hope was that industry would put more stress on waste reduction. And that seems to be happening. Similarly, there is much more demand for product labeling that provides key information to consumers. In California, the passage of Proposition 65 is significant because of its requirement to disclose the presence of toxic chemicals in nearly all consumer products. One of the success stories in California is that no more lead-soldered cans are in use there, even though they are still being used elsewhere.

A FAILED ENVIRONMENTAL STRATEGY CREATES
THE DEMAND FOR AN ALTERNATIVE

The green movement and the increased political attention to environmental issues have happened largely because traditional solutions either failed or seemed burdensome and too costly. Part of the problem was the apparent failure to anticipate and prevent new environmental problems. For example, a headline for a front-page story in the *Washington Post* (1989a) was "Frogs, Toads Vanishing Across Much of World." A government scientist was quoted as saying, "We have to ask whether this is a general indication of environmental degradation." Nearly everyone must ask themselves when reading such frequent types of stories: *What next?* Frogs and toads may not seem essential, but by now, people understand the ecological connectedness among all living species on planet Earth. Somewhere in the complex cycle of interdependent life, frogs and toads are important for humans. Besides, whatever is affecting the frogs and toads may eventually affect humans. Therefore, such observations, like many others in the past, should be seen as warning signs for people, even those who think they can live very well without frogs and toads.

The truth is that environmental problems associated with industrial and household wastes and pollutants *have* become more difficult to predict, understand, and control with conventional pollution control and waste management methods. And nearly everywhere, people oppose facilities, even those using the latest technology, which control pollution or manage waste, such as toxic waste or municipal waste incinerators and landfills. Even though local capacity to manage ordinary household garbage may be running out, communities still oppose new facilities. The same is true when local industry claims that it needs new industrial waste management capacity. Public opposition to all kinds of pollution control and waste management facilities is increasing worldwide, even in poorer nations, because of instant global communication about environmental problems and because of the increasing activities of environmental organizations such as Greenpeace and Friends of the Earth.

Generally speaking, community opposition to waste management facilities is viewed negatively by people in government and industry. The often used acronym NIMBY, standing for "not in my backyard," is commonly used to condemn citizen activists. From a pollution prevention perspective, however, the widespread community opposition to construction of new waste management facilities and expansion of existing ones has been enormously constructive. Companies which produce environmental wastes, government bureaucrats who must manage wastes, and companies that make money from wastes don't like community opposition. Some people in economically depressed areas support new waste management facilities because of the jobs created. But public opposition has helped make pollution prevention more necessary and, therefore, has helped focus much more attention on it by people in government and industry. By using pollution prevention as a positive, pragmatic alternative to waste management, communities have also increased their understanding of their role as consumers as waste reducers. From a pollution prevention perspective, government agencies should require companies that want to build new waste management facilities to demonstrate that waste reduction has been pursued to the maximum degree.

Without strenuous pollution prevention, there will be no end to the building of pollution control and waste management facilities. More people, consumption, economic development, and industrial activity on the planet mean that even controlled, small amounts of waste from individual sources which enter the biosphere can cause problems. And in developing nations, the goal of economic growth to meet survival needs, to achieve a high standard of living, and to manage often rapid population growth has frequently caused government officials to ignore environmental protection, as if it were a luxury affordable by only affluent nations.

But affluent, industrialized nations do more than afford the luxury of environmental protection. There is strong evidence that, over time, affluent industrialized societies produce more waste and pollution as population grows and *on an individual or per capita basis.* Table 1-1 shows data from a number of industrialized nations on the increase in per capita generation of ordinary garbage. Only a few nations show decreases. Overall, aside from more garbage because of population growth, there is about a 1 percent annual increase in municipal waste generation *per person worldwide.* That may not sound like much, but it definitely adds up. From 1960 to 1986, official data from the U.S. Environmental Protection Agency (EPA), which do *not* come from actual measurements of garbage, show a 43 percent increase in the amount of municipal waste produced per person in the United States. The actual amount of U.S. municipal garbage may be nearly twice as large, as explained in a later chapter. Actual data for New York City show an increase in per capita garbage generation of about 100 percent from 1980 to 1989.

As the data in Table 1-2 show, the amount of municipal solid waste being generated in the United States is much higher than that of other industrialized nations with a high standard of living overall, among most industrialized nations in 1985, the United States accounted for over 50 percent (and perhaps as much as two-thirds) of the total amount of municipal solid waste, even though it accounted for only one-third of the total population.

TABLE 1-1.   CHANGES IN MUNICIPAL SOLID WASTE GENERATION
PER CAPITA FOR INDUSTRIALIZED NATIONS

Nation	Percent Change	
	1975–1980	1980 to 1985
Canada		21.2
United States	8.4	5.9
Japan	4.1	−3.2
New Zealand	79.8	
Austria	19.4	3.1
Belgium	5.7	
Denmark		5.8
France		4.6
West Germany	5.3	−8.7
Ireland	7.7	65.1
Italy	−2.0	4.1
Luxembourg	6.4	1.8
Netherlands		3.2
Norway	−1.9	14.0
Portugal		3.6
Spain		28.2
Sweden	3.1	5.1
Switzerland	18.2	9.1
United Kingdom	−1.6	11.4
TOTAL	7.4[a]	5.9[b]

Source: OECD 1989.

Note: Data among nations is only approximate because of wide differences in definitions and methods of obtaining data. U.S. data may seriously underestimate municipal waste generation.
[a]Thirteen nations: 1975 waste, 240.8 million tonnes, population 543 million; 1980
waste, 266.9 million tonnes, population 561 million.
[b]Seventeen nations: 1980 waste, 306.9 million tonnes, population 692 million; 1985
waste, 334.2 million tonnes, population 713 million.

In other words, it is reasonable to believe that pollution prevention actions
can greatly reduce ordinary garbage generation in America without destroying the economy and certainly without degrading the *quality* of life. Waste
reduction is a way for everyone in society to get more value (rather than
garbage) for their money and to improve environmental quality.

All told, with the U.S. population growing at a rate of more than 2 mil-

TABLE 1-2.   U.S. GENERATION OF MUNICIPAL SOLID WASTE
RELATIVE TO THAT OF OTHER INDUSTRIALIZED NATIONS, 1985

Nation	U.S. Per Capita Rate Greater by (Percent)	
	Conservative[a]	Probable[b]
France	174	282
West Germany	134	227
Japan	116	202
United Kingdom	110	192
Switzerland	94	172
Netherlands	65	132

Note: Data among nations is only approximate because of wide differences in definitions and methods of obtaining data. In Chapter 4, the authors show that the official data for the United States seriously underestimates municipal solid waste generation.

[a]Comparison with the estimate by the OECD (1989) for the United States of 196 million tons for 1985, which is higher than the EPA's calculated amount (158 million tons for 1986).

[b]Comparison with a more realistic amount that reflects actual collection data from states is estimated by the authors at approximately 275 million tons for 1985.

lion people a year, an additional 5 billion pounds of municipal solid waste are being produced annually. This is comparable to creating a good-sized city every year, and it is happening at a time when it is more difficult and costly to find a place to put that waste. The combination of increasing per capita rates of waste production with economic development and population growth is devastating in both industrialized and developing nations. The effectiveness of all environmental solutions, including recycling, must be tested against the tyranny of inevitable global economic development, population growth, and, of course, technological progress. Or is it progress?

Technological progress in conjunction with increasing consumption of products often spells trouble. It is not difficult to discern how technological innovations which we think of as progress also produce more waste. Few people anticipated that the computer revolution would produce enormously greater amounts of paper waste or that microwave ovens would produce just as much packaging waste as they do cooked food, but they have done just that. In 1959, when Xerox introduced its dry copier, it was estimated that no more than 5,000 such copiers would be sold in the United States; in 1986 alone, about 200,000 photocopiers were sold, and sales are still rising. The remarkable 62 percent increase in municipal solid waste in Washington D.C., from 1981 to 1988 *on a per capita* basis, as population declined, is largely explained by the increasing wastepaper production in offices. From 1960 to 1982, the amount of paper trash in the United States on a per capita basis increased 35 percent as total paper garbage increased 75 percent nationwide. But photocopying is not all of the problem. Half of all paper made in the United States is used for packaging, which quickly becomes waste.

Social trends have also meant more waste generation. When all family members work and more and more time is spent commuting to work on clogged highways, time constraints motivate "convenience" shopping, products, and services. Materials replace increasingly expensive labor, and convenience produces garbage. Marketing products requires new tactics, and so computerized mailing lists and catalogs produce the convenience of shopping from the home and the inconvenience of piles of junk mail and more garbage to get rid of. In fact, 12.4 *billion* glossy mail order catalogs are mailed each year. This is just a fraction of the 70 *billion* pieces of second- and third-class mail, which also is nearly instaneously turned into garbage. The hidden cost of convenience for a few is a nuisance for most people; it means more waste landfills, incinerators, and the need to spend time lugging out more trash and sorting more household garbage for recycling. A study of small-town post offices in Vermont found that 85 to 95 percent of the substantial amount of paper trash they generated consisted of junk mail thrown away by people emptying their boxes.

Today, hardly anyone has time or is inclined to pay an electric, gas, or telephone bill in person; credit cards have replaced cash payments nearly everywhere. The result is more garbage because paper is produced at the time of purchase when credit cards are used; more paper comes in the mail (bills with advertisements); and still more paper is produced when checks are written and records are maintained by buyers, sellers, credit card companies, and banks.

As another example of the linkage between modernity and waste generation is the growth of terroristic acts, such as product tampering, which have impact because of instant communication systems. The fear of product tampering has resulted in complicated packaging that invariably means more materials that quickly become garbage. Deterrence of shoplifting is one reason for encasing small items in cardboard and plastic packages.

In a society where people are so mobile and businesses change so frequently, it seems necessary to have new residential telephone books and Yellow Pages directories for homes, offices, and public telephones each year. They weigh a lot. But now there is competition among the producers of telephone books because these are profitable advertising media. A single household may receive several books for the same year. Every year probably more than 3 *billion* pounds of "old" telephone books become garbage and end up in landfills.

In addition to new environmental problems being discovered every week or so and old solutions being found to be ineffective, new environmental solutions may be less successful then they first appear to be. With more at stake, companies and industries have become more active, just as environmental organizations have become more grass roots oriented and activist. For example, a lot of rhetoric has appeared about biodegradable plastics, without any convincing information about their effectiveness in landfills or attention to the negative impacts of their decay products. Biodegradable plastics have cleverly been used to reduce public opposition to plastics in general, but they offer few environmental benefits. The one big exception is when plastics in oceans and other waters degrade before harming water species. Similarly, some consumer products said to offer environmental benefits do not, just as advertising and labeling for some food products promoting health benefits are misleading.

GOVERNMENT, A PART OF THE PROBLEM

In the new environmental stuggle, the role of government has become more often questioned, particularly in the United States, which has emphasized the heavy hand of government regulation. The traditional methods of waste management and pollution control have been increasingly regulated by government. In 1985 there were about 7,000 pages of federal environmental laws and regulations; by 1988 there were 10,000 pages. More than any other nation, the United States has used government end-of-pipe regulation embedded in a complex web of legal rules and procedures to try to achieve environmental objectives. It is inevitable that more and more pollution control regulations will be created. Other nations rely more on cooperation between government and industry to achieve national environmental goals. But the U.S. regulatory effort has revealed the limitations of its strategy as much as its achievements. Government regulatory programs are plagued by problems of noncompliance, litigation, loopholes, and slow implementation that are continually revealed by the press and environmental organizations.

Charles L. Grizzle, EPA's Assistant Administrator for Administration, in acknowledging that EPA's employees at its headquarters in Washington, D.C., have suffered health problems because of indoor pollution said: "The irony of this being in the Environmental Protection Agency is very pointed. Perhaps it could lead people in the country to think that if we can't take care of our own employees, how can we take care of the rest of the country?" Many environmental victims would probably say that the environmental problem at EPA headquarters was justice. Photographs have appeared periodically in the press showing Americans holding signs saying that EPA stands for "Every Pollutant Allowed" or "Every Poison Allowed." Clearly, public loss of confidence in government programs in the United States and other nations has been a potent force in stimulating the new green and grass-roots environmental movement.

Moreover, the traditional end-of-pipe regulatory approach has carved up the world of pollution and waste into neat categories like air pollution, water pollution, toxic waste, and municipal waste. But this piecemeal approach has prevented us from seeing the whole environmental problem as it really exists. All too often, pollution control and waste management have only meant shifting harmful substances from one location to another, from one kind of threat to another, and sometimes from one generation to another. All of this results because problems are contained or moved around rather than solved. For example, the start of the environmental movement focused on air and water pollution. But many pollution control technologies that removed pollutants from air and surface waters created solid hazardous wastes which went unregulated for some time. Another example is that of the wood-preserving industry, which responded to limits on discharging contaminated water into surface waters under the Clean Water Act, in part, by using surface impoundments. But some of these surface lakes of toxic waste waters contaminated land and groundwater, which are very difficult and costly to clean up.

Even when hazardous waste became regulated, after air and water pollution, it was still managed in ways, such as injecting liquids into deep wells and burying solid wastes in landfills, which have resulted in pollution of groundwater and contamination of land. Getting the harmful substances out of groundwater and soil is more difficult and more costly than the original pollution control schemes that removed them from air and surface waters. Rarely has pollution control and waste management destroyed harmful substances safely, once and forever, and not merely transferred the problem somewhere else. In making the case for the primacy of waste reduction, the OTA (1986) said:

> Often what is called treatment of waste is simply removal and transfer. For example, evaporation ponds and air stripping columns used for treating liquid wastes purposefully put volatile toxic chemicals into the air, and adsorption materials [like charcoal or carbon] used to remove toxic chemicals from liquids or gases are generally land disposed. Statistics for industrial hazardous pollutants in waste streams sent [legally] to publicly owned water *treatment* plants indicate that only about 50 percent are permanently altered; the rest remain hazardous and are released into the air as volatile emissions, discharged into surface waters, or put into the land as sludge, where hazardous substances can migrate into groundwater. There are concerns about emissions of unregulated toxic chemicals resulting

from incineration; according to EPA more than half of hazardous waste incinerators in 1981 used no air pollution control systems at all.

Even the simplest end-of-pipe technologies fail. In December 1989 the National Research Council reported for the first time the failure to control highly toxic plutonium emissions at Department of Energy facilities. The *Washington Post* (1989b) reported:

> A significant quantity of plutonium has accumulated over the years in ventilator ductwork at a finishing plant in Hanford, Wash., much of it "downstream" from filters installed to protect the public from radioactive emissions. Several experts said the finding suggests that substantial amounts were probably released from the ducts to the outside air.

The day afterward, interestingly enough, another *Washington Post* (1989c) article reported that another study by the National Research Council had found that exposures to low levels of radiation are "more likely to cause fatal cancer than is commonly believed" and that "no radiation exposure can be considered risk free."

As another example of the failure of relatively simple systems, 500,000 gallons of refined oil leaked from an underground pipeline into New York harbor in January 1990. Exxon employees disregarded signals from a leak detection system for almost 6 hours. Why? For a year or more, the system had been giving false alarms weekly. This tertiary prevention system (like a smoke alarm that goes off for no good reason) did not receive preventive maintenance to ensure its usefulness. Nearly all so-called accidents, including Exxon's Alaskan oil spill, are in fact preventable.

Another basic problem is that a lot of hazardous waste is not regulated as such by the government. For example, California has a more comprehensive definition of hazardous waste than the federal EPA. The result is that about half of the regulated hazardous waste in California is state-defined waste. Few other states regulate so much. A chemical called butadiene is used in industrial processes such as the manufacture of synthetic rubber products. In such plants, significantly higher rates of several cancers amont workers have been linked to butadiene. The chemical has been identified, for example, as a major air pollutant in the Houston, Texas, area. Michael Picker (1989) has noted, "According to the EPA, the most serious community toxic health hazard in Los Angeles is the UnoCal facility in La Mirada, which emits cancer-causing butadiene into the air. Is the emission wasteful? Yes. Is it a hazardous waste in either state or federal law? No."

In other words, there are major limitations to traditional technical approaches, which pay little attention to the causes of waste and which try to manage or control waste in safe ways. The key alternative to the old approach that we prefer to call *pollution prevention* encourages a more comprehensive approach. It should be noted that pollution prevention is also frequently called *waste reduction, waste minimization, source reduction,* or *toxic use reduction;* it may also be referred to as the use of clean technology or clean products.

The message of this book is that every one of us has to rethink every work, play, consumer, transportation, and investor activity from the perspective of pollution prevention. Unless we do, the conventional environmental protection systems based on institutions will be increasingly ineffective even as their costs increase. Around the world, more people are beginning to understand that producing less pollution and waste is a matter of collective survival. The concentration of people in cities from Los Angeles to Hong Kong to Cairo to Mexico City has highlighted waste disposal and pollution problems for all to see. Humankind is, indeed, burying itself in its own waste products, although often the burial process may be invisible because of the dispersal and dilution of wastes and pollutants in land, water, and air. Unlike other eras, when the wastes from horses were on streets and black clouds billowed from smokestacks, today's local and global environmental threats are largely invisible to the naked eye but are more threatening. People did not see the 563,000 pounds of uranium dust emitted into the atmosphere over the years from the Department of Energy's Fernald, Ohio, nuclear weapons facility. After years of denial, the federal government agreed to pay the people who live near the plant $73 million.

Pollution control and waste management institutions, laws, and regulations that have been established over several decades are still needed, but they can no longer be relied on to protect people and the planet's fragile ecology effectively and proactively. The end-of-pipe system is literally clogged, backing up, and threatening the global and local health, environment, and welfare. People have literally treated the Earth as a toilet, and it is rebelling. And lastly, there is cause for concern about so-called innovations in regulatory practices euphemistically called *innovative free market incentives.* Such incentives can be completely antithetical to the idea of pollution prevention.

The chief example is the selling of air emissions reduction credits, often called *offsets,* for the six pollutants for which EPA has established ambient air standards. A similar system will probably be used to cope with the sulfur dioxide emissions causing acid rain. Polluters that reduce their air emissions below the values specified in their regulatory permits can sell the difference so that another company can surpass its regulatory limits. In theory, companies have a profit incentive to cut their emissions and sell their valuable "property"—the right to pollute.

For example, Sundance Spas, a manufacturer of fiberglass hot tubs in Chino, California, used the services of AER*X, a firm that brokers rights to pollute. Sundance uses solvents in a laminating process, but in southern California there are tight limits on solvent emissions. With increasing sales, the only way the company could open a new plant a few years ago was to buy pollution credits for $25,000, allowing it to produce another 316 pounds of organic emissions a day.

There are cases where a plant with a regulatory permit could essentially close its operations and make a lot of money by selling its pollution rights. The point is that the environmental regulatory system legally sanctions the right to pollute. Trading pollution rights that does not result in a *net* reduction in pollution is not progress. The sale of pollution rights is easily seen as obscene, immoral, and risky by people who believe that government-allowed levels of pollution—inevitably the result of a political compromise—are too high to begin with. To the extent that some companies have an incentive to cut their pollution, other companies have a way to escape their responsibility to seriously examine and change the sources of their pollution. Moreover, there are few safeguards to assure communities that excessive pollution in their area, obtained by purchasing pollution rights, does not increase health and environmental risks. One solution would be to allow the sale of only a fraction of pollution reductions, and to phase out marketable pollution rights over time as older facilities have more time to implement pollution prevention measures. One innovative twist to help rectify this government device is to force public auctions of pollution rights, giving communities and environmental groups an opportunity to buy the rights and not exercise them. This approach, however, places an economic burden on those at risk rather than on polluters.

POLLUTION PREVENTION LINKS ALL ENVIRONMENTAL PROBLEMS TOGETHER

A large number of problems and threats are classified by most people as environmental. As environmental issues have moved closer to center stage as a threat to planetary survival, as cause for concerns about immediate and local threats to health and environment, and especially as a fundamental component of national and international politics and economics, confusion has also risen. There is a constant flow of information about environmental problems and, less so, about solutions. But experts and exposés seem focused on small pieces of the global environmental puzzle. More and more buzz words enter the language, but technical complexity outpaces understanding for most people. A good example is that people hear about an ozone problem related to air pollution and also about an ozone problem related to global warming.

First, most people do not know what ozone is and how it is produced; it is a molecule with three oxygen atoms instead of the normal two, and it is produced through chemical reactions. Second, few people understand that there are two very different ozone problems, one being too much ozone (in the lower atmosphere that we breathe, which causes smog and respiratory problems) and the other too little ozone (in the upper atmosphere that protects us from harmful radiation, which causes skin cancer). In both cases, consumers play a vital role because they use products that contribute to both problems. Upper atmospheric ozone destruction results from the use of refrigerators, air conditioners, foam cushioning, and virtually every piece of electronic equipment, for example. Smog from too much ozone results largely from products that release volatile organic conpounds present in hair sprays, air fresheners, disinfectants, cosmetics, spray paint, insecticides, all all-purpose cleaners, for example. Every day, California air is polluted from 58,000 pounds of volatile chemicals from paints and 54,000 pounds of discharges from hair sprays. Automobile exhaust contributes to smog.

Nearly everyone is likely to find it increasingly difficult to figure out the causes and effects of environmental problems and how different problems relate to each other. Table 1–3 is a list of generic environmental problems along with general pollution prevention solutions. It is especially important to understand how solutions to environmental problems may have positive or negative effects on other problems. As the list of increasingly complex environmental problems grows, so does the possibility, which many people feel intuitively, that a solution to one problem may only exacerbate another problem. Or people may feel that solving one problem is insignificant because there are so many others for which solutions seem less certain. Additionally, people may feel that individual actions may be ineffective when environmental problems seem so global or ubiquitous.

A lack of public understanding of environmental problems and solutions, and their interrelationships, can lead to public paralysis or apathy. There is a need for people to feel hopeful and not helpless, to feel that they are increasing their understanding and not being buried by information, and to feel that individual actions which are supposed to be good for the environment are really so. Therefore, it is important to examine how pollution prevention relates to other parts of the environmental scene in order to establish the uniqueness of pollution prevention and to energize persons.

The thesis of this book is that pollution prevention offers a universally applicable strategy to solve *all* environmental problems. The trick in identifying pollution prevention solutions is to ask what product, process, or practice, if stopped or greatly reduced, would solve the problem. One must not be deterred by believing that a particular product, process, or practice is indispensable. Successful pollution prevention requires us to reexamine

TABLE 1-3.   SOME ENVIRONMENTAL PROBLEMS
AND ILLUSTRATIVE POLLUTION PREVENTION SOLUTIONS

Problem	Pollution Prevention Solution
Destruction of the protective ozone layer	Quickly replace all ozone depleting chemicals like Freon with safe substitutes
Global warming	1. Replace fossil fuels with energy sources like solar energy 2. Conserve energy in homes and businesses
Toxic waste products	1. Change industrial manufacturing processes and raw materials 2. Buy products whose manufacture produces less toxic waste and that don't contain toxic chemicals
Acid rain	1. Use low-sulfur clean coal in power plants 2. Use renewable energy sources
Smog	1. Use automobiles powered by electricity or alternative fuels 2. Replace many consumer products that release volatile organic chemicals, such as hair sprays, paints, and air fresheners
Too much garbage that is too toxic	1. Buy durable products 2. Buy less toxic products 3. Reuse products 4. Demand less packaging for products
Contamination of groundwater	1. Replace agricultural chemicals with integrated pest management 2. Reduce pesticide use around homes
Lead poisoning of children	Ban lead soldering for plumbing and food containers, and in ceramic cups and plates
Dioxin in consumer paper products	1. Change the bleaching process in papermaking 2. Buy brown, nonbleached products

and rethink the use of products, processes, and practices which may be taken for granted because of their extensive use or because they may seem solely beneficial, as if there are no environmental costs.

Pursuing pollution prevention solutions means asking these kinds of questions about a particular product, process, or practice: Besides the original waste or pollutant and the environmental problem that brought attention to the product, process, or practice, what other waste outputs may be causes of problems? Is it absolutely necessary? Is there an equivalent one available or being used elsewhere? What was used before? Has anyone really tried to find a substitute? Has research and development been started to find an alternative? Who benefits financially from the current use, and will they resist change? If necessity is indeed the mother of invention, then by rejecting dangerous products and the processes that made them, new opportunities for technological innovation and business will be created. Smart, creative people will seize pollution prevention as a business opportunity as much as a solution.

SHIFTING THE FOCUS FROM BACK END TO FRONT END

Population and economic growth can be limited not only by resource constraints, such as food and energy, at the "front end" but also by pollution and waste management problems at the "back end." It is when the back end gets clogged that institutions are forced to target the front end for change. Comprehensive pollution prevention that covers all wastes, discharges, emissions, and pollutants, therefore, closes the circle and refocuses attention on all resources, not simply because supplies may be limited but because resources are used excessively, as well as being processed inefficiently. In other words, pollution prevention and waste reduction force us to question why and how we use raw materials and products and to address our excessive and inefficient use of natural resources. Even though 100 percent efficiency (and complete elimination of wastes) may not be attainable routinely, experience shows us that greater efficiency (and substantial pollution prevention or waste reduction) is nearly always possible.

The giant Ciba-Geigy company reported that in 1979, finished products represented only 30 percent of all outputs and wastes represented 70 percent. In 1980, products, or overall efficiency, increased to 50 percent. In 1988, the efficiency, or fraction of conversion to products, increased to 62.5 percent. True, these data show real progress, but is there any reason to be complacent about an efficiency level of 62.5 percent? Of course not; Ciba-Geigy has a goal of 75 percent efficiency by the year 2000. But is 75 percent efficiency the best? Of course not. There is no law of physics, chemistry, or thermodynamics that requires the production of pollutants and wastes.

Economics should drive the quest for greater industrial efficiency, as should increasing global industrial activity and population because of the limits to the amount of environmental wastes the Earth can tolerate without dying.

Indeed, by understanding the link between pollution prevention and inefficiency, it becomes clear that more than better protection of our environment and health is at stake. If we, as individuals and as a society, practice pollution prevention and waste reduction, we will save money, both individually and collectively. We will use less natural resources, capital, and labor to meet our needs and desires. The real uncertainty is the extent to which we must also reduce consumption. A major goal of this book is to demonstrate that all around us are inefficient and unnecessary uses of materials, for which we pay considerable money, which become or cause waste and pollution and which can be reduced or eliminated. Physical waste is not just a nuisance and an environmental threat. It is literally a waste of money—your money.

Consider the packaging of products. Although there are legitimate functions for packaging materials, things do seem out of control. In 1986 there were at least 53 million tons of packaging garbage in municipal waste. In the same year, it was estimated that $55 *billion* was spent on packaging. This means that over $200 was spent on packaging, on average, for every person in the United States in one year. This is equivalent to spending $1,000 to produce a ton of packaging waste (or 50 cents to produce a pound of packaging garbage). These are the front-end costs. At the back end there is the cost of collecting, transporting, and disposing of packaging garbage; this may amount to an additional $75 or more per ton, or perhaps a total of $5 billion for U.S. waste disposal. Making and getting rid of packaging garbage is big business: $60 *billion* annually and rising. In addition, there are the costs for disposing of the industrial wastes produced in manufacturing the packaging materials; the environmental costs associated with the original manufacture of the packaging materials; and the environmental costs of managing the packaging waste in landfills or incinerators. Counting everything in the life cycle of packaging wastes, the ultimate costs for *1 years's* packaging waste in the United States might total $70 *billion*.

Buying consumer products must, therefore, focus on the product's packaging as well as the product itself. The new interest in "green products" reflects the new emphasis on how individuals can make new environmental and economic choices in the marketplace. But those choices can be made only *if* industry gives consumers real choices and *if* consumers have reliable information about the environmental benefits of specific products. By practicing pollution prevention and waste reduction, we will also spend a lot less money on pollution control and waste management technologies, as well as on the costly government regulatory systems that try to assure their proper functioning. The green movement and pollution prevention, therefore, do not threaten the quality of life, but they do imply changes in life-styles and industrial practices and products. Whether or not our standard of living is reduced depends on how we define standard of living. If the standard of living is mostly a matter of consumption of goods, then it will be reduced if preventive planet care is practiced. But if the standard of living depends more on the satisfaction of human needs and desires with minimum consumption, then it is not threatened. The shift to pollution prevention inevitably feels threatening to many people as changes which are not now fully comprehensible will have to be accommodated. Those changes also provide new business opportunities and new choices for consumers.

A concrete example of new opportunities and choices in the greening marketplace is filters for brewing coffee. Not long ago, information became publicity available that toxic dioxin chemicals produced in white paper by bleaching processes could get into the coffee (or into any product or part of the body in contact with bleached paper). Several companies now manufacture brown, unbleached paper coffee filters that generally sell for the same price as the original white ones. But paper filters are single-use products which quickly become garbage and other companies have entered the marketplace with cloth reusable filters attacking both the single-use and dioxin problems. Cloth filters cost little more than a box of 40 paper filters, a little over $2. By substituting for 365 paper filters, according to the manufacturer, the ultimate savings are significant. Instead of spending over 4 cents per paper filter, the customer will spend less than 1 cent per brewing. This may not seem impressive, but over 10 years, for example, the direct saving adds up to about $150. But using the cloth reusable filter means more work, because it must be washed periodically. Looked at another way, many consumers will continue to pay over 4 cents per brewing for the convenience of paper filters; also, if they didn't switch to the brown ones, they face a health threat from dioxin. Taking a comprehensive environmental view, the cloth filter also means much less use of energy and materials compared to those used to make, package, and transport about nine boxes of paper filters per year. And, of course, the one cloth filter that will become garbage replaces the 365 paper ones and nine boxes disposed of annually; that amounts to well over 300 times as much household waste for a year.

Just when environmentally conscious coffee consumers were given two new choices, however, a much more aggressively promoted alternative entered the marketplace. Ironically, a major coffee manufacturer introduced a new product, made known to many more consumers because of a major advertising campaign, that typifies the historical social trend of selling convenience; its innovation is to sell coffee prepackaged in single-use bleached white paper filters so that the consumer does not have to take the time to place a measured amount of coffee in a filter. Not only will the filter become garbage, but this innovation also deters people from using the spent coffee grinds in home composts.

We do not suggest doing away with all current products, modern industry, or the elaborate legal and technical systems of environmental protection. Doing so is neither practical nor prudent nor necessary. Pollution prevention is *not* antitechnology. But we can no longer unthinkingly believe that traditional products bear no environmental costs or that individual consumers bear no responsibility for environmental impacts. Most importantly, people should not believe that it is possible to pursue genuine advances in pollution prevention and better protection of health and the environment without making significant changes as consumers and workers. Nor can the public depend on traditional government efforts and pretend that they can keep churning out more and more waste from homes, factories, and offices without increasing environmental risks. The current green movement shows that society is at the early stages of a *paradigm shift*—a fundamental change in what people think is the problem and the solutions. Such social and cultural changes occur periodically, but are often difficult to recognize as they first occur and spread incrementally and haltingly among people and social institutions. As a social and cultural change, therefore, pollution prevention poses exactly the same kinds of social problems and institutional challenges that other major changes, such as attempts to achieve racial and gender equality, have caused.

In the case of pollution prevention, because it is so closely connected to industrial activities and products, the inevitable consequence of a successful pollution prevention movement will be a restructuring of industry. Restructuring is a concept used by economists and public policy professionals to describe changes in the core technologies and organizational systems in industries. That is, restructuring of industry because of pollution prevention signifies changes in raw materials usage, changes in basic production technologies, changes in the design and composition of products, changes in the priorities of workers and managers, changes in the use of capital, and changes in the marketing of products. Restructuring of industry, in other words, implies the creation of whole new methods of doing business, producing product lines, and managing industries. Pollution prevention is *not* anti-industry.

In contrast to restructuring, conventional economic growth involves expansion of existing—often polluting—industries and perhaps changes in the ownership of companies through acquisitions, mergers, and other such financial transactions. In the past, industrial restructuring occurred because of major technological innovations such as the invention and widespread use of computers; the shift from peace to war economies an vice versa; and, to a limited degree, the energy crisis of the 1970s and other raw material scarcities. Furthermore, environmental wastes are a real cause of economic damage to especially sensitive industries. Tourism, agriculture, fishing, and real estate, for example, have been hard hit in some areas because of pollution. Overall, pollution prevention does not threaten or impede economic growth and prosperity because declines in some sectors can be offset by health in others and by the creation of new, clean technologies and products. Pollution prevention is *not* antiprosperity.

SUMMARY

It has always been wise to practice prevention instead of cure. Preventive care is one of those core actions that seems inherent in all living species because it is a survival skill. But it also seems to be learned behavior more than instinctual or automatic behavior for humans. With increasing intelligence and cognitive functioning, acting preventively could increase. If humans are the most intelligent of planet Earth's living species, surely they should be the best practitioners of preventive care. And yet, the history of the planet and the current state of affairs reveal that humans have practiced insufficient preventive care on virtually all levels of existence. Personally, socially, and politically, preventive measures are shortchanged, even though prevention is a necessity. The simple truth is that growing population, consumerism, and industrialization cripple current environmental efforts. This makes prevention more critical than ever before.

On the other hand, humans also have shown repeatedly that they can react constructively in the face of great danger. It seems like a case of tempting fate or danger to test our powers. Ignore preventive care, wait for imminent danger, and then react just in time with great technological feats to achieve victory. Hardly anyone really believes that life on planet Earth can continue, with its present tastes and trends, and still prosper or even survive. But many people seem to think that the stresses on our life-sustaining inputs through excessive use and waste generation, as well as direct threats to our health from pollution, can be controlled or redirected before it is too late. They correctly point to endless predictions of global gloom which have never materialized and ignore the countless smaller signs of environmental trouble which other, more pessimistic people focus on. In defense of pollution prevention, we and many other people say: better to be safe than sorry.

Lastly, while we emphasize the critical role of individuals as workers and consumers in gaining control of the production of wastes and pollutants, we definetely recognize the importance of changing institutions and government policies and programs. It is a question of what comes first. We contend that changes in the behavior of many progressive people in the marketplace and workplace must occur first. Once enough people have shown their values, preferences, and skills, then industries, environmental organizations, and governments will follow. It has always been that way. People speak of public leaders, but when it comes to making major social and cultural changes, it is nearly always the actions of a significant number of individuals which spearhead broader changes. Yesterday's radicals are indistinguishable from tommorow's average citizens. Environmentalists correctly remind people that many of "their" issues have become commonly accepted problems years later. When comprehensive *preventive* planet care is shown to be practical by an active, vocal minority, it will be demanded and implemented by the majority. Or else.

2

WHAT POLLUTION PREVENTION IS; WHAT WASTE RECYCLING AND OTHER STRATEGIES ARE NOT

The purpose of this chapter is to shed light on how pollution prevention can receive the attention it merits. If Chapter 1 has achieved its intended goal, then you may feel that too much waste and pollution are being produced and that they threaten humanity. But several decades of little success have shown how difficult it is for people to understand pollution prevention and to practice it with intensity and commitment. All too often, people seek answers elsewhere, almost the same way they seek convenience products: They ask the wrong questions and get the old answers. First, we will define a range of activities that share important characteristics as preventive actions. Distinguishing pollution prevention actions from waste management or pollution control actions is absolutely necessary and requires constant attention by all those interested in preventing planetary destruction.

Some people believe that a waste management or pollution control action, such as land disposal or incineration of hazardous waste or the use of a wastewater treatment plant, prevents pollution. This is because such equipment limits the release of harmful substances compared to uncontrolled raw discharge into the environment. Our perspective, however, is that control and management of environmental wastes already produced cannot guarantee that there will be minimal releases into the environment. Control and management methods are technological countermeasures based on imperfect technology, operated by imperfect people, and regulated and managed by imperfect organizations. The traditional end-of-pipe control strategy has been described as the "strategy of the cork." Technological optimists believe that a cork can be plugged into the places where pollutants are released. Practically speaking, however, the corks are loose and some pollution escapes. Traditionalists believe that improving technology and making more stringent regulations press the corks in tighter. But the corks loosen as technology and people fail, and government regulators are rarely effective at keeping strong pressure on the corks. Also, pollution has a nasty habit of popping out somewhere else, and then the system rushes to stick in another cork.

Like using medication to avoid feeling pain from a headache, infection, or surgical incision, using corks—waste management and pollution control methods—attempts, albeit imperfectly, to minimize the effect or pain of releases of pollutants into the environment. True, some releases and effects are curtailed. But the original toxic or destructive environmental wastes remain hazardous or are transformed into different hazardous substances to some degree. In most cases, environmental releases still occur, either routinely (around loose corks and at places where the government has not required corks) or when control equipment fails (when the corks pop). Significant effects on health and the environment cannot be totally and assuredly eliminated, therefore, with management and control options. That is, there is a false sense of security based on the illusion of effective control. In fact, control is weak, uneven, and often completely absent. Given a choice between preventing a source of personal physical pain and using medication for the pain, surely you would choose avoidance. And so it should be for environmental wastes. Forego pollution control corks and reduce the need for them by producing less of the pressurized pollution to begin with.

When waste management and pollution control can be avoided or eliminated, there will be certainty about avoiding or eliminating health and environmental damage from wastes and pollutants. Prevention offers real power because the source of the problem is addressed. Of course, another possible strategy, other than pollution countermeasures and prevention, is adaptation. Using our health care analogy, adapting to environmental effects is like living with pain or illness without addressing the source of the problem. Examples of adaptation include building dikes to cope with flooding from global warming; getting used to higher temperatures from global warming; routinely using sun-blocking skin lotions to minimize the effects of increased ultraviolet radiation because of ozone layer destruction; using bottled drinking water to avoid contaminated tap water; fencing off land contaminated with toxic chemicals to avoid contact with them; not eating fish caught in contaminated waters; wearing breathing masks to filter out air pollutants; using tanks of oxygen for breathing; and eating synthetic foods free of toxic chemicals. These adaptations sound familiar because they have already appeared. To appreciate the pollution prevention strategy, imagine a world where these and many more such adaptations are an integral part of nearly everyone's daily life. It is a science fiction world of grim penalties paid for believing that pollution is either a necessary price for economic prosperity or is unavoidable.

In contrast to pollution countermeasures and adaptation, we define two basic forms of pollution prevention. *Primary pollution prevention* is avoidance and elimination of the production of wastes and pollutants. *Secondary pollution prevention* is avoidance or elimination of waste management and pollution control actions, even though a waste—or potential waste—is produced. These variations will become clearer in the discussion that follows.

DIVERSE POLLUTION PREVENTION MEASURES ARE THE TOP CHOICES

For some years, nearly everyone who has examined waste management and pollution control has spoken of a hierarchy of preferred actions. A *hierarchy* means an ordered list of preferences. Our conception of the waste management and pollution control hierarchy is shown in Table 2-1. Pollution prevention is a separate class at the top of the hierarchy and is subdivided into several categories. The activities in the most desirable primary pollution prevention category are nearly self-evident, and many examples of them are presented in the next chapter. In the primary pollution prevention category for industries, commercial establishments, farms, military installations, and consumers are actions that eliminate or avoid all or part of the wastes and pollutants where they are first produced. These actions may reduce the volume or toxicity of wastes and pollutants, or both. In this category, it is the character of the original waste-producing situation—the source—that is changed substantially, improved, or made more efficient. The concern in industry about future uncertain liabilities, such as for toxic waste site cleanups and health damages, is best attacked by primary pollution prevention measures and, to a lesser degree, by secondary prevention practices. The same is true for consumer concerns about toxic wastes and dangerous products.

Secondary pollution prevention is less desirable because a material that might become a waste, but doesn't, is produced. It can be thought of as a potential waste because the material would be a waste if it was dealt with differently. But if secondary pollution prevention is practiced, the material does not become a waste. Instead of managing or controlling the material as waste, something else is done. In manufacturing operations, the internal use of a potential waste as a raw material is a key type of secondary prevention. This is internal recycling. For consumers, secondary pollution prevention includes simple repair of products, composting of organic food and yard wastes, and conversion of yard wastes or Christmas trees into mulch. Internal recycling and waste and product reuse, the two important forms of secondary pollution prevention, are discussed later. One way to think of secondary pollution prevention is as a transition between uncompromising pollution prevention and waste management or control.

The key principle of the hierarchy is that aside from pollution prevention (both primary and secondary), all other actions fall into the broad category

TABLE 2-1.   POLLUTION PREVENTION AND WASTE MANAGEMENT:
HIERARCHY OF OPTIONS

CLASS I.   POLLUTION PREVENTION

PRIMARY
In industry and other institutions
Change processes or equipment
Change composition, packaging, or durability of products
Change or reduce raw material inputs
Improve controls of processes
Improve inventory controls
Improve materials handling and cleaning operations
Improve maintenance and repair of equipment

By consumers
Purchase different products
Use less of a product

SECONDARY
In industry and other institutions
Internally recycle waste in process
Reuse waste onsite
Provide waste for reuse offsite

By consumers
Repair products
Reuse products
Provide products for reuse by others
Reuse waste, and compost food and yard wastes
Provide waste for reuse offsite

CLASS II.   WASTE MANAGEMENT AND POLLUTION CONTROL
External waste recycling (which, unlike reuse of waste offsite, involves
manufacturing or processing requiring pollution controls)
Treatment of waste to reduce toxicity
Land disposal of waste
Direct environmental release of waste or pollutant

---

of waste management or pollution control. Giving primacy to pollution prevention means that the best solution is to focus on the production of waste
and on how to cut that production and avoid costly and uncertain waste
management and pollution control. To say that pollution prevention has
primacy means that it is the option of choice and that it is to be considered,

evaluated, and chosen, when feasible, above all other options. No lesser option should be considered acceptable unless the producer of the waste or pollutant has shown convincingly that all forms of pollution prevention have been considered, evaluated, and found impossible to implement, either technically or economically. Some aspects of waste treatment, land disposal, and environmental releases have been discussed in Chapter 1 and are also discussed in Chapter 3. Recycling is analyzed later in this chapter.

There is a very important difference between pollution prevention and traditional waste management and pollution control. This difference is profoundly important but, for most people, difficult to understand because it seems unimportant in terms of solving real environmental problems. Nevertheless, this difference gets to the heart of the *choice* between prevention and reaction. In prevention, *sources* are emphasized; in reactive waste management and pollution control, *effects* receive most attention.

First, consider waste management and pollution control that attempts to control effects. The first step in this approach is to analyze what is to be protected, such as human health and air and water quality. Then it is necessary to determine the nature and level of risks to what is being protected. Then government, supposedly serving the public interest, determines what risks to valued assets are unacceptable. In this way, environmental standards are set for acceptable levels of emissions or releases directly to the environment, or acceptable levels of contaminants in the valued assets (like people, water, and fish) are set. Setting safe or acceptable levels (the two are not necessarily the same) is a very difficult, costly, and slow scientific, political, and bureaucratic process in which economic effects are given equal importance to environmental effects. Environmental protection standards are like safety valves which, theoretically, stop the production of excessive amounts of wastes and pollutants.

Risk assessment and cost:benefit analysis play an increasingly critical role in the traditional reactive, effects-controlling approach. Both methodologies are intentionally cloaked in scientific trappings, even though they are loaded with unscientific human judgments, values, and economic objectives. Moreover, in looking at protection through the lens of effects, controlling effects from some sources of environmental wastes may be balanced against controlling effects from other sources. The idea is that if some outputs from some sources are controlled, other outputs from other sources may not have to be controlled to protect the public or the planet. The inevitable result of the waste management/pollution control approach is that not all environmental wastes or assets are controlled, either at all or effectively enough to truly protect human health and natural planetary assets.

For example, the conventional scrubbers used at coal-burning power plants are designed to take out sulfur dioxide at the end of the pipe to help reduce acid rain. But scrubbers add carbon dioxide to the atmosphere and, therefore, increase global warming. Government policy has cut one pollutant to some extent, but not enough to solve the overall acid rain problem, and increased another, which has not yet become regulated. So many traditional measures are illusory, like building very tall smokestacks—as tall as 500 feet, the height of a 50-floor building. Such tall smokestacks keep the immediate area clean at the expense of more distant communities, which receive the pollutants, especially sulfur dioxide, and the damage. Tall smokestacks illustrate the outcome of cost:benefit, out-of-sight/out-of-mind thinking. These practices externalize or ignore the true costs of pollution in the name of illusionary cost effectiveness.

Now consider the pollution prevention, source-changing approach. It is very different and much simpler. The goal of pollution prevention through source changes and improvements is to eliminate or reduce *all* wastes and pollutants, without exception, without judgments about what is safe, acceptable, allowable, cost effective, and so on. True protection of human health and natural planetary assets is a consequence of practicing comprehensive waste reduction. Environmental quality is maximized when all environmental wastes are eliminated or reduced.

Let us summarize by delineating the comparative advantages of preventive techniques that deal with *all* environmental wastes from *all* sources, including industry, agriculture, transportation, energy supply, building trades, households, recreation, military facilities, research laboratories, schools, hospitals, retail stores, and office buildings. Unlike the back-end pollution control approach, the front-end pollution prevention strategy aims at improving the quality and efficiency of basic industrial practices, as well as the quality and safety of consumer products. Similarly, prevention means that all energy needs are reduced, and the whole production-consumption life cycle of raw materials and energy, as well as all waste outputs at any point during the life cycle, are integrated. These differences result in *economic* benefits from pollution prevention. In contrast, end-of-pipe pollution controls do not contribute to profits at those locations where they are employed (although they produce limited benefits to society), but instead increase costs both directly and indirectly.

Therefore, there is much resistance to back-end, unproductive controls. That resistance has accounted for the sluggish and incomplete progress of the conventional end-of-pipe approach. For example, getting the automobile industry to manage its hazardous paint wastes in order to minimize health and environmental effects adds to its costs. Regulations which attempt to force automobile makers to use better but more expensive forms of waste management are resisted. But when automobile manufacturers install robots to spray precise amounts of paint on cars within sealed booths, paint waste is eliminated. By getting 75 percent of the paint on the cars and recovering and reusing the remainder, the automobile manufacturer has cut costs, improved quality, and increased profits. *And* the manufacturer has provided a better environmental solution than any form of waste management. However, note that paint companies may sell less paint (and produce less pollution) and waste management companies may have less business, so that they must strive for business success by concentrating on making money without counting on waste generation. For such reasons, even regulations that in one way or another require economically beneficial pollution prevention are likely to be resisted, just as end-of-pipe regulations are resisted.

To summarize, in comparison to back-end control techniques, front-end source changes and improvements are to be preferred because:

- there is clearer responsibility for environmental consequences at the source;
- there is greater ability to manage, with reliability and predictability at the source;
- by addressing the sources of *all* environmental wastes, no loopholes are created for some health and environmental effects;
- international cooperation is seen as more necessary because prevention in one nation (like a ban on a chemical) is easily negated by a lack of it elsewhere;
- scientific uncertainties about causes and effects do not justify inaction;
- surprises from newly perceived health and environmental problems, and the delays associated with setting acceptable levels of pollution and risk, are minimized because *all* environmental wastes are subject to elimination and reduction proactively; and
- from a systems perspective in which all direct and indirect, short- and long-term costs to all parties are accounted for, prevention is less expensive than reaction and, therefore, makes all economic systems more efficient.

There is also a tertiary level of pollution prevention, which we have not included in the hierarchy and do not discuss in this book. In this category are actions that do not deal directly with wastes and pollutants, but rather with detecting, at the earliest possible time, the *effects* of wastes and pollutants which already exist. Early detection of human health and ecological effects can be considered tertiary pollution prevention to the extent that detection prevents, reduces, or minimizes significant adverse effects and helps primary and secondary pollution preventives to be implemented. Health screening tests may, for example, detect levels of toxic chemicals in people's bodies. *Environmental monitoring* refers to the testing of air and water for chemical contamination and to measurements that reveal specific environmental effects such as destruction of the ozone layer, global warming, and damage resulting from acid rain. Environmental monitoring also examines various living organisms, like fish and birds, as they generally show early signs of environmental pollution and degradation. A common example of tertiary prevention is a home smoke alarm. It is supposed to prevent injury from a fire, but of course, this depends on its effectiveness for enough of a fire to produce smoke. The smoke alarm does not, therefore, prevent a fire from careless cigarette smoking as well as not smoking in the first place. A sprinkler system is an example of a control response to a fire, and not a preventive device, because its function is to quench the fire. Similarly, emergency toxic chemical spill teams are not preventive but reactive.

INTERNAL RECYCLING AND REUSE
OF PRODUCTS AND WASTES
(SECONDARY POLLUTION PREVENTION)

When a material that has served its original purpose, and could become a waste, can be used within the same process or operation as its original production, then the *internal* recycling does not pose the environmental risks or uncertainties characteristic of external recycling, which requires transport of the waste to another location where it undergoes processing. It is by avoiding waste handling, transport, and processing that internal recycling offers a controlled and certain option. Internal recycling in industrial facilities is often called *closed-loop* or *inprocess recycling;* this is common in large-scale chemical processing plants in which a by-product or waste from one process may be piped to another part of the plant where it is used directly. Small industrial facilities can also find ways to recycle internally some materials that would otherwise become waste, such as recovering and using volatile solvents that formerly were vented into the atmosphere.

Internal recycling is also possible in the home. For example, composting waste in backyards is becoming increasingly popular; through biological processes, it converts organic waste into heat and usable soil additives that can replace commercial mulch and fertilizers. The key characteristic of internal recycling is that, unlike external recycling, the material is controlled by the waste producer. It is this attribute that makes internal recycling a form of pollution prevention. The material is a waste if it is *not* internally recycled, but it is a raw material when it *is.*

A recent important example of internal recycling in automobiles is the use of charcoal canisters to capture gasoline fumes. Canisters large enough to capture all fuel vapors can prevent the release of enormous quantities of hydrocarbons which contribute to smog. The captured fuel is routed to the engine. Thus the system is completely internal to the automobile. In late 1989, EPA proposed requiring all automobiles to have large enough canisters to capture all fuel vapors. The agency estimated that the additional national cost of $100 million a year would be offset by fuel savings of $600 million a year plus avoided environmental effects.

The difference between internal recycling and onsite waste reuse is subtle but significant. Internal recycling can be done, and usually is, with liquid and gaseous materials. Beneficial waste reuse is usually reserved for liquid and solid materials. With internal recycling there is no consequential movement of waste as waste. With waste reuse, however, waste is physically moved to another process or operation at the location where the waste was produced. Waste reuse may also happen by transporting the waste to another company facility, another company, or a place serving as a transfer point, such as a government or commercial waste exchange. Any form of waste reuse, as for internal recycling, must not include chemical processing or treatment which is or resembles either basic industrial processes or waste treatments that produce environmental wastes. It would be unusual to speak of a waste reuse facility or operation as one would refer to a recycling facility or company. Waste reuse should be a simple movement of waste from the place where it was first produced to a location where it can be used in its original physical and chemical forms. Some pollution prevention purists may assert that waste reuse should not be considered pollution prevention because of the risks associated with the handling and transport of waste, which indeed can be significant. Offsite reuse of hazardous materials is definitely the lowest option for industry and for households within the secondary pollution prevention category.

Another significant concern about waste reuse, both onsite and offsite, is the use of waste as a fuel in boilers and similar equipment. In such cases, the waste substitutes for ordinary fuel usage, which itself produces emissions. Should the waste being reused contain hazardous substances, there is an environmental risk, just as with hazardous waste incineration, and there may be substances in the waste which, although not toxic themselves, may cause new toxic substances to be produced in the burning equipment. Burning of waste is, therefore, *not* considered a secondary form of pollution prevention, but rather a kind of waste management in which the substantial processing converts the latent heat value of the material into usable thermal energy.

A particularly interesting case of reuse of garbage produced by individuals as well as businesses is the reuse of old automobile tires in Cape May County, New Jersey. Since 1986 the county has used over 150,000 tires to make artificial reef in the ocean. This reuse keeps tires out of the landfill and helps the area's sport fishing industry. The tire reefs are made at a state prison. Combined with concrete slabs to anchor them, the tire reefs make excellent fish habitats. Direct observation has confirmed the attractiveness of the reefs to fish. The goal is to use all collected tires for reefs. Generators also have been given an economic incentive to provide tires for the reefs rather than take them to the landfill. For reef use, generators pay $25 per ton instead of charging a $60 tipping fee at the landfill. Nationwide, disposal of tires remains a big problem. Many people want to pursue recycling, which, however, involves substantial processing, such as burning to produce energy or shredding to produce road material. Such processing may produce pollution. Also, some disposal sites have enormous quantities of tires stored above ground; periodically, these sites burn and release air pollutants. Reusing tires, therefore, as tires, such as for ocean reefs, offers environmental benefits.

The more typical reuse of waste in industry is illustrated by the practices at the Mehoopany plant of the Procter and Gamble Paper Products Company. The plant won a 1989 Governor's Waste Minimization Award, employs 3,000 persons, and manufactures sanitary tissues, kitchen towels, and disposable diapers. Major plant activities include wood receipt and preparation, sulfite pulping, papermaking, and conversion into finished products. After various waste prevention actions, 635,800 pounds of dry solid waste are produced each day. The largest operational sources of waste are, in descending order, woodyard (30.4 percent), product conversion (26.3 percent), sulfite pulpmaking (24.6 percent), papermaking (8.3 percent), steam generation (5.2 percent), general services and maintenance (4.8 percent), and wastewater treatment (0.3 percent). Reuse as commodity and raw material onsite and offsite is 26 percent of the total and gives the company a value of $4.4 million annually. Specific examples of waste reuse include wood bark for nursery sales, cellulose fines for cattle feed, wastewater sludge for fertilizer, felts from papermaking for plant road base, air system baghouse dust as a chemical raw material, and ash from the steam generator for a potting soil additive. Plant waste management practices consist of burning of waste (44.3 percent), inplant recycling (17.6 percent), and landfilling (12.1 percent). About half of the landfilled waste is from air and water pollution control facilities. Use of landfilling decreased from 67 percent of the total in 1972 to 12.1 percent for 1989.

Product reuse by consumers has some similarities to waste reuse, which pertains mostly to industry. The attributes of control and avoidance of waste production and management make the reuse of products that would otherwise become garbage an important form of secondary pollution prevention. The critical difference between product reuse and external recycling is that product reuse—like using returnable bottles—involves very little industrial activity that is costly, environmentally threatening, uncertain, and outside the control of the waste producer. Product reuse generally necessitates relatively simple cleaning or repair instead of the major industrial activity of materials recycling. Unlike external waste recycling, therefore, product reuse involves much less uncertainty about the desired outcome.

However, there is some uncertainty because the product may not ultimately be reused. And of course, eventually even reused products break or wear out.

Historically, product reuse generally declined because of competition from low-cost, newly manufactured products, the relatively low cost of waste disposal, and the desire for convenience. The convenience factor is significant because, as with recycling, the consumer must cooperate and take special actions. For many products, all three of these factors have changed over time to favor reuse. But for many other products, the desire for convenience and the low cost of single-use products have crowded out reuse. Interestingly, the advent of recycling has not yet refocused attention on product reuse, which, for households, would require a similar special cooperative effort.

On the other hand, because product reuse does not involve fundamental industrial processes, people can sometimes do it on their own. Many people find creative ways to reuse products; this is particularly true of poorer people, both in industrialized and in developing nations. Even in the United States, people have sometimes adopted extreme and creative measures to reuse products which ordinarily are garbage. Houses have been built largely from used automobile tires and empty metal beverage cans. Some 70 houses in the Taos, New Mexico, area, for example, have been built from tires filled with dirt. Walls of empty beverage cans, in combination with cement, can offer insulation, heat release at night, and low cost. A more prosaic example of reuse in building construction is the reuse of bricks from demolished buildings.

WHY RECYCLING IS GOOD, BUT NOT AS GOOD AS POLLUTION PREVENTION

As the best of the waste management options, recycling deserves support and implementation. In the past few years, recycling of various waste materials, such as newspapers, aluminum cans, plastics, and glass bottles, has received enormous attention and public support. One way to look at recycling is to realize that because the materials are recycled and not thrown away as garbage, those materials are by definition not garbage. But another perspective is that even products containing recyclable materials which have served their intended purpose are garbage once discarded, and they remain waste until they have been converted into a new product. Our point is that a lot of uncertainty separates the original creation of waste and the final use of recycled material.

The difference between being recyclable and recycled is like the difference between being employable and employed, because it is easy to be the former but not the latter. Recyclable materials have to be collected and separated (or kept separate from other materials), and then transported to a facility which is essentially like a basic industrial materials processing plant. In other words, recycling is external to the original waste production location. Finally, to complete the cycle, the new material made from the original waste must actually be used; otherwise, it remains a waste. At any point along the recycling path, the process may fail. For example, a bag of discarded newspaper or cans from a household may not reach the recycling facility, or recycled material may find no buyer and be discarded. Those who produce the original waste are not normally in control of the entire process, nor can they know or be assured that the desired outcome has been attained.

The waste producer, household, or company that participates in recycling is bound to feel good about helping to recycle because it seems more valuable and sensible than disposing of waste. And it is. It is when recycling is compared to pollution prevention that its limits become apparent. Cooperating at the beginning of the recycling process by separating and setting out waste materials for collection *is* commendable, but there is no personal responsibility for carrying out the whole complicated recycling process. If there is a significant chance that the recycling process will not be completed successfully, the cooperative waste producer's efforts have been wasted and the cooperative individual may have been deceived or deluded. If the recycling process is not completed successfully, then the personal effort expended in separating waste materials and assisting in their collection is for naught. This possibility is not unrealistic. In fact, the practical and economic limits as well as the environmental failures of waste recycling are well known. Here are some important examples.

Because recycling depends on having fairly immediate commercial markets for recycled materials, its ultimate outcome is uncertain. There have been many instances, for example, where waste newspaper has been separated, collected, and transported to central collection places but not recycled. In August 1989 it was reported that there was a glut of 1 million tons of collected newsprint nationwide. Sometimes collected newspapers have been disposed of in landfills instead of being remanufactured into "new" paper. Such events may result from having too much wastepaper relative to available capacity at recycling plants. Instead of recycling plants paying for waste, they may require payment for taking it. Or the demand for recycled paper may be low because of low prices for virgin paper, less need for paper by end users, or inferior properties of recycled paper, at least for some uses. Some industries which produce virgin materials may, indeed, be threatened by waste recycling and may cut their prices to offer stiffer competition.

In addition, over the longer term, technical problems may arise with the continued recycling of previously recycled materials. That is, over time, the properties of some recycled materials may degrade, especially relative to those of virgin materials. For example, the fibers in paper become shorter during recycling and, over time, recycled paper becomes weak. In recycling ferrous scrap materials, engineers have been concerned about the buildup of impurities which can weaken and embrittle steel made from scrap. Overcoming such problems is not impossible, but it typically requires more processing and increases costs. The idea of blending virgin and waste materials to help alleviate these problems sounds reasonable, but it may require a mix of technologies which makes recycling significantly more costly.

A more subtle and growing technical problem with recycling is that advances in materials and in the design of products and packaging have resulted in complex combinations of very different materials. A package, for example, may consist of layers of paper, metal, and several kinds of plastics. Recycling of household garbage is simple for simple products and discards which are clearly in a single category, like glass, aluminum, or plastic. But with complex products and packages, the consumer being asked to separate wastes, or even machines separating waste in central garbage collection or management facilities, may not be able to do so. Moreover, if assumptions are made about what a piece of garbage is, and these assumptions are wrong, then recycling operations may suffer serious upsets and inefficiencies. Wastes sent into the recycling system may emerge later as waste.

Another problem is that recycling facilities may themselves be a source of environmental problems, because, unlike waste reuse, material is processed in physical or chemical ways. For example, New York State put its pilot battery recycling project on hold while it awaited the outcome of investigations into pollution around the Mercury Refining Plant in the city of Latham. High levels of mercury were found in the soil around the plant.

One significant historical problem has been facilities which collect large quantities of industrial toxic waste for recycling but let them deteriorate, leak, and contaminate soil and groundwater instead of recycling them. Many federal Superfund sites slated for expensive cleanup started as recycling facilities. Other cleanup sites were created when recycled wastes were placed directly into the environment to serve some purpose. Such waste recycling has sometimes been discovered—all too late—to be the sources of toxic contamination. For example, there have been cases where so-called recycled oil has been used as a dust suppressant on roads, only to be recognized later as a source of dioxin, which requires very expensive cleanup. This was the origin of the Missouri dioxin Superfund sites, which involved extensive relocation of people and government buyouts of their property. Ash from incinerators and cement kilns has been used as aggregate for road construction. But such ash can contain significant amounts of toxic metals which may, with the help of rain—especially acid rain—leach out over time and contaminate groundwater.

Even facilities that recycle all the waste materials they receive may create new pollution and hazardous wastes, just as do ordinary industrial facilities that make materials from virgin raw feedstocks. Toxic chemicals may be needed to reduce recycled materials to usable forms. Sludge, air emissions, and water discharges from processing paper can contain toxic metals from inks or dioxin from previous bleaching operations. A 1982 EPA study found toxic pollutants in paper recycling processes that were not in virgin processing; 12 pollutants were found in both processes, but 11 occurred in higher levels in recycling than in virgin processing. Some recycling facilities take used oil and burn it, which may produce air pollution. This is also true of resource recovery facilities that are really incinerators using the energy content of municipal waste. Recycling facilities which process and purify waste motor oil produce sludges that are hazardous waste when they contain toxic metals.

Automobile shredding facilities have become cleanup sites because of extensive contamination of land by toxic metals and PCBs. Secondary lead smelters emit enormous amounts of lead into the atmosphere and send a lot to landfills. In December 1989, the Lancaster Battery Company in Pennsylvania was ordered to pay a $250,000 fine for pumping lead-contaminated waste into a nearby field. In early 1990, EPA selected the C&R Battery Superfund site near Richmond, Virginia, for a $15 million dollar cleanup. C&R Battery reclaimed lead and plastic from car and truck batteries from 1970 to 1985 and contaminated soil by dumping lead-laced battery acid into ponds. Electric steelmaking furnaces using ferrous scrap produce major amounts of hazardous waste. The lesson to be learned is that even though recycling may reduce pollution compared to making products from virgin materials, recycling may still produce a lot of environmental wastes which true pollution prevention that reduces material and product use could cut.

Producing energy to make virgin materials or process recycled materials creates pollution. And while it is true that recycling often saves energy compared with making the same materials from scratch, especially for recycled aluminum, the comparison should include feasible pollution prevention alternatives. A good example is a beverage bottle. The glass can be recycled and made into a new bottle. That clearly saves energy over manufacturing a new one from scratch. Pollution prevention consists of using a returnable bottle and is better yet. One analysis, for example, found that the use of 10-time refillable glass bottles is over 20 percent more energy efficient than recycling glass into new containers. The use of just a few standard refillable glass bottles for all beverages in the United States would save an enormous amount of money and waste. Companies could put their own labels on standard bottles, as is done in Denmark.

It should also be noted that recycling of materials from wastes of households and offices requires a collection system much like that used for taking wastes to disposal facilities. That collection system typically accounts for about 80 percent of the total waste disposal costs and generates significant pollution from the large-scale use of conventionally fueled trucks.

In general, recycling facilities should be considered like other industrial facilities and regulated appropriately to safeguard health and the environment. However, recycling industries often oppose being regulated by government. One reason is that regulation imposes costs, and recycling often sits precariously on the edge of economic competitiveness.

So many people and organizations have become enamored of recycling that it may seem backward and brainless to criticize it. Perhaps the most serious problem with the current emphasis on recycling is that it takes attention away from the even better solution of pollution prevention. There is enormous popular praise for recycling aluminum beverage cans, but hardly a mention of or question about the use of aluminum cans in the first place. The aluminum can industry started recycling, in large part, to fend off the use of returnable glass bottles.

Although recycling is less advantageous than pollution prevention, compared to other waste management options it is at the top of the list of options. To the extent that people and organizations are enraptured with recycling and committed to it, pollution prevention receives less interest and support than it deserves. Indeed, advocates of recycling may see no immediate or long-term need for pollution prevention, although other proponents seem to understand that recycling is not a panacea.

Recycling of household waste has received much of its backing, particularly from industry, because it seemingly offers the best of two worlds. On the one hand, recycling seems to address the well-publicized problems of municipal waste management capacity shortfalls, such as declining space in landfills. Moreover, opponents to the siting of new landfills and incinerators see recycling as a key practical alternative. By supporting recycling, the opponents to landfills and incinerators argue that new waste management capacity will not be needed. Also, economically, recycling of valuable materials seems to make a lot of sense. But, on the other hand, by stressing recycling of wastes, it may seem that the production and consumption of all the products which eventually make up the garbage to be recycled can go on their merry way.

Manufacturers count on the public's enthusiasm for recycling and to maintain their historic partiality for—or unconcern about—garbage-intensive products and packaging. Consumers who have become addicted to a lifestyle that produces more and more garbage every year can feel good about participating in recycling programs. In other words, by staying with an end-of-pipe solution, it may seem as if our society can continue its historic patterns of production and consumption without any major inconvenience, cost, or disruption. Recycling seems like a solution without much sacrifice. However, an especially important omission of the recycling advocates is their lack of attention to the production of industrial pollution and toxic waste associated with making consumer products, even if those manufacturing processes use considerable amounts of recycled materials. This omission is a lack of consideration of the full life cycle of products, from their initial manufacture to their final disposal.

The point here is not to oppose all recycling but to put recycling in a context which recognizes that, without giving the appropriate priority to pollution prevention, environmental and waste management problems will persist over the long term. Excessive, avoidable garbage is caused by excessive, unnecessary production and consumption. Recycling does nothing to slow those processes. Landfill capacity is likely to continue to decline, and siting new landfills and incinerators will likely remain difficult. Moreover, the increasing use of municipal waste incinerators still requires substantial landfill capacity for the enormous quantities of ash residues produced. Meanwhile, the generation of garbage—without a shift in strategy to pollution prevention—will keep increasing on a per capita basis and with population growth. To these factors should be added the uncertainties of markets for recycled materials and the environmental issues of recycling discussed above. The inevitable conclusion of a comprehensive view of both waste production and management in modern society is that recycling major parts of household garbage and other wastes, such as office paper, is not a panacea for either environmental or waste management problems.

An interesting proof of this viewpoint is what is happening in Japan. Advocates of municipal waste recycling usually point to Japan as a great example of successful recycling. A 1989 *Newsweek* story on garbage concluded: "Yet Japan has not conquered garbage. The overall recycling rate peaked at about 50 percent; it has dropped during the 1980s. . . . Tokyo and three neighboring prefectures will have an excess of 3.43 million tons of garbage by 2005, and might have to ship it elsewhere." Might this be a consequence of the Americanization of Japanese consumerism? It seems so. It should also be noted that Japan has close to 2,000 waste incinerators, while the United States has fewer than 200 large municipal waste incinerators.

A more subtle negative consequence of the emphasis on household waste recycling in affluent societies is that many developing nations will continue to aspire to the American consumption-intensive lifestyle. Many aspects of American culture are exported, as any traveler knows. Although the United States has lost its dominant role in world manufacturing and trade, it remains a potent influence on global tastes and consumer demand—not the least of which is the demand for waste-intensive products and services. However, developing nations may not implement massive recycling programs. A lot depends on the level of wealth. In nations with very poor people in large cities, recycling of municipal solid wastes (but not industrial wastes and air and water pollutants) will probably remain extensive because it is a primary means of work for some people. Other places may landfill or burn large amounts of municipal wastes. Nor are environmental issues always effective in shaping public opinion and driving recycling programs, as they are in the United States.

Moreover, in resource-rich nations, the production of virgin raw materials is an integral part of developing economies. Therefore, there may be significant opposition to recycling programs that take away business from primary industries (especially after recycling in advanced nations has put a dent in primary industries' exports and revenues). Furthermore, if advanced industralized nations do not emphasize pollution prevention and motivate the production of new "green" products for the largest consumer markets, then the world's manufacturers will have little reason to stop producing waste-intensive products. Unless pollution prevention becomes a global strategy, the net result of this global demand-supply condition will be that garbage generation will rise with rising economic development, population growth, affluence, and consumption, and pollution associated with producing consumer goods will worsen.

HOUSEHOLD HAZARDOUS WASTE COLLECTION PROGRAMS

One of the fastest-growing innovations in waste management during the past few years has been the increasing number of community household hazardous waste collection programs. They started in 1981 in a handful of states. By 1989 more than 2,000 programs in 45 states were operating. The basic idea is that households identify, collect, and take hazardous household wastes to a central collection place. Local governments like the idea of diverting hazardous waste from their landfills. Even though, on average, household hazardous waste has been found to represent only 0.3 to 0.6 percent of household trash, equating to perhaps 30 pounds *per year* per average household, it adds up to a significant problem in large urban areas. For example, measurements for New Orleans indicated that about 700 tons of household hazardous waste ends up in landfills there and about 300 tons in Marin County, CA. In Los Angeles, 10 tons or more per day of household hazardous waste might be disposed of in the area's landfills.

Some people think that collection programs have been successful. But the cost of the programs on a household or ton of waste basis is normally very high. Costs per household participant can total $100 to $200. In most cases, less than 1 percent of households have participated, although one program in Connecticut has a 5 percent participation rate and one in San Bernardino, California, reported a 10 percent rate. Significant money must be spent on extensive public education so that people can identify household hazardous wastes, which typically have been set aside for some years but still remain in homes. Such wastes include discarded paints, varnishes, solvents, car batteries, and used motor oil. When households are well informed and motivated enough to partake in a community program, relatively large amounts of accumulated hazardous wastes are collected. Because of accumulation, the amounts collected can be high, often 50 to 100 pounds per household.

The issue we raise is this: What happens to such collected wastes? They are exempt from federal regulation and so, unlike regulated industrial hazardous wastes, they can be disposed of as normal household garbage. Of course, it would make no sense to go to the trouble of collecting household hazardous waste merely to put it into a normal municipal landfill. But what if this waste is managed by a commercial waste management firm and is put into a hazardous waste landfill? To nearly everyone, land disposal of hazardous waste just delays the inevitable contamination of land and groundwater. Indeed, some household hazardous waste collected has been landfilled: 75 percent of Florida's waste from its first ''Amnesty Days'' was landfilled, as well as 23 percent of the waste in San Bernardino County. The best management option would be recycling, but little of that appears to be going on, and it would be difficult because of the complex mixtures of relatively small amounts of wastes.

The next best solution is destruction of the collected household hazardous waste, which today means incineration. There is no other practical waste treatment for the diverse mixtures of toxic wastes collected in such programs. But, here too, most people have serious concerns about the safety of commercial hazardous waste incineration facilities. Proponents of household hazardous waste collection programs like to say that the programs will help people appreciate pollution prevention. On the other hand, the fact that a number of large corporations support these programs suggests that, as with recycling, the real purpose is to help make consumers feel good without threatening their current waste-producing consumer habits. Commercial waste management firms also make money from the programs.

ENERGY CONSERVATION FIGHTS GLOBAL WARMING AND COMPLEMENTS POLLUTION PREVENTION

The pollution prevention theme of this book does not cover all human follies or all threats to people and natural resources. There are a number of serious planetary problems which are not discussed. These include deforestation, desertification, soil erosion, loss of biological diversity, and depletion of nonrenewable natural resources. True, these are human-made problems, and they seem to be environmental in character because the word *environmental* can be interpreted very broadly. But these problems of deterioration and wasting of the planet are not caused by the production of wastes and pollutants and their effect on our ecosystem. Such problems are a direct consequence of misguided, shortsighted, and often greedy human behavior, and they too demand solutions. These problems are best defined as resource conservation issues and not environmental ones, although some of them have environmental consequences, such as global warming. The solution to resource problems is obvious: stop the human practices which are the direct cause of the problems and practice conservation. Theoretically, of course, we can contend that conservation is by definition a form of preventive behavior and a contributor to preventive planetary care.

Conservation and pollution prevention can be distinguished in a technical way. Divide the system up into two parts: human existence inputs from the planet and outputs to the planet. Inputs are all those things that are required directly or indirectly for living, such as food, oxygen, sunlight, water, minerals, plants, coal, petroleum, and agricultural soil. Anthony Lawlor (1990), in a letter to *Time,* gave an elegant description of our relationship to inputs:

> The importance of the planet's health hits home when we realize that there is no clear boundary defining the edges of an individual's life and the environment that sustains that life. We breathe oxygen produced by trees on the other side of the globe, drink water that may have evaporated from the Indian Ocean and eat food grown as far away as Chile and Australia. All this is activated by the sun from a distance of 93 million miles. When, with pollutants and toxic waste, we poison the natural systems that allow us to live, we engage in the worst kind of self-abuse.

Outputs from human existence are a combination of inevitable and preventable wastes or by-products of human life and the complete array of social, commercial, and industrial activities used to support life. Outputs include carbon dioxide, human wastes, heat and myriad industrial, agricultural, and municipal wastes and pollutants. Human life is threatened when there are insufficient necessary inputs and when outputs are harmful to inputs or human health. Therefore, from a long-term perspective, it is critical to conserve inputs, especially ones that are nonrenewable; and from a short-term perspective, it is critical to minimize outputs that are harmful or not converted into inputs.

There are strong connections between energy conservation and pollution prevention. First, conserving energy inevitably means that less pollution is produced, because pollution results whenever energy is mass-produced from coal, petroleum, natural gas, or nuclear reactors. Of particular concern is the direct relationship between global warming and emissions of waste gases from the burning of fossil fuels, principally coal. Atmospheric carbon dioxide, the major greenhouse gas in the Earth's atmosphere, has increased by about 20 to 30 percent in the past century. Worldwide, emissions of carbon dioxide from burning fossil fuels have almost quadrupled since 1950. Fossil fuel–produced carbon dioxide accounts for 70 percent of all human-produced carbon dioxide and is projected to cause over half of the global warming for the next 100 years. Globally, production of carbon dioxide from fossil fuel combustion is over 1 ton per capita.

About 20 percent of global carbon dioxide, however, is produced by the United States. In other words, releases of fossil fuel–derived carbon dioxide in the United States equate to about 4 tons per person, about twice the level of Japan. The developing world produces fossil fuel carbon dioxide at a rate of much less than 1 ton per capita. Increasing world population and economic growth, as well as abundant resources of coal, mean a future of more total global energy use and enormous amounts of greenhouse pollutants. For example, China depends on coal for 75 percent of its energy and expects to double its use of abundant coal reserves by 2000. At present rates of increase, global carbon dioxide will probably increase another 10 percent or more by early in the next century, at which time it is likely that the Earth will be warmer than it has been in the past 100,000 years. Increasing atmospheric concentrations of global-warming gases are certain, but the actual level of global warming remains uncertain. One large uncertainty, for example, is how the oceans may mediate global warming.

Deforestation is generally recognized as a very serious cause of excessive carbon dioxide buildup. Actually, it currently accounts for 20 percent of anthropogenic global carbon dioxide emissions. Removing forests means removing an essential part of the original natural cycle consisting of carbon dioxide formers and, in the case of forests, destroyers. Deforestation in Brazil and some other countries attracts considerable attention. However, at a more local level, consumers and businesses can reduce carbon dioxide by mowing their lawns less often (producing less carbon dioxide from lawn mowers and letting more grass consume carbon dioxide). People and government agencies can also plant more trees and shrubs, not to mention practicing a multitude of energy conservation techniques.

But of course, the degree of pollution varies enormously among different energy generation systems. From a pollution prevention perspective, it is not just a question of how much energy is produced but also of how it is produced. For example, burning natural gas produces less carbon dioxide than burning oil, which produces less than burning coal. In fact, coal releases about twice as much carbon dioxide as does natural gas per unit of energy obtained, with petroleum about midway in between. There is evidence that burning natural gas in new kinds of turbines, a form of waste reduction technology, could increase its carbon dioxide emissions advantage over coal to a factor of 3. Other examples of true prevention approaches to dealing with air pollutants from fossil fuel burning include more use of solar energy for heating and electricity generation, as well as wind and tidal power—approaches which produce no greenhouse gases.

Aside from primary energy production substitutions, which are resisted by the coal and utilities industries, there are other important applications of pollution prevention principles in energy production. A good example is replacement of high-sulfur coal with low-sulfur, or clean, coal, or using different coal-burning processes (such as gasification and fluidized bed combustion), so that air pollution associated with acid rain is greatly reduced when the coal is burned. (But the greenhouse gas carbon dioxide is still produced.) Eastern European industries burn vast quantities of high-sulfur brown coal and require twice as much energy as Western nations to produce a dollar's worth of goods; the result is that Eastern Europe accounts for two-thirds of Europe's sulfur dioxide, which causes smog and acid rain, but only one-third of its gross domestic product. Likewise, in the United States, it has been difficult to overcome the economic interests that benefit from the use of high-sulfur coal. The result has been a less than enthusiastic movement to true pollution prevention alternatives and more interest in end-of-pipe techniques to capture sulfur dioxide and reduce emissions.

In the critical area of transportation, it is now estimated that the global fleet of automobiles will triple to over 1 billion vehicles over the next 30 years. America's 150 million motor vehicles, driven 2 trillion miles annually, account for about 44 percent of the primary ingredient in smog—hydrocarbons. Change is necessary. For automobiles, pollution prevention means less automobile use and more purchases of the highest-efficiency cars by individuals. It also helps to have government requirements which force manufacturers to make cars that have higher fuel efficiencies and use alternative fuels, such as methanol, which, although less polluting than gasoline, produce large amounts of the carcinogen formaldehyde, and ethanol, which is cleaner and is added to gasoline to produce gasohol. However, the American automobile industry resists government regulations requiring cars to get more miles per gallon of gasoline, and the petroleum industry resists alternative fuel requirements. Consider some pollution prevention measures that affect fuel use and pollution:

- Natural gas is the cleanest alternative fuel; the United States has 30,000 natural gas–fueled vehicles compared to 300,000 in Italy. Using natural gas, however, requires significant changes to automobiles and fuel delivery systems.
- A front-end change is the new gasoline commercialized by the ARCO Chemical Company. By using methyl tertiary butyl ether, which in-

creases the oxygen content of gasoline, less carbon monoxide and volatile chemicals are emitted to the atmosphere. This type of octane enhancer is better environmentally than the more often used aromatics, such as benzene, toluene, and xylene. The latter are major toxic air pollutants that contribute to smog and toxic air emissions; they replaced leaded gasoline—a good example of unsafe substitution resulting from a government ban. Aromatics constituted about 35 percent by volume of American gasoline products in 1989, compared to 22 percent in 1971, 1 year after Congress passed the law that resulted in the banning of leaded gasoline. The ARCO product has 50 percent less benzene and one-third less of other aromatics than conventional gasoline. Its introduction into southern California was driven by the state's push to clean up its heavily polluted air. The technology used by ARCO has been available for several years. Although most American consumers have no convenient fuel alternative, they can choose to use lower-octane gasoline that contains a lower level of aromatics, with little if any effect on engine performance.

- For operators of diesel vehicles, the Harrier Diesel Cleaner illustrates an equipment change that can increase fuel efficiency and cut emissions of carbon monoxide, hydrocarbons, nitrogen oxides, and smoke by large amounts (about 50 percent) by electronically metering a water emulsion into the engine.
- Consumers can correct a widespread problem: underinflated tires. It has been estimated that 2 billion gallons of gasoline are wasted each year because of underinflated tires that cause poor mileage efficiency. Soft tires also mean more tread wear, which causes more small particles of tire material to enter the environment.
- Employers and office employees can switch to telecommuting. Commuting to work by car is replaced, at least partly, by working at home and communicating with the office through computer modems, fax machines, and sophisticated telephone message machines. In southern California, curbs on air pollution are pushing companies and workers to adopt this option.

For consumers, an example of combined energy conservation and waste reduction in the home is the use of fluorescent light bulbs to replace traditional incandescent ones. An 18-watt fluorescent bulb produces as much light as an ordinary 75-watt one but uses about one-quarter to one-third less energy. Using the fluorescent bulb prevents the burning of 400 pounds of coal, forestalling the release of an enormous quantity of air pollutants, including about 700 pounds of carbon dioxide and 12 pounds of sulfur dioxide. National use of such fluorescent bulbs could cut several hundred million tons of carbon dioxide emissions annually. Such new types of fluorescent bulbs last about nine times longer than regular bulbs. The consumer who uses the fluorescent bulb will also save money, about $10 to $20 a year per bulb. Note, however, that fluorescent bulbs contain more hazardous material and, therefore, produce a more hazardous solid waste upon disposal.

Another example of meaningful energy conservation that reduces pollution is improving the efficiency of household refrigerators. If 100 million U.S. refrigerator owners cut their electricity use by 10 percent, consumers would save $700 million annually and the emission of 8 million tons of carbon dioxide and 60,000 tons of sulfur dioxide would be eliminated. Such an improvement is a near-term possibility.

Pollution prevention leads to energy conservation because it nearly always means that raw materials have been used more efficiently. This, in turn, usually means that less energy has been used to produce raw materials and products. Later, when less pollutants and waste are produced, less energy is used to operate pollution control and waste management systems or to clean up environmental problems. A study by the Organization for Economic Cooperation and Development of 200 French companies that practiced waste reduction found that 51 percent also saved energy and 47 percent saved raw materials, which meant more energy savings overall. Often waste reduction projects consist of internal plant recycling of materials with enough heating value to substitute for regularly used fuels. For example, a 3M plant manufacturing industrial tapes installed a pipe to catch solvent vapors being released to the atmosphere. The solvent is sent to plant boilers as supplemental fuel. Modifying the boiler facilities cost $270,000 but saved $155,000 in energy costs alone in the first year.

We conclude, therefore, that energy conservation is a close counterpart to pollution prevention. Promotion of energy conservation by citizens, industry, and government assists pollution prevention, and vice versa. Dealing with global warming definitely requires massive global energy conservation and a switch to lower carbon content fuels, because 70 percent of carbon dioxide emissions result from energy production. Technologically, it is feasible to achieve all the energy conservation that is necessary to cope with global warming, even with population and economic growth. The U.S. use of energy per dollar of GNP is about twice as high as that of Sweden, France, and Japan. But then again, the People's Republic of China and Hungary are using energy at twice the rate of the United States.

After examining a global warming prevention program, G.R. Davis of the Shell International Petroleum Company concluded (1989) that its "resource allocation could be handled without crippling economic growth." Furthermore, "the costs for prevention programmes could appear high but the benefits of effective joint action may well be higher. If nations can forge a common vision and purpose to tackle this problem, then a basis for progress on a wider agenda may have been laid.''

However, it is critically important to also emphasize that pollution prevention will not be maximized by relying on energy conservation efforts. Pollution prevention must receive primacy for its own environmental and economic benefits because energy conservation for its own sake may not always have government, industrial, or individual support, as history has shown. Nor can energy conservation address all environmental problems. History has shown that low market prices for energy and abundant fossil resources in some parts of the world pose serious problems for energy conservation.

Also, a large fraction of pollution prevention measures are likely to have rather small energy conservation benefits relative to primary environmental ones. Such pollution prevention actions would not, therefore, be taken if only energy conservation guided decisions. The energy conservation benefits of pollution prevention actions are best captured as a specific economic benefit and as a way to explain full life cycle benefits, such as for manufacturing products. Indeed, it is the reduction of energy and materials use and the pollution associated with their production that must be invoked for reducing the production of ordinary household garbage, even if it is free of toxic chemicals.

STEWARDSHIP AS MORAL RESPONSIBILITY

The strategy called *stewardship* or *conservatorship* receives relatively little attention. Some environmentalists, industrialists, and others speak about the need for people and institutions to accept their role as stewards of the planet. What does this mean? Being a steward means being a custodian, guardian, or trustee of the planet and of its ecology and natural resources. The key point is that being a steward does not imply the means by which the role of custodian, guardian, or trustee will be implemented. Stewardship is a statement about ecological and moral responsibility with which few people would probably disagree.

The flexibility to interpretations of what stewardship means in itself makes it attractive, especially for those threatened by activist environmentalism. Indeed, one can argue that the traditional end-of-pipe environmental approach could be used as part of the "good management" that this approach signifies, because stewardship focuses on conservation of inputs and not production of outputs. People who advocate the old approach to environmental protection can believe that making pollution control and waste management more effective is completely consistent with the goal of stewardship. Stephen Muratore (1985) has described the inadequacy of contemporary views of stewardship:

> Even at its highest spiritual expression, our ecology movement and our understanding of stewardship has very little to do with traditional stewardship as it has been revealed through the ages in various cultures and societies. We compliment ourselves on thinking that we practice stewardship when we put some bandaids on problems such as attaching pollution control devices to our cars. We think that's stewardship. Or when the Environmental Protection Agency comes out with a statement that the proposed new dam we're going to build has a "tolerable" environmental impact, we say, "good stewardship!" None of these approaches actually addresses the fact that the methods, technologies, and ways of thinking that we are employing, as a matter of course, are themselves destructive at the core. So we put on control devices and control legislation to try to channel the effect, but we have never addressed the root. If we want verification or if we want to examine how we are doing with our stewardship, it seems the most obvious place to look is to Nature herself. . . . Nature is showing us that there is a disease rampant upon the face of the earth, of incredible magnitude, and it is not a minor sickness but has all the signs of a terminal illness for the planet.

Another issue is that stewardship opens the door to arguing about the exact impact of pollution and wastes, if any, on inputs necessary for human existence. Proponents of stewardship are more likely than green environmentalists and other environmentalists, for example, to offer analyses about effects, costs, and benefits. And such analyses are used to support the proposition that it is feasible and correct to maintain current products, processes, and practices because of an adequate supply of inputs and the ingenuity of scientists and engineers. As the director of the OTA, John H. Gibbons (1988a), explained:

> I think energy can enhance both [standard of living and quality of life], if we apply wisdom and thoughtfulness to how we produce, convert, and use it. That's what we call the conservator society . . . it's quite possible to get what we want from our resources by using them more elegantly, in ways that minimize their net flow through society. We don't need to have all the waste. We *do* need to look at *total* cost . . . future generations aren't going to have the rich inheritance of natural resources that we've known. But we can leave them rich technological capabilities to offset that loss.

Gibbons (1988b) has also said, "There is no fundamental dichotomy between conservation and growth. Indeed, both our own economic future and development in the Third World will depend on careful stewardship of resources."

Stewardship is, in other words, amenable to interpretations that make it less threatening than pollution prevention to existing institutions and economic interests. Consider the following two excerpts from advertisements: the first by a company in the consumer product business and the second by a major producer of industrial chemicals. First, from Dow Chemical Company:

> You get the commitment of ChemAware product stewardship, a systematic program to help you inform your employees how to properly buy, process, handle and dispose of our products. ChemAware is part of the Chemicals and Metals Department's ongoing product stewardship program to help customers respond to the rapidly changing safety, waste disposal and regulatory environment. The costs of proper disposal can be controlled. Call . . . to get more information on ChemAware. Before you have a problem. Before your business goes down the drain.

Second, from Anheuser-Busch Companies:

> As our nation moves into the 21st century, American indsutry will be called upon to meet ever higher standards of environmental stewardship. All companies must accept the challenge—being mindful not only of their regulatory obligations and controls, but also of promoting environmental awareness in everything they do . . . so must we adopt an uncompromising stance when it comes to protecting the environment. . . . People deserve a quality environment now, and in the future.

The word *stewardship* appears in both advertisements, but the first message does not contain useful information about the meaning, goals, or effects of "environmental stewardship." In fact, Dow Chemical is one of the most progressive and committed U.S. companies when it comes to waste reduction within its own facilities, but the advertisement is aimed at its customers. The second advertisement defines "product stewardship" only in terms of buying, using, and disposing of products, but there is not one mention of the concepts of conservation, prevention, or reduction. At about the same time that the Anheuser-Busch advertisement appeared, the following news report (*Pollution Engineering* 1989) of a classic end-of-pipe illegal act by Anheuser-Busch also turned up:

> New York Attorney General Robert Abrams has announced that Anheuser-Busch, Inc. has pleaded guilty to a series of water pollution violations at the brewery's water treatment plant in Baldwinsville, NY and is required to pay $1 million in criminal fines and civil penalties. Reportedly, A-B employees discharged phosphorus-contaminated water into the Seneca River and provided de-

ceptive sampling data to the N.Y. Department of Environmental Conservation. According to the State, Anheuser-Busch employees kept secret from DEC a hose which illegally bypassed critical portions of the water treatment system.

True believers in stewardship may also concentrate on population growth as the key problem and foster birth control or other means to limit population growth, perhaps only in some parts of the world. Like a focus on energy conservation, and probably more so, requiring population control can make it difficult to achieve broad public support for and cooperation with other actions. Finally, as with energy conservation, any strategy based on conservation faces stiff opposition because there have always been scholars and experts who maintain that cries of resource depletion have been made for many decades and have never come true. Technological optimism in this area means that more reserves of resources will be found, that material substitutions will reduce demand, and that recycling will increase as reserves decrease and prices rise.

WHAT IS SUSTAINABLE
ECONOMIC DEVELOPMENT?

We have saved for last the newest grand strategy, and one that has quickly received remarkable support worldwide: *sustainable economic development*. Some of the world's most prestigious organizations and individuals have embraced and endorsed this strategy. Many governments have formally adopted it as public policy. There have been few if any public arguments against sustainable economic development. Our aim is not to discredit it but to explain how it differs from the pollution prevention strategy. Sustainable economic development is likely to prove ineffective because it will not necessarily result in near-term pollution prevention or energy and materials conservation in either industrialized or developing nations.

An important problem with sustainable development is nailing down its meaning. Consider the following three views:

1. The World Commission on Environment and Development's 1987 report *Our Common Future* (often referred to as the *Brundtland report*) defined sustainable development as the set of changes that "meet the needs of the present without compromising the ability of future generations to meet their own needs."
2. William D. Ruckelshaus—more noted as a past administrator of the EPA than for his current position as chief executive officer of Browning Ferris Industries, Inc., one of the world's largest commercial

waste management companies—said (1989) that "economic growth and development must take place, and be maintained over time, within the limits set by ecology in the broadest sense—by the interrelationships of human beings and their works, the biosphere and the physical and chemical laws that govern it."

3. The New Zealand government said in 1989: "Sustainable management means managing our use of the environment so we don't end up with species extinction, over-exploitation of resources and expensive pollution clean-ups. . . . It means looking after the environment so it can continue to support us and provide us with economic and social benefits. . . . This doesn't mean we have to shut up shop and stop using resources. But what we have to do is think more carefully about the impacts of our decisions and what they will mean in the long term."

These definitions are difficult to disagree with because they are statements of idealistic philosophy and morality rather than pragmatic plans for comprehensive action. But, historically, a philosophy of concern for others and a moral valuing of others have rarely controlled enough human behavior to curb immediate human appetites which translate into consumption of manufactured goods and natural planetary assets. This is not merely a result of a complete lack of concern for others in societies, but also a reflection of how various kinds of thinking lead to decisions which seem to make sense for the short term but are lethal in the long term. Such thinking includes:

- a reductionist view of ecology which divides the world into parts as if they do not affect each other;
- technological optimism about future increases in resources and solutions to health and environmental problems;
- a conservative view of what constitutes scientific evidence of environmental effects that asks for research and studies instead of action;
- a utilitarian view of resources which sees "common" resources, such as air and water, as available for use by anyone; only counts animals, plants, minerals, and other natural resources consumed directly by people; and ignores effects on "unnecessary" ones or those for which substitutes are conceivable.

The difficulty of shifting decisions from a short- to a long-term basis that is an essential component of sustainability has also been recognized by Mike Flux, environmental adviser to the giant ICI chemicals company in England. "There's an inherent conflict in time-scales. The idea of sustainable development is something I can accept philosophically, but the time-scale can be measured in decades, even generations. Even oil, for all the scares, is a resource which will prove finite outside our lifetimes. The facts are that most people in industry operate on a two year time-span."

It is not only big industries in affluent nations that have limited vision. Poorer nations, which often have considerable foreign debt and a population that is struggling to survive and prosper, often use natural resources with little regard to long-term effects or even short-term environmental ones. If decisions throughout the world are made with a limited view of the future, and if there are abundant reserves of many natural resources, such as fossil fuels, then will decisions compatible with sustainability really be made before global environmental catastrophe occurs?

A reason to be concerned about the current frenzy of support for sustainable economic development is the malleability of the meaning of *sustainable development*. John Pezzey, a consultant at the World Resources Institute, has observed (1989) that "even ardent advocates of sustainable development admit that the concept has numerous, wide-ranging and often rather vague definitions—when I stopped counting in August I had reached over 60." Pezzey goes on to say, "It is a beguiling type of jargon that is enabling people to jump on a bandwagon without really changing their style at all. Technically, sustainable development is a broad concept and it's hard to prove the specific policy required to do it."

Many people in American industry interpret sustainable development as something for developing nations but not for the United States. For example, Robert A. Anderson (1989), of the Weyerhaeuser forest products firm said, "I frankly relate it to developing countries. If you look in terms of priority, major problems are in Third World countries."

The views of other industry people reveal the historic resistance to major changes, especially about alternative energy sources necessary to address global warming. James C. Leathers (1989), of the Duke Power Company, said, "In applying the term to us, sustainable development says we would generate more power from solar, wind and the like. But in the near term, those things are not technically and economically where they must be to be competitive." Similarly, Betsy Ancker-Johnson (1989) of General Motors said that alternative renewable energy sources were either "not economically feasible" or "not practical to supply now."

Many environmentalists and some professionals have a much more positive evaluation of energy alternatives. A 1990 report by the Union of Concerned Scientists concluded not only that increased use of renewable energy sources is the solution to the global warming problem, but also that such alternatives as solar and wind energy cost the same as new fossil fuel power plants. Michael Brower (1990) said, "Although some technical issues remain to be solved, there appear to be no insurmountable barriers to prevent renewable energy sources from eventually meeting most, if not all, of US and world energy needs."

Another, industry viewpoint, from Philip X. Masciantonio (1989) of USX Corporation, is: "In contrast to the belief that industrialization was the pump that primed environmental degradation, the Brundtland report predicts that economic development (industrialization) is the engine that drives global environmental improvement." In suggesting that future economic growth is necessary to pay for environmental improvement, it is crucial to also commit to a redesign of industrial technology in all its forms.

Sustainability advocates seem to be sending two key messages: (1) that industry can keep on expanding and (2) that consumers can keep on consuming because industry will be cleaner and will provide green products. To be cleaner should mean adoption of pollution prevention. If that is the case, then industry needs to articulate its pollution prevention actions and impacts and demonstrate its commitment. The danger is that industrial growth may not be based on the cleanest possible technologies and products.

As for so-called green products, they may have little benefit for long-term sustainability, just as food products with a minuscule amount of oat bran do not really provide health benefits. There are, in other words, good reasons for environmental activists and members of green organizations to be skeptical about the movement behind sustainable economic growth. Sandy Irvine (1989), a member of the Gosforth Green Party in northeast England, has made these insightful observations about the sustainable economic development movement: "belief in sustainable growth is no different to a belief in perpetual motion. *Sustainable growth is a contradiction in terms.* In practice, 'sustainable' production has been responsible for displacement of human communities, intolerable health hazards, much cruelty to animals, and destruction of other species in general." Irvine also notes that the " 'scientific management' of America's forests, for example, justified the replacement of old woods by uniform 'tree factories' in which toxic spraying has threatened wildlife and humans alike."

In other words, the commitment of industrialists and politicians to economic sustainability may reveal a very different set of environmental values and priorities than those of activist environmentalists. And so, we ask the next question.

CAN SUSTAINABILITY BE TRANSLATED INTO EFFECTIVE POLICIES?

In the prevalent definitions of sustainable development that speak of concern for the future, there is little to disagree with. In fact, it is fair to say that such noble thoughts have been conceived by many people throughout recorded history. American Indians, for example, certainly expressed and tried to live by the philosophy of sustainable economic development because they were in harmony with nature and had an inordinate concern for future generations.

One of the more remarkable prophetic statements was made by Chief Seattle, leader of the Suquamish tribe in the Washington territory, in a 1854 speech given at the time Indian lands were transferred to the federal government: "We know that the white man does not understand our ways. . . . He treats his mother, the earth, and his brother, the sky, as things to be bought, plundered, sold like sheep or bright beads. His appetite will devour the earth and leave behind only a desert. . . . Continue to contaminate your bed, and you will one night suffocate in your own waste." No doubt other cultures also tried to live within their own limits without jeopardizing their future or that of others.

Sadly, the obvious lesson of history is that societies which understood and tried to implement sustainable development have been replaced by societies with an unsustainable lifestyle. Indeed, Ruckelshaus (1989) acknowledged exactly this ugly truth: "In the industrialized world, unsustainable development has generated wealth and relative comfort for about one fifth of humankind, and among the populations of the industrialized nations the consciousness supporting the unsustainable economy is nearly universal." Ruckelshaus goes on to link the traditional approach to environmental protection with unsustainability:

> With a few important exceptions, the environmental protection movement in those nations, despite its major achievements in passing legislation and mandating pollution-control measures, has not had a substantial effect on the lives of most people. Environmentalism has been ameliorative and corrective—not a restructuring force. It is encompassed within the consciousness of unsustainability.

This is a remarkable statement of the ineffectiveness of the dominant environmental protection strategy of today. It also helps explain that the dominant use of unsustainable development creates illusory wealth at the expense of people in other places and other times.

To make sustainability an effective strategy, it is necessary to bring current environmental destruction and long-term costs into decisions. This has been a major failure of the traditional and still prevailing end-of-pipe strategy. If, historically, the multi-decade-long success of the industrialized modern world has been dependent on *unsustainable economic development,* then is it realistic to seek a near-term shift to sustainable economic development? The thesis of this book is that a shift to a pollution prevention environmental protection strategy is far more feasible and implementable than an overt attempt to change Western and developing nations to a sustainable consciousness which does not pinpoint implementation. This is so because pollution prevention is a concrete set of actions. A commitment to pollution prevention is less grandiose and ambitious than an attempt to restructure global societies along the lines of sustainability. Sustainability is more noble philosophy and goal than near-term, pragmatic means. To take sustainability seriously and to implement it *immediately* is literally like turning industralized societies upside down and inside out. Pollution prevention can start the conversion of polluting, consuming industrialized nations into far leaner and cleaner societies, and it can ensure that industrialization in developing nations does not follow the unsustainable style.

In sum, sustainability lacks short-term feasibility and may be a ploy that allows the "haves" to keep what they have, without necessarily practicing pollution prevention, and prevents the "have-nots" from becoming "haves." It is the longer-term nature of sustainability and its easier application to less developed nations that appeals to the establishment more than to environmental activists. Its attractiveness is that politicians can mold sustainability to suit many different interests without calling for immediate actions that threaten the status quo. Conversely, there is considerable evidence, which will be discussed later in this book, for the near-term feasibility and practicality of pollution prevention.

Advocates of sustainability talk about specific measures. But when they do, they do not speak of the shift to pollution prevention. To the contrary, there is considerable talk of using market and regulatory incentives to achieve better environmental protection, and not of government actions such as banning of toxic chemicals and products and a requirement of waste reduction. One such favored government measure is called *emissions trading*. This allows industrial producers of pollution and waste to sell their "pollution rights." That is, pollution below some level would be marketable. The theory is that those industrial polluters which can profitably reduce pollution will do so, and those which cannot can buy the rights from those which can. Some problems with this market approach seem evident, other than the fact that it explicitly denies the principle of pollution prevention that *all* environmental wastes should be eliminated or reduced.

For example, the scheme seems more like a strategy to redistribute the generation of a given amount of pollution, without concern about unacceptable effects from high-pollution producers. Also, there would be an economic incentive for firms to shut down industrial activity or move it to another country and gain the profit from selling their pollution rights asset. Another issue is whether waste generators who, theoretically, cannot afford to reduce pollution can afford to buy the right to pollute from others. More fundamentally, what would be the critical levels of pollution below which some producers could sell pollution rights and above which other producers would need to buy them? This issue seems like a classic pollution control regulatory dilemma. There is every reason to think, based on the implementation of current end-of-pipe regulatory programs, that the critical levels for individual pollutants would be too high and that levels would not be set for many pollutants.

We end this discussion of sustainable economic development by citing the environmental statement of Pope John Paul II made at the end of 1989. Advocates of sustainability would do well to heed these wise words in designing specific implementation measures and evaluating progress:

> Faced with the widespread destruction of the environment, people everywhere are coming to understand that we cannot continue to use the goods of the earth as we have in the past. . . . Simplicity, moderation and discipline, as well as a spirit of sacrifice, must become a part of everyday life, lest all suffer the negative consequences of the careless habits of a few. . . . The newly industrialized States cannot . . . be asked to apply restrictive environmental standards to their emerging industries unless the industrialized States first apply them within their own boundaries. At the same time, countries in the process of industrialization are not morally free to repeat the errors made in the past by others, and recklessly continue to damage the environment through industrial pollutants, radical deforestation or unlimited exploitation of non-renewable resources.

SUMMARY

For those who believe in the need for new environmental solutions, different strategies have been conceived of and compete with each other to some extent for the attention and commitment of people, organizations, and governments. Nearly all the main strategies, however, can be seen as complementing each other. For example, stewardship is a good statement of moral responsibility. Energy and materials conservation are fitting strategies to safeguard life-sustaining inputs. Pollution prevention can avoid problems resulting from excessive loading of our planet with wastes and pollutants. Sustainable economic development is an admirable general statement about a desirable equilibrium state with considerable focus on the long term.

While all of these approaches are consistent with preventive planetary care, the goal of this book is to make the case for immediate pollution prevention as the key pragmatic approach. People and organizations that emphasize one of the other strategies cannot be assured, or assure others, that their efforts, even if successful, will result in maximum pollution prevention. This is a particularly important concern about the sustainable economic development movement.

The Dutch government has shown that it is possible to accept the goal of sustainable development and to recognize fully the unique benefits of pollution prevention. The Dutch have distinguished between two paths to sustainable development—traditional economic growth based on end-of-pipe and pollution control techniques versus economic growth stressing the use of environmentally friendly or clean technologies and products. Moreover, they have committed themselves to a shift to pollution prevention and have reached an important conclusion. "[W]ithout structural changes in consumption and production processes, sustainable development is out of reach" (Lander and Maas 1990). The United States and other nations could benefit greatly from the advanced thinking and work in the Netherlands.

The green movement is hard to define, but generally, green organizations throughout the world have a number of environmental, conservation, social, and political objectives. In the United States, traditional environmental organizations dominate and generally have narrower interests. The pollution prevention approach of this book is compatible with most green and environmentalist views, but it does not tackle a host of green concerns outside of basic environmental problems associated with the effects of wastes and pollutants. Many products are being given the green label by manufacturers, environmental groups, and government agencies, but a good fraction of them are rejected by authentic green activists. For example, green organizations may have certain social or political criteria for what constitutes a green product, such as work conditions in the original manufacture of the product or the political system in the country of manufacture. A company that manufactures a product may have policies or activities that disqualify its products from being considered green, regardless of facts about the product's relationship to wastes and pollutants. The use of certain animals or plants may be deemed unacceptable as well. Although, from a pollution prevention perspective, a product may have produced little waste in its manufacture, contain no toxic chemicals, and produce a minimum of postconsumer household waste, non-environmental factors may be used by zealots to refuse green status.

But a larger problem is that products labeled or identified as green may not have resulted from the maximum application of pollution prevention principles. For example, a so-called green product may itself contain no toxic chemicals, but its manufacture may have required the use of toxic chemicals and may have produced hazardous wastes or pollutants, including greenhouse gases which get relatively little attention when consumer products are discussed. In other words, the full life-cycle environmental aspects of products must be but generally are not considered in detail. Additionally, such misleading green products may eventually lead to the production of a relatively large amount of household waste compared to alternative products. Another problem is that a seal of approval may evaluate the product but not its packaging.

3

ACHIEVING SUCCESS
BY OVERCOMING OBSTACLES

No matter how good an idea is, many obstacles have to be overcome to obtain all feasible and readily attainable benefits. Ignoring obstacles to pollution prevention only reduces the likelihood of wide-scale success. A February 1990 report on waste reduction by the National Research Council recognized the presence of "important institutional and behavioral barriers to waste reduction. Evne when presented with information purporting to show benefits to self-interest, however, many waste generators evince surprisingly little interest in change."

Much of this chapter discusses industrial practices, including many specific examples showing how obstacles to pollution prevention can be overcome. This will benefit people who work in all kinds of industries and commercial or institutional settings. However, people who do *not* work in industry can also profit from learning about the industrial situation. For consumers, for participants in public interest organizations, schools, and the political system, and even for investors, the knowledge presented in this chapter can affect consumption of products, expectations of industry, and the ability to evaluate information from industry. The chapter ends with a description of an unusual and exciting case of citizen activism that resulted in major industrial waste reduction of benefit to the local community. People who work in industry can learn a lot from this story because it illustrates the new grass-roots environmentalism, especially as it applies to obtaining greater pollution prevention instead of traditional end-of-pipe techniques.

THE IMPORTANCE
OF UNDERSTANDING OBSTACLES

We focus on obstacles to waste reduction not because we see them as insurmountable but because they can be overcome *if they are understood and confronted.* For decades, progressive people have thought and written about pollution prevention, often using other names for it. But the massive amounts of wastes being produced today (described in the next chapter) and the escalating environmental problems reveal how ineffective past pollution prevention efforts have been. There has been more than talk. What seems, and is in fact, a lack of waste reduction success on a broad scale obscures smaller pockets of success. There have been years of successful experience with pollution prevention in industry and by consumers because there have always been individuals who, for a variety of reasons, have persevered.

Hundreds of examples of successful waste reduction help illustrate the technical feasibility and economic benefits which can be obtained on a much larger scale. To illustrate the breadth of industrial waste reduction efforts see Table 3-1, which is a summary of a study of some 500 cases by Donald Huisingh, one of the true pioneers in pollution prevention. We can hypothesize that perhaps the world will slowly move toward optimum pollution prevention. But why wait 50 years when the benefits of pollution prevention are needed now? In a world of industrial growth, population growth, changing industrial technologies and products, and newly discovered interrelated environmental problems, small successes and incremental progress are not likely to prevent global environmental threats. Resistance to pollution prevention in industry often takes the form of arguing for a slow enough pace to avoid disruptions, to allow the use of invested capital, and to allow more time for research. While there is merit to many of these considerations in specific cases, waiting for pollution prevention is the same as ignoring the host of environmental problems that need urgent attention. The better position to take is that *wastes can be eliminated or reduced until proven otherwise.* Part of knowing whether a waste producer has thoroughly examined pollution prevention is to understand the full range of obstacles to it and the means used to overcome those obstacles.

People who come to believe in the urgent need for pollution prevention must, therefore, understand that in a world in which the *opposite* of pollution prevention is the dominant practice for workers and consumers, it is necessary to identify and understand obstacles to pollution prevention in order to keep them from stopping, slowing, or limiting progress. That is an important objective of this chapter. Based on a decade of studying both industrial and consumer pollution prevention, we have constructed a framework for describing how people and organizations encounter and overcome obstacles. What we have done is to model actual practice and simplify the stages that people go through. Our model necessarily simplifies complex industrial and consumer actions, but this approach helps us identify and discuss the practical problems facing consumers and workers.

The basic idea underlying our approach is that all pollution prevention efforts start out easy and become progressively more difficult over time. In fact, the first rule of pollution prevention practice is that it is easy to accomplish a little but remarkably more difficult to maximize results. This characteristic is both positive and negative. It is positive because it is easy to ac-

TABLE 3-1. EXAMPLES OF SUCCESSFUL INDUSTRIAL WASTE REDUCTION

Industry	Method	Percent Reduction	Payback time
Pesticide production	Separation of plant wastes	100% of dust	10 months
Farm equipment	proprietary	80% of sludge	2.5 years
Automotive equipment	Pneumatic cleaning process	100% of sludge, water	2 years
Microelectronics	Vibratory cleaning process	100% of sludge	3 years
Paint, coatings	Pneumatic cleaning process	100% of solvent, paint	< 1 year
Leather making	Ion-exchange, adsorption	99% of chromium	2.5 years
Pharmaceutical production	Water-base solvent replaced organic solvent	100% of solvent	< 1 year
Organic chemical manufacturing	Adsorption, scrap condenser, conservation vent, floating roof	95% of cumene	1 month
Photographic processing	Electrolytic recovery, ion exchange, adsorption	85% of developer; 95% of fixer, silver, solvent	< 1 year
Equipment manufacturing	Ultrafiltration	100% of solvent, oil; 98% of paint	2 years

Source: Huisigh (1989).

complish something at first, and people can verify the feasibility, practicality, and benefits of pollution prevention techniques. No amount of waste is too small or insignificant. The accretion of small successes is part of the secret of successful pollution prevention. The negative aspect of easy early success is that after positive results are obtained, there is uncertainty about whether preventive efforts will continue. After achieving some benefits and feeling good about their results, people may, and often do, stop their efforts because they feel blocked and become unable or unwilling to prevent more waste and pollution. Some people simply feel that they have done enough at that point. Continued pollution prevention progress entails more commitment, discipline, and work; taking more risks; changing more old habits; and perhaps spending more money before realizing longer-term economic benefits. There is, in fact, no end to pollution prevention, just as preventive health cannot stop on the assumption that the benefits will keep accruing automatically. With changing products, technological innovations, and new circumstances, it is necessary to continue pollution prevention. Otherwise, the whole system will slip back into its historic wasteful, polluting ways and environmental problems will persist.

LEARNING FROM SUCCESS AND LEARNING TO QUESTION DATA

A remarkable record of long-term waste reduction is shown in Figure 3-1. For more than 80 years production of sulfuric acid at the Leverkusen plant

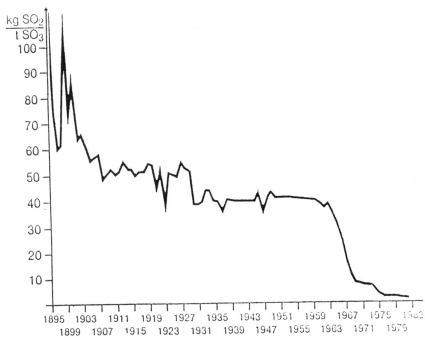

Figure 3-1. Reduction in sulfur dioxide emissions from sulfuric acid production plant in West Germany. (*Courtesy of Bayer-AG*)

of Bayer AG in West Germany, emissions of sulfur dioxide per unit of output have decreased. The point is that, at any time, engineers may have believed that they had optimized the process. But major waste reductions went on and on for a variety of technological, economic, and regulatory reasons. Overall, waste generation dropped from about 100 to about 2 kilograms of sulfur dioxide per ton (kg/t) of sulfuric acid, for an aggregate 98 percent reduction over some 80 years. A 50 percent reduction in waste output sounds impressive. For the Bayer plant, there were actually four times when a 50 percent reduction relative to past waste generation took place.

The first 50 percent reduction, from about 100 to 50 kg/t, took about 10 years; the second, from 50 to 25 kg/t, took about 50 years; the third, in the early 1960's, from 12 to 6 kg/t, took only a few years; and the last, from 6 to 3kg/t, took about 10 years. Is another 50 percent reduction possible? Why not? Certainly, whatever obstacles existed in the past were overcome.

Citizens who demand more waste reduction from industry and technical people who work in industry need to be aware that continuing waste reduction progress cannot be ruled out, regardless of past accomplishments. Indeed, the burden of proof should be on the generators of environmental wastes to show that they are continuing to pursue pollution prevention. It is fairly common today for waste producers to have waste reduction goals that seem very high, like 50 percent. However, a 50 percent reduction relative to some base year does not rule out another major reduction in a relatively short time.

It is also important that the data reported be calculated and presented as a quantity of waste per some measure of output from the facility. Waste quantity may be in units such as pounds, tons, kilograms, or cubic feet. Output is much more complicated because facilities which produce environmental wastes also produce a very broad range of products. Output, therefore, may range from something sinple, like pounds, tons, or kilgrams of product, to units such as number of automobiles, number of television sets, number of airplanes serviced, square feet of surfaces electroplated, and so on. *Warning: any kind of data that purports to describe waste reduction or pollution prevention only in terms of changes in waste outputs, such as pounds or tons of wastes or pollutants, cannot be relied on.*

Many different factors can cause a decrease or increase in waste outputs which masks what really has happened from a pollution prevention perspective. The chief problem is a change in the level of production output, such as the quantity of chemicals produced at a plant. If production of a valuable product decreases, then total facility or company waste generation will decrease, even if no change in process technology or operations have been intentionally made in order to reduce the amount of waste per unit of product. Or a large company might close a plant or end a specific production line, change the kind of product made, or take advantage of a regulatory loophole and remove a waste from the category being described. Such events decrease total company or plant waste generation and may seem beneficial but may only transfer waste somewhere else. To cite an extreme example of a change in the industrial base that is not pollution prevention, even though it does lead to reduced pollution, if all manufacturing and use of automobiles in the United States stopped, there would be a gigantic waste reduction. But that is not what pollution prevention is about, although some radical green activists may believe so. Such a change is not a restructuring and restyling of American industry; it is a destruction of it—and the latter is *not* what we are suggesting.

Although changing or eliminating products is sometimes necessary, the more important way to think of waste reduction is that the efficiency of activities is increased by reducing the *relative* amount of environmental waste generated. *The general objective of pollution prevention is to reduce the unit amount of waste generation so much that even if the level of facility product output increases, the total facility generation of environmental waste decreases.* Likewise, at a national level, pollution prevention will be a genuine success if all levels of wastes and pollutants decline even if population, personal consumption of goods, and industrial output increase. A token or inadequate amount of pollution prevention does not stop increases of total company, industry, national, or global waste generation.

STAGE 1:   ATTENTION AND COMMON SENSE GET FAST RESULTS

The key aspect of the first stage of pollution prevention is that workers and consumers can exploit readily visible, easily implemented, low-cost, and low-risk opportunities. Over the past several years industry has gained considerable experience in promoting waste reduction. One of the most important lessons learned is how easy waste reduction is at first. Having waste reduction in mind, workers and managers can walk through industrial operations and spot pollution prevention problems that can be corrected in days and weeks, without any doubt about the economic benefits. In industry and commercial establishments, neither technology nor capital stands in the way. Studies, engineering analyses, and testing are not necessary. Early actions generally involve changing procedures rather than core production technology, major equipment, or products.

Much of the time, all that has to be done is to stop using waste-intensive practices which, upon reflection, are unnecessary in whole or in part. For example, people in industry have described many cases of reducing the flow of water in cleaning operations involving toxic chemicals, covering vessels containing volatile chemicals, improving inventory controls to eliminate outdated chemicals which become hazardous wastes, reusing off-specification products (like paints) in the original process, and replacing water or solvent cleaning of equipment with other methods. There have been instances where workers showed no restraint in throwing away shop rags contaminated with solvents. But when it was realized that a drum of such rags was costing hundreds of dollars to dispose of as hazardous waste, practices changed quickly. The Adolph Coors Company replaced many chemical solvents with biodegradable cleaners, but not until after the solvent trichloroethane leaked in 1981 from its plant in Golden, Colorado, and contaminated two of its groundwater wells used for brewing.

Stanadyne, a manufacturer of plumbing equipment, was producing hazardous waste from an electroplating line. A significant part of the waste resulted from plating parts which were found to be defective after plating. The simple waste reduction solution: inspect parts *before* plating rather than afterward. A Department of Defense facility in southern California was generating over 30,000 gallons of cyanide-based wastewaters annually from plating operations. With a little attention, it was realized that the hazardous waste was created because considerable plating solution clung to parts being removed from the plating bath (this is called *dragout*) and contaminated the rinse water. The waste reduction solution was to install drain boards to keep the plating solution in the plating bath. Cyanide waste was cut 90 percent, with a capital cost of less than $900 and a monthly savings of $784, yielding a payback period of just over 1 month.

Dow Chemical had plans to install a coagulation system to remove suspended solids in large volumes of "whitewater" in a latex process. The system was estimated to cost $250,000 in capital and $70,000 a year in annual landfilling costs. The waste reduction solution was simply to implement an intensive maintenance program to improve seals and close off leaks. Only $10,000 was spent for a tank to hold the remaining latex leakages for recycling back into the process. Thomas H. Lafferre (1989), an executive with Monsanto Company, has said, "Our experience indicates that substantial reduction in waste generation can be accomplished without a lot of capital expenditures simply by paying closer attention to the operating conditions."

Consider the office situation. GSD&M is an advertising firm in Austin, Texas, with 200 employees. Here are some things they did to reduce waste production: they stopped ordering over 600 Styrofoam cups a week and asked employees to bring their own reusable mugs; substituted china at conferences and meetings for throwaway plates and plastic utensils; and stopped delivery of drinks in six-packs with plastic yokes and switched to supplies in larger containers. There are countless other waste reduction opportunities for offices, especially cutting down on paper use. Replacing paper towels in restrooms with either air blowers or cloth towels makes a lot of sense. Using paper towels in a typical restroom in an office building can cost several hundred dollars a year and produce nearly a ton of trash. A most impressive product innovation, simple yet elegant, is the introduction by 3 M of a small slip of adhesive note paper that serves as a fax transmittal memo. Over the last several years of exploding use of fax machines, nearly every user has used a whole page to identify the receiver and sender. 3M's product, one of its Post-It brand of products, eliminates a whole piece of expensive fax paper that nearly always quickly becomes paper waste.

One idea applicable nearly everywhere is to make two-sided copying the default option for photocopying machines. The Human Services Department of Itasca County, Minnesota, reached 50 percent for two-sided copying and was striving for 80 percent, a level that would save about 270 reams of paper and more than $900 per year, according to a February 1990 report. Another idea is to use the back side of sheets of paper already used once (such as for a draft) for still more routine, uncritical printing before recycling the paper. Alternatively, nearly everyone could use the back side of scrap sheets of paper for pads of memo paper and telephone messages before recycling the paper.

Changing the default settings for printers to reduce margins should also be examined. Not sending mail copies of faxed letters, except in certain exceptional situations, is another good waste reduction practice. And a very smart idea is the use of two-way envelopes instead of enclosing a return envelope. The Illinois Secretary of State's office used two-way envelopes to mail license plate renewal applications. In 1989 it saved $57,547 and 2 million envelopes by doing so. The Geico insurance company has also switched to two-way envelopes in its billing of customers.

There has been less experience with large numbers of consumers. But there is indirect evidence that consumers have the potential to change their behavior quickly. Consumers change their food preferences quickly when health considerations enter the picture. News stories about apples contaminated with the pesticide Alar stopped apple consumption nearly instantaneously across the United States. Information about the benefits of certain foods in preventing heart disease has resulted in skyrocketing consumption of oat bran products and fish with corresponding declines in consumption of eggs and red meats. In terms of household behavior, garbage recycling programs across the nation have gained widespread public cooperation. Even special programs by local governments to collect household hazardous wastes, which generally require people to take toxic wastes to collection places, have succeeded.

Compared to early successful industrial waste reduction efforts, there are important parallels for consumers. After all, household garbage is visible on a daily, almost an hourly, basis. People might immediately cut down on their use of garbage-intensive practices like routine use of paper towels, purchase of single-use, disposable products, and subscriptions to newspapers or magazines they do not read. A shopper can say "no" to a smallish bag and put the purchased product in another, larger shopping bag, or even in a pocket or handbag if that is practical. Moreover, the enormous number of salespeople and checkout clerks in America can suggest to shoppers that a bag is not really necessary! It would also be easy for most people to think of ways to make available old clothes and magazines for reuse by others. Pieces of paper or cardboard which ordinarily would be thrown into the garbage receptacle, for example, could replace new note paper. A plastic container from a microwave food package could be washed and used again and again to heat up foods in the microwave oven. People can stop buying Styrofoam cups and plates and plastic forks, spoons, and knives and go back to using washable utensils. Many packing materials could be used directly or after being cut up as soil additives for gardening and indoor planting, instead of buying commercial products or expensive potting soil to keep soil porous.

The First Obstacle

The key to commonsense waste reduction is getting people to pay attention to reducing waste generation when and where it first occurs. The first major obstacle to pollution prevention is inattention to the need, the benefits, and the immediate opportunities. A key part of the process of overcoming ignorance about pollution prevention and just plain lethargy, apathy, and inertia is to accept a basic principle of pollution prevention: absolutely everyone in every industrial, commercial, office, military, school, or household setting can contribute to the effort. The challenge, therefore, to managers in industry and citizen advocates of pollution prevention is to bring waste reduction to the attention of everyone, educate people about the benefits of waste reduction, and provide simple information about the successes achieved elsewhere.

Companies with organized waste reduction programs, like Dow Chemical's "Waste Reduction Always Pays," Chevron Corporation's SMART program—"Save Money and Reduce Toxics," and 3M's "Pollution Prevention Pays" programs, have consistently succeeded because key people in their organizations act as advocates or champions of waste reduction, spreading the gospel of pollution prevention and making converts because of their zeal and commitment. And the same is probably necessary in offices, retail stores, schools, and households. One individual can act like a catalyst, bringing pollution prevention possibilities to the attention of other people and motivating them to change their behavior.

Using people-to-people contact is an *active* tactic, compared to *passive* approaches such as relying on published company policy statements, publications, conferences away from the workplace, information centers, and telephone hotlines. Passive approaches do not work as well as active ones because they depend on getting people's attention or participation, a difficult task, as professionals in sales and marketing are well aware. The success rate will be higher for personal contacts, such as informal and formal presentations from advocates who are friends or fellow workers, and in industry from onsite technical assistance teams from government or universities, followed up by passive measures.

Getting the attention of people outside the workplace is more difficult. Even with all the recent talk about environmental problems, the ugly truth is that many environmental problems get little sustained attention by major news media. Without that, households and consumers may be unaware of a particular environmental problem and, particularly, their relationship to it. To an even greater extent, there is little big news coverage of necessary and successful pollution prevention in industry and everywhere else. Good environmental news is not considered newsworthy by the press and television. True pollution prevention, for example, has received little news attention compared to waste recycling. Public attention to environmental problems ebbs and flows with the episodic "crisis" character of news coverage. Behind the "quick-hit" news coverage that helps build a crisis atmosphere may be a new report from some government agency or an environmental group that documents a particular problem. But a majority of the public, in their role as consumers, may miss such "hit-and-run" news coverage, and hardly anyone reads the detailed reports.

The environmental fad or crisis-of-the-moment condition has also preempted a comprehensive approach to *all* environmental wastes and pollutants. This is critical because another principle of pollution prevention is that there is no single waste or pollutant that causes any or all environmental problems. Nor, therefore, is there a single solution. For long-term, comprehensive protection of health and the environment, the public must want to eliminate or minimize *all* wastes and pollutants. Only pollution prevention intrinsically addresses *all* environmental wastes.

The lack of sustained public and news media attention to the full range of environmental problems and wastes is not the only problem. Contamination of food products aside, the American custom has been to focus on waste and pollutant production by industry and, more recently, government, but not by the general public and smaller-scale commercial activities. It is particularly difficult to focus attention on an invisible waste that appears to cost the waste producer nothing significant. For example, cooling towers used in ventilation and air conditioning systems in hospitals, hotels, shopping malls, and other large commercial buildings emit water vapor. However, the water vapor contains chemicals used to treat the water to prevent various problems in the equipment. Some water treatment chemicals contain hexavalent chromium, a carcinogen. Only at the end of 1989 did the EPA ban the use of hexavalent chromium compounds in such applications because of the high risk of lung cancer. Hundreds of Americans may have already gotten lung cancer because no one paid attention to the direct discharge of the chemically contaminated emissions. Clearly, such health effects were preventable.

The building construction industry has received relatively little attention as a preventable source of environmental pollution. But Charles Howell, a former construction company executive and Tennessee Commissioner of Conservation, had a bold, creative idea. In November 1989 he started The Tennessee Initiative for Environmental Sensitivity in Construction. The program addresses "the presence and continued use of environmentally insensitive construction methods and materials." Some of the problems noted by Howell include formaldehyde-treated wood products that give off odorless toxic gas; asbestos products; use of chloroflurocarbon (CFC) blown polystyrene materials which, in their manufacture, help destroy the Earth's ozone layer; and the use of pesticides. The new organization will recommend environmentally sensitive alternatives. In other words, the effort is really a pollution prevention one. And the formation of the group was designed to overcome inattention by nearly everyone—architects, builders, manufacturers, home purchasers, and government officials—to the construction industry's contribution to major environmental problems. To that point, note that in 1989 the Construction Industry Institute developed a design evaluation matrix that contains no criterion related to waste or environmental effects.

For consumers in general, a key initial obstacle to pollution prevention is inattention to the problem. A good example of untapped potential are a host of consumer actions which contribute significantly to the chemical contamination of underground water supplies. Anthony Earl (1989), former governor of Wisconsin, described this phenomenon:

> Groundwater is an out-of-sight, out-of-mind problem for most people. The problem is that groundwater pollution is so incremental that we don't see it happening. If a lake or a stream begins to go bad, if we can't catch a fish, if there are algae blooms, we recognize it. But because groundwater pollution is so incremental, we don't notice it, and we continue to ignore it. Everyday we continue to destroy it by the application of fertilizers and pesticides, by urban development that is thoughtless and reckless, by failing septic tanks, by a whole host of topics, which we simply won't come to grips with. If there is any water problem that calls for a prevention rather than a reaction and clean-up approach, it is the groundwater problem. But it requires a protection strategy. We have got to break away from our old notion of not addressing a problem until we have a crisis on our hands.

In the United States, about 100 million people depend on groundwater (the rest make use of surface waters). True, thousands of U.S. citizens have had the unfortunate experience of discovering that their groundwater is contaminated. But most Americans are oblivious to their use of products which are directly responsible for the slow contamination of groundwater. Consumers buy pesticides and fertilizers, support more local commercial development, and pay little attention to maintaining their septic tanks or eliminating their need by using municipal sewer systems. Much can be done by government and environmental organizations to inform the public in the strongest terms of the connection between daily activities and groundwater contamination. Moreover, doing so can educate people that being an environmentalist, which 75 percent of Americans consider themselves, means more than being angry about big chemical companies and the government. It means accepting responsibility and taking control.

The Second Obstacle

The second early obstacle to waste reduction is the lack of immediately obvious economic benefits even though such benefits exist. Saving and making money is an immutable motivation for all people. But because the true costs of waste have either been at the end of the pipe, instead of upstream where wastes are produced, and because many costs have been externalized to society as a whole, workers in industry may not see the economic benefits of pollution prevention. Managers in industry have learned that changing accounting procedures to allocate waste costs to profit centers where wastes are produced is critical. Also, providing rewards for successful efforts is a powerful tool. After all, workers who cut the production of wastes make money for the company. Being successful at waste reduction is like producing more product, saving energy, using less raw material, or working faster. An office worker who convinces management, for example, that the routine practice (or default option) should be to set office copying machines to copy on both sides of paper has made money for the company.

In other words, waste reduction should be a criterion for deciding employee salaries, promotions, bonuses, and special rewards and awards. The need is to ensure that economic benefits for the organization (private and public) derived from successful pollution prevention are distributed directly to workers. Otherwise, workers may not carry out feasible and beneficial waste reduction of the easiest kinds, especially since workers will focus on achieving things for which they are rewarded. Here are two stories of industrial experiences, which could apply to many other situations, that clearly illustrate several things: first stage, commonsense successes in industry and how a lack of a perceived benefit keeps workers from reducing waste.

For several years, a medium-sized plant paid about $100,000 per year to dispose of partly full barrels of dried glue that were being generated at the rate of 40 to 50 barrels per month. Each barrel was one-quarter to one-third full of dried, unusable glue because the normal practice was for the glue operator at the beginning of each shift to switch to a full barrel of glue that would last throughout his shift. The operators did this to stay competitive. Otherwise, they would be held responsible for stopping the operation while they switched barrels, and their productivity would be reduced. The operator's improved productivity was achieved by waste generation—whose disposal costs, however, were not directly connected to the operator's behavior.

The waste reduction solution ultimately used by the plant manager was to buy a $200 glue pump that locked in place and could only be removed by the shift foreman after the pumping mechanism reached the bottom of the barrel. That prevented partially full barrels from being thrown away as garbage. This solution cut the cost of glue hauling, and waste disposal was reduced by about 80 percent. That is, the $200 investment and the change in procedures saved $80,000 every year. It also saved raw material costs because less glue became waste. Another solution might have been to instruct the glue operators simply to stop their practice of beginning their shifts by getting rid of the partially full barrels. Management could have tried to show the operators that their normal behavior was in fact *not* saving the company money because the waste disposal costs and increased glue costs outweighed the benefit of less frequently interrupted operations.

The second example is a large-scale spray painting operation using several different colors. The original practice was to paint products in a spraying booth according to the order in which the products were received. Therefore, one product would very likely be followed by another requiring a different color paint. After painting a product, it was therefore necessary to flush the apparatus with solvent to prepare for the next item, which required a different paint. The flushed solvent containing paint entered the wastewater treatment unit. Forty percent of the paint that missed the product entered the waste stream as well. Production line people had no direct economic incentive to change their practices. Eventually, management decided to look at the source of waste. The waste reduction solution was simply to reprogram the computer to change the way the order of painting was established. The new criterion was color. Products with the same color would be painted in a group each day. The waste management costs were reduced by about 20 percent, with no up-front capital cost for the change.

For consumers, direct economic benefits and incentives may be more elusive. Managers of industrial and commercial establishments have the advantage of knowing or being able to find out what their waste management and environmental controls cost. Few consumers and households, however, are conscious of their direct waste costs and, even less so, of the untold indirect costs hidden in the prices of goods and services and taxes or of the inevitable health and environmental costs which they may ultimately bear personally. Linda W. Little (1988), Executive Director of North Carolina's

Governor's Waste Management Board, summed up the problem of lack of readily seen incentives for consumers this way:

> For the general public about the only incentive is that "virtue is its own reward." As a rule, the cost to the homeowner of garbage pickup is unrelated to the amount produced. There are no regulations requiring waste minimization. As for disincentives, it is usually attractive to buy products [producing hazardous waste] in the giant economy size, and fancy and excessive packaging is used to lure customers. . . . When the public becomes personally involved in pollution prevention and waste reduction, when the public understands where wastes come from, and when the public understands the need for waste management facilities, I believe that the whole waste management program in this country will benefit.

Little goes on to draw attention to the lack of public focus on nonindustrial and consumer-related wastes and pollutant generation, which also helps to explain why the system has created few economic incentives for consumer, household, and small business waste reduction:

> The fish in a river cannot distinguish between a regulated [industrial] hazardous waste and a non-regulated [consumer] waste, between the lead in an industrial effluent and that in a gasoline spill, between the pesticide in a pesticide manufacturer's wastes and that running off a cornfield, so we might as well move beyond this artificial and simplistic categorization wherein we scare the wits out of people in regard to one [industrial] source and for all practical purposes ignore other [consumer] sources.

The Third Obstacle

A third important initial obstacle to pollution prevention in industry and elsewhere is that attention may only be directed to officially recognized, government-regulated wastes and pollutants. Except for a limited number of chemicals banned by the government, producing wastes and pollutants is not illegal. At best, government regulations limit flexibility for managing wastes and pollutants and, in some cases, limit the amounts that can be released in untreated form to air, water, and land. Even though (as discussed in the next chapter) the data add up to an enormous quantity of wastes and pollutants, there are several reasons to conclude that even more wastes and pollutants are being produced. Every federal environmental statute and regulatory program has exemptions and loopholes. The General Accounting Office once estimated that only 10 percent of hazardous wastes might be regulated under the federal hazardous waste program. Relatively small industrial generators of hazardous waste can dispose of that waste as if it were harmless. Household hazardous waste is exempted from regulation. Only a handful of the hundreds of toxic chemicals are regulated under the Clean Air Act program.

The pollution prevention issue is whether people in industry who are accustomed to dealing with regulated materials will devote waste reduction efforts to unregulated wastes and pollutants. The latter offer little economic incentive for reduction because relatively little money may be spent on their management and control. Consumers may simply take the position that producing garbage of any type is not against the law and, therefore, they need pay no attention to reducing its generation. The basic way to deal with this obstacle is to acknowledge the shortcomings of government efforts and emphasize the certain and potential damages to health, the environment, and welfare that *all* wastes and pollutants pose.

The Fourth Obstacle

A fourth initial obstacle is that people may see no difference between an end-of-pipe pollution control approach and a front-end pollution prevention solution. In households, people may see no reason to cut down on the types of garbage that are or may be recycled after the materials are placed curbside. If their community has an incinerator for burning municipal waste and generating some electricity, they may even believe that it would be wrong—and certainly unnecessary—to cut down on their garbage production. Most people may not perceive that the biggest environmental effects of garbage incineration, toxic air emissions, and burial of hazardous ash affect them directly, even though air pollution and groundwater contamination may eventually do just that. High fractions of the toxic metal mercury from batteries, for example, are in garbage being burned, and it is difficult to trap mercury in air pollution control equipment. Research has found that even after recycling all yard waste and 50 percent of paper and plastics, 83 percent of the mercury remained behind.

Moreover, although there has been a lot of talk about insufficient landfill capacity for municipal waste, except for a few unusual situations, most Americans have not encountered problems in disposing of their daily garbage. Nor do most Americans suffer any direct adverse consequences from landfills. In fact, because about 80 percent of household waste on average, and often much more locally, is disposed of in landfills, which indisputably pose environmental problems, most Americans should be motivated to reduce the volume and toxicity of their own trash. Yet most people do not seem to appreciate the extensive, documented environmental problems of common municipal waste landfills. These problems are summarized in Table 3–2. It is imperative that Americans understand that current legal methods of waste management do not prevent pollution and do not offer the certain protection that waste reduction does. And with pollution of all kinds existing everywhere, the hope of getting out of harm's way is naive. When

TABLE 3-2.   EVIDENCE OF ENVIRONMENTAL THREATS
FROM COMMON MUNICIPAL WASTE LANDFILLS

Ambient environmental monitoring: As of November 1986, only about
25% monitored groundwater, 15% monitored surface water, 7% monitored methane gas, and 3% monitored other air emissions.

Leachate: Typically contains many hazardous chemicals. Over 100 potentially harmful substances have been identified in municipal solid waste
(MSW) landfill leachate. The ratio of median measured contaminant concentrations to federal regulatory standards is very high for some chemicals:
chromium, over 3 times higher; lead, over 3 times higher; manganese,
nearly 200 times higher; nickel, over 4 times higher; thallium, over 4
times higher; benzene, about 44 times higher; 1,1-dichloroethane, over
200 times higher; methylene chloride, over 1,000 times higher; methyl
ethyl ketone, over 2 times higher; phenol, over 2 times higher; tetrachloro-
ethylene, nearly 20 times higher; 1,1,1-trichloroethane, over 4 times higher;
trichloroethylene, about 20 times higher; vinyl chloride, nearly 20 times
higher.

Leachate collection: Only 11% of existing landfills have any type of leachate collection system, and the presence of a leachate collection system
is not necessarily sufficient to prevent groundwater contamination. The
EPA has identified MSW landfills equipped with leachate collection systems that failed to prevent such contamination because of inadequate design and/or construction.

Leachate treatment: Once collected, leachate is rarely treated to permanently render harmless the many hazardous chemicals it contains.

Risks from leachate: It is impossible to determine the actual risks posed
by leachate from most landfills because groundwater monitoring data
are rare. This lack of monitoring is alarming because down-gradient
drinking water wells exist within 1 mile of an estimated 46 percent of all
MSW landfills. An EPA model that evaluates the environmental effects
of only eight pollutants in landfill leachate found that over 17% of landfills pose lifetime cancer risks that would qualify them for cleanup under
Superfund regulations. There are over 7,500 operating MSW landfills in
the United States.

Cleanups because of contamination: In May 1986, 184 (22%) of the 850
sites proposed for the Superfund National Priorities List (NPL) were
municipal landfills. The general lack of engineering controls at most exist-

TABLE 3–2. CONTINUED

ing landfills, combined with the fact that many landfills have accepted or continue to accept hazardous wastes or industrial nonhazardous wastes, suggests that additional municipal landfills may eventually require remedial action. The EPA identified 135 of 163 landfills studied that constitute a threat to human health or the environment. Moreover, for the 184 MSW landfills on the NPL, approximately 72% were associated with releases into groundwater, 44% experienced surface water contamination, and 17% experienced air emission problems. By June 1988 the number of municipal landfills on the NPL had increased to 249 (21%), and 42 of them were still operating.

Gas emissions: MSW landfills generate several gases that pose risks to human health and the environment. The primary gases are methane and carbon dioxide, but numerous organic chemicals in gaseous forms are emitted as well. The EPA estimated that about 200,000 metric tons of nonmethane volatile organic compounds (VOCs) are emitted from landfills each year. In one study in the San Francisco Bay area, VOCs were present in the gas at 47 of 60 landfills. VOC emissions can affect ozone concentrations because many of these emissions are ozone precursors. Some researchers suggest that substantial releases of hazardous substances will occur from MSW landfills even where no regulated hazardous wastes have been accepted.

Source: Adapted from Office of Technology Assessment (1989).

that lesson is learned, people will lose faith in traditional pollution control and embrace prevention.

In industry, incineration of hazardous waste and sending wastes offsite for recycling are often believed to offer the same benefits as true preventive actions. But, as discussed in Chapter 1, all forms of waste management and pollution control are inferior to pollution prevention because there are always risks when waste is handled and processed. Technology, people, and systems inevitably fail. The enormous liabilities and costs facing industry from the Superfund toxic waste cleanup program are proof of the failures of traditional end-of-pipe techniques. Convincing people that prevention is always the first choice is critical if they are to take even the simplest pollution prevention actions. Industrial workers and managers need to know that they must think and practice prevention. A key way to institutionalize this practice is to require that every corporate proposal for investment in end-of-pipe equipment contain a detailed account of how pollution prevention options were identified and evaluated.

A major problem during the past several years is that American industries have invested enormous amounts of capital—hundreds of millions of dollars—in constructing hazardous waste incinerators on their industrial plant sites. To achieve lower costs through economy of scale, engineers typically oversize the capacities of these incinerators by assuming *increases* in future production of hazardous wastes. Indeed, a consulting firm, the Freedonia Group, has projected an average annual growth rate in incineration of hazardous waste of 18.5 percent from 1988 to 1993. Many companies also burn wastes, such as gaseous chemicals, not defined for regulatory purposes as hazardous waste.

Dow Chemical had an experience that more companies need to have. At its Baton Rouge, Louisiana, plant, 1,100 pounds per hour of unreacted byproducts from an ethylene dichloride process had been incinerated. Although useful steam was produced, maintenance costs were high because of the presence of hydrochloric acid. After an analysis, the waste stream was piped as feed to another process in the plant, a relatively simple action, and the incinerator was shut down. The payback period for this change was less than 1 year, even after the lost steam production was taken into account.

The pattern of investment in onsite incineration has been spurred by government restrictions on land disposal of hazardous waste, increasing costs of commercial offsite incineration, concerns about commercial capacity, concerns about cleanup liabilities from using commercial companies, and the perception that more and more industrial wastes will be regulated by the government as hazardous. All these concerns are valid and could, of course, also be addressed by waste reduction. Capital investment in incinerators, however, creates a powerful disincentive for waste reduction and instills an attitude among engineers that waste reduction is unnecessary and inconsistent with company policy. Executives who have signed off on these massive capital projects may fear looking like failures should future successful waste reduction verify that the incinerators were a poor choice. To the extent that the costs for onsite, or captive, incineration are less than the costs for offsite, or commercial, incineration, there is less economic incentive for waste reduction because there is less waste management cost avoidance.

One way for companies to offset the negative impacts of expanded incineration capacity on waste reduction is to charge those company profit centers that use company incinerators a surcharge per ton of waste burnt. This extra cost will boost the incentive for waste reduction, and the money collected can be put into a company fund designated for waste reduction research and development projects. Proposals can then be made to get support from the waste reduction fund. The same principle can be applied, of course, to any company-operated waste management facility. For this scheme to work, it is necessary for companies to have cost accounting systems that fully charge specific waste producers within a company for waste management costs. Without such internal charges, there is no economic incentive for waste reduction at the microlevels where wastes are produced from individual production lines.

Similarly, every proposal for investment in new plant or equipment should be accompanied by an evaluation of how new, clean technology has been considered and used. For example, senior company managers need to have independent information on the availability and use of alternative low-waste technologies elsewhere in the company, in the industry domestically, and in foreign industries. Senior management cannot accept at face value the conclusions or assertions of engineers that a new process or new plant has been absolutely optimized from a pollution prevention perspective.

The Fifth Obstacle

A fifth obstacle in the first stage is psychological resistance to change. Joseph E. Harwood (1988), environmental affairs manager for Duke Power Company, summed up the problem this way:

> Now is the time to look at production processes. Can they be fine tuned to minimize waste production? Are waste management practice costs approaching the value of the product being produced? Cultural changes are slow because they are also philosophical changes. There is also psychological opposition to change. Many like the old ways of doing things and there is often an attitude of "if it isn't broke, don't fix it." These attitudes and fear of new approaches must be overcome before effective waste reduction can be implemented.

Engineers in industry may also feel that they have already optimized "their" processes, operations, and products. It takes effort to show experienced engineers that they have not necessarily focused enough attention on *all* wastes and pollutants being produced and that they have not been aware of the full short- and long-term costs of generating wastes and pollutants. An example of the subtleties of industrial waste reduction is that most large-scale industrial processes consist of chains of subprocesses which precede or follow each other. If, for example, each subprocess has an efficiency of 70 percent for materials and energy conversion into useful material, then for four steps the overall efficiency is only 24 percent. Even if each step had an efficiency of 90 percent, the overall process efficiency would be only 66 percent for four steps and 53 percent for six steps.

It is easy, therefore, for engineers who are responsible for specific steps of a long, complicated industrial process to feel good about their low waste generation, even though substantial waste outputs are produced overall. In early 1990, ICI in the United Kingdom announced its intention to build an agricultural chemical intermediates plant that would use new process technology to reduce waste generation. The technological improvement was said to result from reducing the number of process stages. Ryan Delcambre (1988) of Dow Chemical, one of the top waste reduction experts in American industry, expressed the best attitude toward waste reduction and process efficiency: "We have a very aggressive program, operating on the premise that any waste generation whatsoever represents an inefficiency in manufacturing. From a base year of 1982, our waste generation [in 1987] has gone down by 37 percent on a pound-waste/pound production basis."

Some engineers may also believe that past energy conservation efforts have also done the trick for wastes and pollutants. That belief is incorrect. Moreover, whatever things were done to conserve energy may not have relevance or the same importance for pollution prevention. The better approach is to learn from energy conservation experiences. According to Dan Steinmeyer (1989) of Monsanto:

> For the last 15 years in *energy,* we've attacked with enthusiasm and capital. The result is an improvement in efficiency, which measured in dollars, is a large fraction of current profits. The opportunity in *preventing* our other *wastes* isn't as easy to recognize and we're only beginning the battle. It's going to be a tougher struggle, but we have at least had the advantage of the energy example to learn from . . . there was a 7 year gap between our capital spending peak and the price spurt which triggered it. . . . If we don't move more quickly than we did in energy, there will be a lot of incinerators and costly biological treatment plants built—if the products survive the impact on profitability. In energy a project carried almost the same net present value in 1989 as in 1985. In contrast a waste prevention project derives part of its value from treatment capital avoidance. A "good idea" that eliminates a process waste is too late if it comes after the incinerator to burn the waste is already built.

There is proof that experienced engineers have not practiced pollution prevention to the maximum extent practical. A few major corporations with intensive waste reduction programs have achieved substantial waste reductions in older plants and processes. They have done so by taking advantage of new technology, as well as by providing new motivation and sometimes new engineers. Industrial and other workers may fear adverse consequences if they now identify waste reduction opportunities that could have been implemented years earlier because past actions were inefficient and waste-intensive and contributed to environmental problems. Managers may fear that starting pollution prevention efforts may set them on a path that will require them to make major changes in their products, risking their existing success in the marketplace. A few managers may view environmental issues as fads that will disappear, and they may also not show interest in pollution prevention because they are not legally required to do.

Consumers may believe that waste reduction is not their problem and that they have enough problems to deal with. For example, Paul Ruffins, the executive director of Black Networking News, said (1989) that "most black people simply don't consume as much energy or produce as much pollution as wealthier whites. Being less of the problem, many black folks may feel less obliged to be part of the solution." This attitude may characterize many poorer people in the United States and elsewhere; however, Ruffins also notes that "many black communities have made the connection between pollution, disease, cancer and birth defects."

All mental attitudes which impede the initial interest in pollution prevention have a kernal of legitimacy and should not be derided, ignored, or underestimated. Effective preventive planetary care requires participation by all classes of consumers and workers. Psychological obstacles can be overcome by respecting psychological blocks, helping people to start practicing pollution prevention, and making them feel good about their new successes by giving them appropriate rewards and recognition. Moreover, current and potential health effects need to be explained in terms of everyday habits, practices, and consumption to demonstrate personal benefits.

People working in industry may not see the connection between what they are doing in the workplace and their lives. A story is told of an environmental coordinator for a major aircraft manufacturer. Solvents were found in the plant's storm drains. The source was discovered to be several floor drains in a shop area. Workers were pouring used solvent down the drains— a practice that probably occurs in many places. Signs were put up that read "NEXT STOP IS YOUR FAUCET AT HOME." Technically, this was not correct, but the idea that "what goes around comes around," which is a fundamental precept of pollution prevention, was sound. This violation of company policy stopped.

To sum up, the way to start easily accomplished pollution prevention practices is by marketing the concept, spreading the word, and helping people develop a feeling of success and accomplishment. Corporate policy statements on the importance of waste reduction focus the attention of workers on waste reduction. Government efforts can do the same for consumers. Slogans, campaigns, speeches, buttons, and all the other paraphernalia for motivating and selling to people are critical. For example, at a Chevron chemical plant, a series of posters was created on the theme "Stop Waste; Make Money." One of the posters was used to sponsor a children's coloring contest; several videos have been produced; and recognition awards and prizes have been given to employees who came forward with waste-cutting ideas. Chevron started its SMART program in 1987; in the first year it reduced the amount of hazardous waste generation by 53 percent and saved $3.4 million companywide. But Chevron's SMART program was initiated originally only for hazardous waste. Chevron said that it would establish similar programs for air and water at a later time. Bill Mulligan (1989), Chevron's environmental affairs manager, told employees: "We all generate waste, which makes us all part of the problem. But we are all part of a solution. The creative talents of all of us can lead to a better, cleaner environment."

For early success in pollution prevention, people who never have seen waste generation as their responsibility—that is, most people—need to understand that waste will no longer be taken care of by others, such as environmental engineers or garbage truck drivers. Moving from the end-of-the-pipe mentality to the pollution prevention paradigm means making waste reduction part of everyone's everyday thinking and responsibility—just as preventive health care is an individual responsibility that more and more people accept and benefit from.

STAGE 2:   INFORMATION NEEDED ON WASTES, SOURCES, AND REDUCTION TECHNIQUES

The danger is that people and companies may not go beyond the first stage of waste reduction, in which opportunities spring from direct experience with naked-eye observations of waste generation. To maintain progress and to attain systematic, comprehensive waste reduction, it is necessary to have detailed information to assess reduction opportunities which become more subtle and sophisticated. However, as with stage one, in stage two there are no major technological obstacles or major capital investments, and quantum reductions in waste generation are possible. The problem is to discover exactly where to focus efforts and what available technology and products to use.

For people in industry, information is needed on all wastes. Why is a waste produced from a process or activity? What is its linkage to raw materials and product quality? What are the quantity, chemical composition, hazards and liabilities, relationship between generation and production levels, and regulatory status of wastes? Industrial companies need to establish permanent information collection and tracking systems. Just as companies recognize the need to maintain constant, reliable information on inventories and sales of products, costs, and financial standing, so too must they maintain sophisticated databases on exactly what wastes their facilities are producing.

Ideally, at any moment, plant managers should have the capability of determining how waste production is being affected by everything going on within the company's operations. Changes in products, work shifts, raw materials, specific workers and managers, weather, level of capacity usage, and many other factors can inform top management exactly how the production of *all* waste outputs varies over time. Traditional engineering material balances must include detailed data on the use of natural resources, like air and water, and the production of even minute quantities of wastes and how they are being handled. With good information, management can measure waste reduction progress for specific operations and products, implement employee award and reward systems, and set or adjust plant and company waste reduction goals. Nearly all large generators of environmental wastes have found it useful to set numerical goals to spur worker interest and to have a benchmark to measure progress and success. It is crucial that such numerical goals be ambitious. Otherwise, workers and outsiders, such as community members, may not take top management's commitment seriously.

A good example of ambitious numerical goals presented as forcefully to the general public as to its own workers and managers is that of General Dynamics. In 1984 the company initiated its waste reduction program with a goal of zero waste generation. In 1988, General Dynamics produced 7,784 tons of hazardous waste compared to 28,099 tons in 1984. Over that time, sales increased by 30 percent. Even with that impressive accomplishment, the company has had to shift its target for zero generation to 1991. And the company had not yet mounted an attack on reducing air and water pollution.

Another ambitious goal was announced by Monsanto. In 1988, Monsanto said that its goal is to reduce toxic air emissions by 90 percent by the end of 1992. In more general terms, Monsanto (Lafferre 1989) has said, "Our goal should be zero spills. Zero releases. Zero incidents. And zero excuses. That goes for us, our suppliers, our contractors and our shipping firms." Monsanto must now translate its rhetoric into reality. Management has put pressure on its workforce.

There are generic and all too common problems with waste reduction goals. They often speak about reductions in the use of land disposal of hazardous wastes, but that does not mean true waste reduction because waste treatment such as incineration could be used. Alternatively, only a small portion of the total waste universe may be covered by a specific goal. Indeed, very few large corporations with waste reduction programs address all waste outputs, including nonregulated ones. Companies with very broad waste coverage include Polaroid and 3M.

A substantial problem for consumers is knowing what wastes are toxic, in whole or in part, and what has happened during the manufacture of the products they use. Available product information is very limited. At this time, product labels and information that come with products do not inform consumers very well, if at all, about the nature of the waste produced after product use or during product manufacture.

Information on waste reduction techniques from many external sources is also needed. For example, information about new raw materials or manufacturing techniques from vendors of materials used by industrial companies is important. Typically, companies buy chemicals and other materials from other companies. Feedstocks for a company may be the cause of waste generation, and the seller of those feedstocks may not assist in finding substitutes for them. Getting information on substitutes for feedstocks is difficult for small manufacturers, commercial establishments, and institutions such as hospitals and schools. Large corporations, which are very large producers of environmental wastes, have found it necessary to establish corporatewide waste reduction programs in order to ensure that individual facilities become aware of and evaluate many different waste reduction techniques. Consumers must examine the options in stores and seek different stores with different options. A good example of a general opportunity is that more and more people in industry and individual consumers are discovering that they can replace traditional chemical solvents with water or biological ones, and new painting materials are being sold.

Although costs and benefits are self-evident or easily calculated in this second stage, consumers must establish new behavior and consumption patterns and companies must build an organization for implementing waste reduction, including getting and distributing information and measuring progress. Getting detailed information on waste generation and reduction techniques can cost millions of dollars for large industrial facilities. Large companies are typically more able to handle this second stage than small and medium-sized firms, which may find it difficult to devote people as well as money to this effort. Even in some large companies, maintaining interest in waste reduction may be hampered because waste management and pollution control costs seem relatively low. For example, automobile, aerospace, and electronics companies have lower environmental costs than chemical companies.

When business is good, profits are high, and waste quantities are high, companies whose waste management and pollution control costs are relatively low may have little interest in waste reduction. Even companies with high waste costs may see waste reduction as less important when their business cycle is in an upswing. Conversely, when business is bad, profits are low (or there are losses), and waste quantities are low, companies may not be able to devote attention, people, or money to waste reduction projects, even though waste reduction could convert the bad condition into a better one. This is why it is necessary to have an ingrained pollution prevention ethic, a permanent management commitment, and an effective support organization to ensure that waste reduction always receives attention.

The role of government becomes more evident in the second stage. Government agencies can distribute information about successful waste reduction techniques in many different industries and for major classes of consumer products; state agencies can provide onsite technical assistance for industry, which has been shown to be very effective and low in cost. For example, Ventura County, California, has run a very successful inplant technical assistance program and, unlike other government programs, has measured the program's impact. Over two years during which about 100 companies received onsite visits from county personnel, 15 percent of the hazardous waste was cut. Some government requirements for information on waste generation, such as those promulgated under the Resource Conservation and Recovery Act, help drive companies to obtain detailed information on waste generation.

Information required from companies for the Toxic Release Inventory, under Title III of the Superfund Amendments and Reauthorization Act, provides powerful information to consumers about what environmental wastes the manufacture of products causes and provides—and, therefore, a strong incentive to industry to focus attention on waste reduction. Many of these government information reporting requirements, however, do not apply to small businesses and also may not cover the full range of wastes and pollutants. Considerable attention is being focused on the need for new kinds of information on product labels. Inevitably, most national governments should and probably will establish legal requirements and systems for helping concerned consumers identify environmentally beneficial products. Environmental organizations have also stepped in with information designed to help consumers make new choices.

STAGE 3:   ANALYSES MUST REVEAL COSTS, RISKS, AND BENEFITS

We will consider industrial waste reduction first because of the greater experience with it. Passing through the first two stages for a particular process or waste may take from 1 to 5 years in individual industrial plants, commercial establishments, or companies. The next major obstacle to large-scale waste reduction is economic uncertainty because of substantial changes in technology and equipment. Changes in core process technologies and interruption of production have to be considered. For example, Merck and Company went to a second-generation manufacturing process to make a cholesterol-lowering drug produced by fermentation. The first-generation process involved the use of toluene solvent to extract the active ingredient from biomass. This generated wastes with toluene. In the new process, toluene was replaced with an environmentally more acceptable solvent, isopropyl acetate. Over 90 percent of the new solvent is being recovered and reused internally. This is also the case with another solvent, ethyl acetate, used in a later step in the process.

Changing fundamental industrial processes is never easy, especially when it also involves a major change in a critical raw material input to the manufacturing process. Rhone Poulenc in France converted a carbon monoxide production unit from coke to natural gas. Waste effluents were cut 95 percent, and operating costs, worker conditions, and productivity also improved.

Other aspects of stage three waste reduction include the following: greater involvement of senior production people is probably necessary; the environmental impacts of changes made for waste reduction purposes have to be analyzed; and major capital investment becomes necessary and risk increases. Moreover, investment payback periods become longer, and capital needs compete with more traditional uses for capital. Testing and development needs increase. There is increasing need to consider changes in products, either to facilitate reduction of production waste or to reduce the quantity or toxicity of postconsumer waste. In other words, waste reduction is no longer simple and self-evidently feasible or profitable. But quantitative analysis can overcome obstacles and provide management with exactly the kind of understanding necessary to sustain and strengthen its commitment to pollution prevention. The U.S. Air Force has had a strong waste reduction program. Its spending of over $62 million from 1984 to 1988 to reach its goal of 50 percent reduction of hazardous waste generation by 1992 is estimated to save $42 million annually by 1992.

All of this leads to the need for formal project and facility analyses, which are called *waste reduction audits* or *assessments*. These analyses must capture and identify costs, benefits, uncertainties, risks, schedules, and relationships to other company plans and programs, such as R&D, expansion, diversification, and marketing of new products. There are hundreds of successful examples in the literature on waste reduction in virtually every industry. Here are some examples of third stage waste reduction:

- General Electric Medical Systems replaced a paint stripping operation using methylene chloride with sand blasting and mechanical sanding; it had found that methylene chloride material and waste management costs were $2,525 annually, but that the sand blasting replacement would cost only $2,000, offered a payback of 0.8 years, and lowered the company's liability.
- Emerson Electric, a power-tool manufacturer, automated its electroplating system. The new system increases productivity, saves elec-

tricity, and reduces waste, partly through more efficient rinsing of the plated parts. Carrying less electroplating solution into rinsing baths generates less hazardous waste. Plating wastes decreased from 450 to 360 pounds per day. The $158,000 investment paid for itself in little more than a year.

- 3M stopped cleaning flexible metal electronic circuits with toxic chemicals and started scrubbing them with pumice. It reduced hazardous waste by 40,000 pounds annually and saved $15,000 annually on raw materials, labor, and waste disposal. The $59,000 cleaning machine paid for itself in only 3 years.
- 3M used to wash reactor vessels with large quantities of water, creating considerable amounts of hazardous waste. It bought sonic cleaning equipment for $36,000 to vibrate residue off instead, and saved $575,000 in the first year.
- Clairol of Camarillo, California, cut waste generation by 70 percent from 1985 to 1989 and saved more than $500 million with its zero waste program. One project developed a new system to flush pipes used for making hair-care products. The old system used large quantities of water. The new system uses a foam ball propelled through the pipes by air to collect and reclaim residue from the pipe walls. The system cost $50,000 to install and has reduced waste by 395 gallons per day, saving $240,000 annually.

Without formal analyses, people may incorrectly conclude that they have exhausted their waste reduction opportunities or that the costs of implementing waste reduction are too high, or they may even pursue projects which are either technically, economically, or environmentally ill advised. In the last category, for example, a facility invested more than $100,000 in equipment to recover chrome from plating rinse water for reuse, but when the equipment did not seem to be operating properly, it belatedly determined that only $100 worth of chrome annually was being lost in the wastestream. Without formal, comprehensive analyses, people in responsible positions may miss opportunities to reduce nonregulated wastes or relatively small wastestreams which nevertheless involve substantial costs and liabilities. Formal analyses performed by inhouse staff or outside consultants can overcome unintended prejudices of key professionals against change, such as the psychological attitudes of engineers discussed earlier.

However, a continuing problem, even when formal analyses are done, is that many economic benefits of a waste reduction option are not captured because they are difficult to quantify. Examples are reductions in future liabilities associated with any form of hazardous waste management, spin-off technological innovations and businesses, and improvement in the public image of a company which reduces public opposition to new company activities. But although these factors are difficult to quantify, the task is not impossible. Westinghouse, for example, has estimated that its waste reduction projects have reduced future liabilities by $700,000 to $2 million. When avoided waste management costs are assessed, inevitable future increases in them are frequently ignored. Thomas E. Higgins (1989), a consulting engineer, described this situation: "Benefits include the value of any marketed good that might be produced, the value of nonmarketed goods (such as annualized savings of future liabilities or current disposal costs), and the value of any secondary benefits (i.e, public relations, community relations, or some other 'public' benefit). These benefits are then summed up. Determining the time benefit of nonmarketed goods and secondary benefits is extremely difficult and is *typically pursued by academicians rather than private sector analysts*" (emphasis added). In other words, Higgins acknowledges that people in industry do not capture the full range of economic benefits and, therefore, may not pursue waste reduction projects even though these would result in net economic benefits.

Analyses may have to go beyond the traditional economic ones associated with project evaluation and capital investment. Merck and Company has, for example, developed a computer expert system called the *environmental assessment system*. It assesses the potential environmental impacts and implications of manufacturing processes. Merck says that the analysis offers flexibility and the opportunity for making process changes and modifications aimed principally at achieving true reduction in waste generation. The system estimates the extent of waste reduction possible by assessing candidate wastes for internal solvent recycling. In one case, total waste disposal costs per unit of final product decreased from $291 to $67 with solvent recovery.

Small businesses definitely find the third stage difficult because it requires much more time and money than the previous two stages and because it is a continuing activity, at least for the next decade or two. The use of outside consultants becomes more necessary. But even large companies may find this stage so burdensome that interest in waste reduction may wane. At the highest levels of corporate management, there may be less interest in pursuing uncertain, high-cost activities even if they are designed for waste reduction. Seasoned technical professionals and managers may feel that they have reached the limits of improving or fine-tuning processes after accomplishing a lot of waste reduction. Moreover, the results of analyses will inevitably reveal the need for large amounts of capital for large waste reduction projects, so that waste reduction competes with other traditional needs of company resources, such as expansion, modernization, diversification, and acquisitions.

Regardless of the economic benefits, waste reduction projects may not get funded. This is the time when corporate policy statements about commitment to pollution prevention are tested. One solution, used by Dow Chemical, has been to establish special internal capital funds for waste reduction projects. Another good approach is to lower the usual company financial decision-making hurdles for waste reduction projects, such as accepting a longer time for recouping investment costs. This can be justified by acknowledging that analyses are likely not to capture fully the long-term economic benefits of waste reduction projects.

It is precisely because of the potential for this stage to bring to an end the company's or plant's waste reduction effort that the role of government becomes more critical. Government policies, national goals, jawboning, and performance requirements can keep the pressure on companies to maintain their commitment to waste reduction. The use of special economic incentives such as tax breaks, for example, may spur capital investment in waste reduction, which may seem less attractive than other uses of capital (e.g., expansion and diversification). Government small business loans for waste reduction can be given special preference. And much more attention is needed to offering flexibility to comply with current regulations, so that companies can channel their capital investment into pollution prevention instead of more pollution control facilities.

Much effort and good analysis by people in industry will be needed to understand the benefits of using a pollution prevention solution instead of a traditional end-of-pipe one driven by regulatory requirements. A good example of this new kind of thinking happened at a 3M plant that makes magnetic oxides for recording products. Ammonium sulfate was passing through the plant's wastewater treatment plant untouched and entering a local river. The company's analysis showed that it would have to invest $1 million for pollution control equipment to control the chemical's concentrations in its effluent. 3M also determined that it would take $1.5 million to introduce a material reclamation system so that ammonium sulfate could be reused as fertilizer. However, even though the waste reduction solution meant more capital investment, 3M adopted it because the change would also produce annual revenues of $150,000 and management looked at that figure relative to the *incremental* increase in capital investment. In this way, the additional capital investment was recouped in about 3 years.

The discussion about this stage has been for industry and other organizations, such as government agencies, offices, and commercial activities. What about people in their role of consumer? Is it necessary and feasible for consumers to perform detailed, quantitative analyses to evaluate their waste reduction efforts? Practically speaking, it is expecting a lot of people to take the time to do such analyses, assuming that they have the skills to do so, which is certainly not the case for a large fraction of Americans. A more practical view is that consumers can use the analytical results of others, especially environmental and public interest groups which they may belong to, as well as the findings of government agencies. Also, most homes are not as complicated as factories. This reduces the need for formal analysis.

STAGE 4:   ULTIMATELY, NEW TECHNOLOGY AND PRODUCTS ARE NEEDED

When speaking of R&D, it is self-evident that individual consumers are not capable of creating new industrial processes and products. The role of the individual consumer gives way to institutional actions, but consumer interests can shape the agenda. Indeed, consumers have enormous power in the marketplace because a relatively small shift in market share of a major product can spell success or failure. And in the wings are countless entrepreneurs and companies with small market shares which eagerly seek new products. Although the rapidly emerging green consumerism is filled with problems, such as deceptive claims and unnecessarily high premium prices, it is rich in opportunities for consumers to define their values, preferences, and demands, which will, in some measure, help set short- and long-term R&D priorities in industry. This same basic argument applies to individual workers in industry, offices, schools, and government. Workers who identify sources of waste and pollutant production can help set R&D needs, and workers often can suggest approaches to waste reduction which need R&D to apply successfully.

Along with technical R&D, there is another similar type of investment in the future: educational programs. American society has hardly scratched the surface of researching and developing new courses and products at all levels of education. Pollution prevention is exactly the kind of fundamental, simple, and easily understood concept that can be introduced into primary schooling, highly specialized college curricula, and worker training.

Industry is getting a stronger message: eventually, for both manufacturing processes and products for industry, government, and consumers, new technical solutions based on pollution prevention must be sought through R&D. Indeed, during the three previously discussed stages, many waste reduction needs should be identified. Completely new manufacturing processes and products can be considered, with waste reduction a primary goal. Designing, making, and marketing new consumer products pose the greatest challenges. The idea of gaining a competitive advantage through selling products that appeal because they offer environmental benefits could be the major marketing breakthrough of the 1990s. Support for this perspective includes the following:

- A 1989 survey by the Michael Peters Group, a consulting firm, found that 89 percent of 1,000 Americans interviewed expressed concern about the environmental impact of their purchased products.

- Referring to new companies marketing environmentally beneficial products, Prof. Ray A. Goldberg (1989) of Harvard's School of Business Administration said: "This is not just a small market niche of people who believe in the 'greening' of America. It is becoming a major segment of the consuming public."
- In November 1989, Stephen Garey formed an advertising firm in Los Angeles that is probably the first one devoted to environmentally sensitive products and services. According to the *Wall Street Journal* (1990), a newspaper ad by Stephen Garey & Associates said: "Despite overwhelming evidence of serious damage to the Earth, tens of thousands of companies, large and small, continue to make products and profits in ways that can only lead to further ill health and tragedy." The firm said that it would employ its skills on behalf of "any firm whose product, service and method of manufacturing brings no harm" to the planet.
- David A. Nichol (1989), president of Loblaw supermarkets, the largest chain in Canada, which had an immediate success with environmentally friendly products, said: "From a marketing point of view, it's potentially the most profound change we've seen in the consumer goods business for a number of years and perhaps decades." Because Loblaw is owned by a Canadian food producer, its market success with existing products will likely affect R&D for new products.
- In the United States, the Wal-Mart chain of 1,347 stores has initiated a campaign of identifying products with environmental improvements, and attempts to influence both manufacturers and consumers.

To put current optimism in historical perspective, however, consider the following finding about public opinion in the United States, published in the June 7, 1974, issue of *Beverage Industry:* "Overwhelmingly, those interviewed said 'ban them [nonreturnable beverage bottles] by law' rather than increasing the cost of products to cover disposal costs or putting up with the pollution problems they cause." The public preference for returnable bottles and for having government ban nonreturnable bottles was just as much a sign of green consumerism as any that exists now.

Moreover, John R. Quarles, Jr., Deputy Administrator of the EPA, said *15 years ago:*

By waste reduction we mean to include every change in our production and consumption practices that will result in less waste of raw materials or energy and will reduce disposal problems. Shifting back to the returnable bottle is one example and a good one, though unfortunately far too few people recognize that it is only one of literally hundreds of good opportunities to achieve waste reduction.

When was the last time anyone used a returnable beverage bottle? The lesson: don't ignore the obstacles to waste reduction.

Products free of toxic chemicals and products which generate little household waste could have the same kind of appeal to consumers as foods which help prevent disease and products which have higher quality. Conversely, more conventional products which contain hazardous substances and generate a lot of garbage could be seen increasingly as dangerous as cigarettes and as unattractive as defective and short-lived products. U.S. manufacturers need to investigate international market opportunities for what are being called *safe substitutes, toxin-free products,* and *green products.* Otherwise, the U.S. trade deficit could become much worse as foreign companies move faster to capitalize on global environmental consciousness.

But large-scale product change will require wholehearted R&D programs by manufacturers of consumer products, and eventually these efforts will affect producers of primary chemicals and the materials they use. Some examples of significant efforts, sometimes benefiting just one company and sometimes many others, are as follows:

- Polaroid Corporation spent years developing a battery for its film packs which does not contain mercury; this is a real environmental benefit for municipal wastestreams and helps Polaroid reduce its own hazardous waste.
- Cleo Wrap, a printing company, spent 6 years developing water-based inks to replace traditional solvent-based ones in its production of wrapping papers. The company was able to virtually eliminate hazardous waste generation. It saved $35,000 in annual waste disposal costs, avoided new regulatory burdens, lowered its fire insurance premiums, and got a lot of good publicity for its efforts.
- Reynolds Metals Company developed technology to use water-based inks to replace solvent-based inks at two printing plants of its flexible packaging division. It was facing the requirement to reduce air emissions of volatile chemicals by 65 percent. The plant changes necessary to continue using solvent-based inks would have cost about $30 million. Instead, the company developed the alternative technology, reduced its ink costs and its insurance bill, and now wants to license its new technology to others.
- Union Carbide found a way to use carbon dioxide to replace current organic solvents for spraying paints, particularly in large industrial operations. Between 30 and 70 percent of the volatile solvents released to the atmosphere, which contribute to smog, could be eliminated. It took 4 years of research and millions of dollars for Union Carbide to develop this innovative process.
- Cosmosol, Ltd., has a new hair spray called New Idea that contains

no alcohol, a volatile chemical that contributes to smog. It replaces alcohol with a safe substitute, a water-soluble formulation using dimethyloxide, and each spritz releases only 20 to 40 percent as much vapor as other aerosol sprays.

- The Westinghouse R&D Center developed a process that uses ultraviolet light instead of hazardous, volatile solvents to dry and set colored paints.

- Du Pont has been spending large sums of money to find alternatives for CFCs that deplete the ozone layer. In 1988 it spent more than $30 million for process development, market research, applications testing, and small-lot production of CFC alternatives. Spending in 1989 was about $45 million for R&D. Du Pont has planned to introduce new products starting in 1990.

- Ciba-Geigy developed a new process for producing dyes for cotton without the use of a mercury catalyst, cutting its consumption and loss of mercury by thousands of pounds a year.

A specific focus on waste reduction is not the only role for R&D. Normal R&D projects for new products must also include a new emphasis on reducing waste generation during production. A good example of this comes from Merck and Company, a manufacturer of pharmaceuticals. During a 5-year period from research through scale-up in a pilot plant and initial manufacturing, a focused waste reduction effort resulted in elimination of 50 percent of the waste that would have otherwise been produced from a complicated multistep synthesis process to make a new broad-spectrum antibiotic. Moreover, the waste reduction effort was applied to full production of the product. Impressive accomplishments include an 82 percent reduction in methylene chloride consumption and an 80 percent reduction in acetone use. Although Merck spent a lot of money on this waste reduction, its economic benefits have outweighed its costs. One analysis by Merck revealed that incineration of waste solvents would cost 70 cents per pound of solvent which cost only 21 cents a pound to buy originally. Avoiding waste production and treatment, therefore, presented a substantial economic opportunity.

Clearly, many small, medium-sized, and large companies will face problems in committing resources to R&D. Some industries already have problems with low levels of R&D, and others use R&D for other objectives which have little to do with concerns about waste or pollution generation. Government could play a major role at this stage by funding R&D programs which could benefit large segments of industry, both by giving tax breaks for company R&D and by working with industries to establish R&D priorities to benefit all companies within them. And of course, government has sometimes applied the greatest pressure of all by banning chemicals or products. This is a potent tool to spur R&D and one which may be used with more frequency. The rapid apparent success in finding substitutes for CFCs is impressive. To spur development of new consumer products, the government could help develop special labeling to identify environmentally beneficial products for consumers; this is already happening in Canada and Europe.

HOW ONE COMMUNITY OVERCAME OBSTACLES

The Chevron USA petroleum refinery in Richmond, California, was identified in the early 1980s as a serious source of pollution of San Francisco Bay. As the largest of six bayside refineries, Chevron contributed 35 percent of the chromium and more than half of the nickel discharged from all major point sources into the northern part of the bay in 1986. Scientific evidence showed that these and other metal wastes were toxic to sensitive aquatic life even in very low concentrations. Weak federal standards did not prevent substantial discharges into the bay. The San Francisco office of Citizens for a Better Environment (CBE) led an effort to appeal Chevron's discharge permit from the state's water resources control board. The goal was to get more stringent standards that would force the company to reduce discharges, preferably by pollution prevention. In 1985, the effort succeeded. Chevron was required to analyze its processes and operations to reduce or eliminate pollutant discharges. As part of a settlement process between CBE and Chevron, reduction of discharges would be based on either pollution prevention or treatment.

The plant had been using conventional water treatment technology built in the 1960s and 1970s, but Chevron chose not to add more of this kind of technology to meet the new objectives. Chevron pinpointed specific sources of waste generation by conducting a waste reduction audit, including a mass balance favored by CBE. The *mass balance technique* provides precise data on the flow of a specific substance, such as chromium, through all the components of the plant's processes. The audit revealed that more than 90 percent of the chromium and nickel discharges to the bay came from several sources in two relatively small areas of the very large 2,900-acre refinery. More specifically, the major source of chromium was the zinc chromate added to cooling tower waters to inhibit corrosion, scale, and slime. Nickel came from a catalyst used in petroleum refining. A substitute was found for the zinc chromate that cut chromium discharges, and the reformulation of gasoline to eliminate lead dramatically reduced lead discharges.

At first, nickel discharges were reduced by installing mechanical seals on washer pumps and piping to segregate high-nickel wastewater and by recycling some nickel internally. Water use was reduced in many plant areas to cut effluent discharges by an average of 15 million gallons per day. Some additional water treatment was also introduced, but this produced solid, hazardous waste for disposal. By late 1988, toxic metals discharges to the bay had been reduced by 67 to 97 percent. Chromium discharges to the bay had decreased by 1,200 pounds per year, for a 67 percent reduction; and nickel discharges had decreased by 7,600 pounds per year, for an 86 percent reduction. Lead discharges were reduced by 97 percent. No reductions were achieved for selenium and some other toxic pollutants.

The experience with chromium illustrated the inadequacy of government environmental regulations under the Clean Water Act. By using pollution prevention techniques, chromium discharges were reduced to 600 pounds per year even though the federal government's best available technology standard allowed 7,000 pounds annually. The conclusion drawn was that the standard was at least 10 times too lax. Moreover, the Chevron experience showed that reducing chromium alone does not necessarily reduce discharges of selenium and other toxics, as the federal standards assume.

Greg Karras of CBE said the following (1989) about the Chevron experience:

> By encouraging manufacturers to prevent pollution before it starts, citizens can help industry break out of the waste management culture that promotes continued use and disposal of hundreds of millions of pounds of toxics every year. This Chevron case study provides a prototype for citizen activism and industry action. . . . An audit identifies fruitful alternatives to current pollution control practices. To take advantage of these alternatives, citizens should get involved in toxics-use decisions. Direct public involvement in local decisions can be effective in the face of government regulatory programs that heavily favor pollution control instead of pollution prevention . . . citizens should demand and share in these audits; and government should require them. . . . Chevron's resistance to [using a mass balance] turned to cooperation the moment the in-plant audit information, particularly the mass balance results, was made public. . . . But the refinery . . . still resists calls to reduce air emissions of toxic petrochemicals.

At other places in the United States, communities and grass-roots environmental groups continue to try to execute "good neighbor agreements" like that achieved by CBE with Chevron. The goal of most community-oriented environmental groups is pollution prevention. Some of the success of this Chevron story must also be related to the introduction of the company's SMART program in 1987. However, there may be some truth to the belief that it was the CBE effort that helped educate Chevron about the benefits of a corporate waste reduction program. Most companies are likely to resist grass-roots activism. For example, a community organization named BayCap in the Houston, Texas, area formulated a good neighbor agreement to address its concerns about the high incidence of cancer and 7,000 tons of air emissions from the Exxon Chemical Americas plant in

Baytown. But Exxon turned it down even though the company had helped form BayCap to improve its communications with the community and had provided detailed information about its emissions to BayCap for a year previously.

We conclude that there are several limits to this general approach of citizen intervention. First, it is difficult to obtain coverage of all wastes and pollutants. Second, the community needs a lot of resources, especially technical expertise, to participate effectively. Third, current government environmental regulations stand in the way of promoting pollution prevention solutions and obtaining lower levels of discharges than allowed by law. The effect of government regulations is twofold. First, companies may be disinclined to pay attention to wastes within the limits allowed in the permits granted to them by state or federal agencies. Second, citizens may only be drawn to regulated wastes covered by permits because there is some public data on those wastes. Citizen activism may get excellent results in a limited number of places, but it is no substitute for improved corporate and government policies and programs to affect all of American industry.

MONEY MADE FROM WASTES, NOT POLLUTION PREVENTION

A major, but often overlooked, generic impediment to nationwide reduction is the existence of a very large U.S. waste management and pollution control industry. This industry has experienced phenomenal growth and profits in the past decade. For example, the Freedonia Group has reported that total revenues for handling hazardous waste products (e.g., incineration equipment, treatment chemicals, analytical instruments) and providing services (e.g., consulting/engineering, cleanup, incineration and landfill services, transportation) totaled $505 million in 1977. This had increased to $9.7 *billion* in 1988 and is projected to increase to $20.2 *billion* in 1993 and $43 *billion* in 2000. The average annual compound growth rate was 30.8 percent from 1977 to 1988 and is projected to be 15.8 percent from 1988 to 2000, a decrease but still a very high industry growth rate.

The point is that this hazardous waste industry lives off of the wastes and pollutants produced by American industries. In fact, this industry is just a small part of the larger environmental products and services industry that makes money from air and water pollution control, industrial solid waste, and municipal solid waste activities. Data reported by the *Environmental Business Journal* indicates that the entire environmental industry had revenues of about $400 *billion* in 1989.

No such industry sits back and responds passively to developments and changes as if they have no influence on them. To the contrary, the environmental industry is becoming increasingly well organized and politically active. True pollution prevention activities by industrial waste producers and by government are not in the financial interests of the environmental products and services industry. The environmental end-of-pipe industry markets their services and products aggressively, and is a potent force undermining waste reduction initiatives in the public and private sectors. However, it should be noted that rarely do such end-of-pipe financial interests publicly and explicitly oppose pollution prevention. The natural appeal of pollution prevention for the public good would make that kind of public opposition appear crass and virtually unpatriotic.

The result is that resistance to pollution prevention takes other, more sophisticated and subtle forms. For example, major waste companies have moved quickly to support waste recycling, because they can make money from recycling activities and because public support of recycling diverts interest from pollution prevention. They also support strong traditional regulatory programs to maintain requirements for strict end-of-pipe controls and equipment. Pollution prevention is often painted as a long-term goal but not an immediately effective strategy. This is ironic because in so many cases it now takes a long time to locate, get government approval for, and construct major new waste management facilities, particularly municipal and hazardous waste incinerators. Considerable effort is given to maintaining that essentially voluntary industrial waste reduction is effective and that strong government programs, possibly including legal requirements to practice pollution prevention, are unnecessary.

Sometimes it may also appear that a company has been attracted to the waste reduction area because its marketing literature uses terms such as *waste reduction* and *waste minimization*. However, for the most part, little business is generated by pursuing genuine waste avoidance and elimination. Consulting firms often gain entry with rhetoric about waste reduction but are predisposed to sell traditional end-of-pipe products and services. They continue to market those services instead of reduction and prevention ones. It is classic bait and switch marketing. An important principle for waste producers to remember is that the greatest expertise about waste production sources is possessed by the people, engineers and plant workers alike, who work with those sources. Outside consultants with a proven commitment to pollution prevention over many years, therefore, are best seen as traditional management consulting services.

The need is not for detailed technical expertise but for help in dealing with the obstacles discussed above, which are more closely related to motivating people and changing company procedures and organization (e.g., information collection systems, waste reduction audit procedures, and employee recognition criteria) to meet pollution prevention needs. The strategic need is to redirect and channel the internal company expertise about processes, operations, and products to attain well-defined corporate pollution prevention objectives. Only a true believer in pollution prevention can really help a company obtain maximum benefits.

A note of caution: in many companies there are environmental engineers and regulatory compliance staff which may be predisposed to accept the advice of and respond positively to the marketing of environmental products and services companies. The same is true for owners and managers of small businesses and commercial establishments who turn to environmental firms for assistance. Getting help that is dedicated to the pollution prevention interests of a waste generator requires care and vigilance by the company.

HOW MUCH PROGESS IS BEING MADE?

Identification of obstacles to pollution prevention would not be important if waste producers had already learned to overcome them. On the consumers' side, there is no reason to believe that there has yet been a significant amount of waste reduction, despite the enthusiasm for recycling. Many companies are accomplishing much more waste reduction than ever before, and some companies are beginning to receive recognition for doing so. For example, as reported in *Fortune* in February 1990, Franklin Research and Development of Boston rated 25 companies for their environmental performance, and 2 got high ratings because of their waste reduction programs. In the oil industry category, Amoco got the high rating and Exxon the low one. In the photo equipment category, Polaroid got the high rating and Eastman Kodak the low one. Six companies got high ratings because of their recycling efforts.

Overall, however, there is no reliable data which documents a major turndown across all U.S. industries in the production of all wastes and pollutants over time. Gregory Hollod (1990) of CONOCO, a leading industrial waste reduction expert, said that "industry needs to move more aggressively towards reducing waste" and that "there is a need for industry to act with more urgency." Two academic specialists (Vig and Kraft 1990) in environmental policy observed that "it has often been cheaper [for industry] to go to court to avoid compliance than to purchase and install expensive equipment, or simply to keep old polluting plants in operation rather than build new, more efficient ones for which standards are higher." Blatant noncompliance with government regulations is also pervasive. EPA data reveals that compliance with hazardous waste regulations, for example, was only 52 percent for industry and only 40 percent for federal government agencies in fiscal year 1988. The popular belief that endless government regulations and increasing waste management and pollution control costs automatically result in pollution prevention ignores alternative responses. Some companies do, of course, respond with pollution prevention, but many do not.

When some companies report percentage decreases in waste generation, the information is often misleading for the following reasons. Only a narrow definition of waste is usually used, mostly federally defined hazardous waste. Waste reduction sometimes results from the successful effort by a waste producer to *delist* a waste from the hazardous waste regulatory system by convincing a government agency that the waste does not meet certain definitions or have certain characteristics. Meanwhile, the waste is still being produced. Reductions in the release of pollutants to the environment don't always result from true pollution prevention actions, but rather from the use of end-of-pipe treatment which may simply transfer pollutants somewhere else. This is especially true for reported reductions in air emissions. Wastes sent offsite for recycling or burning are often counted as wastes reduced. Reductions are not usually related to production outputs. Therefore, other factors may explain waste reductions, such as plant closings or product changes. Sometimes a large company may "export" its waste generation by using another company to perform a waste-intensive operation.

Not only are enormous quantities of wastes and pollutants still being produced by industry, but there are signs that the level of hazardous waste generation (wastes managed in facilities regulated under the Resource Conservation and Recovery Act) for the United States as a whole will continue to increase. For example, the Freedonia Group has projected that U.S. hazardous waste generation will increase from 291 million tons in 1988 to 510 million tons in 2000 in spite of industrial waste reduction efforts. Other categories of industrial wastes, such as air and water pollutants, might similarly increase. Should this happen, today's most skeptical environmentalists who say that industry is not doing enough waste reduction will have been vindicated.

From a public policy perspective, the dilemma is whether to wait for the real data to come in or to work on the assumption that the current voluntary waste reduction system is not accomplishing enough and, therefore, that more aggressive government efforts are necessary.

The following are additional pieces of evidence for concluding that much more waste reduction is needed:

* Analyses of projects and professional papers being labeled as *waste reduction, waste minimization,* and similar terms actually consist of large numbers—typically 75 percent—of end-of-pipe techniques rather than true pollution prevention. The public opposition to land disposal of hazardous wastes, in particular, has driven industry to so-called waste treatment techniques, such as incineration, but not necessarily to pollution prevention.
* A study by the State of Illinois of 275 companies concluded that in

1985 over 50 percent of hazardous waste generators had not yet begun serious waste reduction.

- A study by a California public interest group of 100 small metal electroplaters found that 75 percent were not implementing waste reduction plans in 1986.
- In a study by the State of New Jersey of 22 firms, only 41 percent said that they had implemented some waste reduction from 1981 to 1985, and 36 percent said they would do so in the future.
- Data reported by the Chemical Manufacturers Association (CMA) on hazardous waste trends has been interpreted by the group to show major waste reduction from 1983 to 1987. However, the data represents only about 12 percent of the U.S. chemical industry. Participation by companies in annual surveys is voluntary, and the sample is not likely to be representative of the whole industry, because companies which have not done significant waste reduction are unlikely to participate. The CMA data also does not relate waste generation figures to production outputs. Moreover, the data presented on a year-to-year basis discloses that waste reduction has leveled off since 1983.

EPA's information on waste reduction, collected through the requirements for reporting to the Toxic Release Inventory, is not especially impressive. Only 11 percent of facilities (2,090 out of 19,278) reporting voluntarily gave information on their successful waste reduction from 1986 to 1987. Actually, only 6 percent of chemical specific forms (4,352 out of 74,152) filed gave waste reduction results. Only 802 of the 4,352 forms had sufficient correct data to allow analysis of the actual amount of waste reduction. For that subset of voluntarily reported information, waste generation decreased from 121 million pounds in 1986 to 73 million pounds in 1987 (not counting offsite recycling). Because only those companies that have reduced waste generation are inclined to report this information voluntarily, the results undoubtedly overestimate recent waste reduction. The subset is so small that only a few large facilities could account for most of the reduction reported. The results are somewhat better if production of useful outputs in reporting facilities *increased* from 1986 to 1987, which it did for 76 percent of facilities; this doubles the effective reduction to about 100 million pounds. Compared to the total amount of waste reported in 1987, 22.5 *billion* pounds, the reduction of even 100 million pounds is only 0.4 percent of the national wastestream reported to this database, which itself is only a subset of all environmental wastes.

A major study by the Organization for Economic Development and Cooperation in 1985 concluded that neither U.S. nor European industries had widely implemented pollution prevention. In fact, a French study found that in the industries which account for most waste and pollutant generation, such as chemical manufacturing and steelmaking, less than 3 percent of the facilities could be classified as "clean."

SUMMARY

This is just the beginning of a pollution prevention social experiment. The four basic stages of waste reduction—(1) getting people's attention and applying common sense, (2) getting information, (3) analysis, and (4) R & D— will go on forever as more and more people get involved, new circumstances require new efforts, and people try hard to achieve complete pollution prevention.

Public policy commitment to pollution prevention and serious government programs grow more important as industry and consumers face increasingly stubborn obstacles, but they have barely begun. Weak as current data collection is, we collect more than enough data to know in 5 years whether we are making progress, and certainly to know in 20 years whether we—industry, government, and consumers—have made a serious commitment to pollution prevention. It is not necessary to have precise figures, but rather to have qualitative proof of whether or not substantial waste reduction has occurred.

Waste reduction by consumers has hardly begun. But the question for many people today is whether industry has already turned the corner and made a serious, long-term commitment to pollution prevention. If it has, then there is less reason for public concern and for new government initiatives such as regulatory requirements for waste reduction audits and maximum allowable levels of waste generation.

We strongly believe that the hundreds of examples of successful industrial waste reduction that we and many others cite do not demonstrate a full pollution prevention effort across all American industries, especially if wastes and pollutants are given the broadest definitions.

Many studies of waste reduction have reached the same conclusion: technology is not the key obstacle to a massive reduction in waste and pollutant production. Given the wide range of obstacles to success, future pollution prevention is more dependent on leadership than on technology. Leadership is needed *now* to communicate the benefits of preventive planetary care and to overcome the many obstacles discussed above. Inevitably, many people will have anxiety and resist change as they correctly sense that there will be winners and losers as industrial processes and consumer products change to prevent pollution. However, the necessary public good outweighs narrow, selfish financial interests.

4

DATA TELLS THE STORY: TOO MUCH WASTE

In this chapter, we present detailed information on the complex diversity of environmental wastes and on the enormous amounts of them being produced by all parts of American society. Our purpose is not to present a compendium of data on environmental wastes but to use selected data to show that the production of wastes in the United States is very large, especially compared to that of other nations with high standards of living. Information is also given to support the proposition that in advanced industrialized nations like the United States, the tendency is to produce more wastes every year. And in developing nations, waste production is also likely to increase and become more toxic in nature. These trends result from increasing population, increasing industrialization, and increasing use of waste-intensive and toxic products and services.

THE ENVIRONMENTAL WASTE UNIVERSE

There is no complete account of the whole environmental waste universe. Up-to-date, accurate data on nearly any component of the world of waste is also a constant problem. No government agency has ever been given the responsibility to obtain routine, periodic data on all wastes produced by society. Part of the reason is that many wastes are not regulated by the government; therefore, there appears to be no official need for the information. Another complexity is the historic disposition to think of environmental wastes in reference to air, water, or land.

Releases of wastes to the air is a particularly difficult area to get good information on, except for a few types of substances that the EPA has tracked through a monitoring program. For wastes discharged into water, there was no national data reporting or collection until 1987. Wastes that end up on land have gotten the most attention but that has been sporadic, although the data is more comprehensive than for air or water.

A very good source of air data is the annual *National Air Quality and Emissions Trends Report*. In March 1989 the report for 1987 emissions was published, the 15th in the series. Of particular interest in this report is the trend over 15 years of the generation of six major environmental wastes for which there are National Ambient Air Quality Standards. In 1987, 102 million Americans lived in counties in which one or more of these standards was exceeded. About 150 million tons of the six pollutants were released into the atmosphere in 1987.

Figure 4-1 shows 15-year (1978 to 1987) trends for five key wastes released into the air: sulfur dioxide, for which most of the 17 percent decline occurred prior to 1983; carbon monoxide, for which there has been a steady decline, but not much from 1986 to 1987; nitrogen dioxide, for which there was a 3 percent increase from 1983 to 1987; volatile organic compounds (VOCs), which have been holding steady for several years; and lead, for which the remarkable decline of 94 percent over the 15 years is accounted for by government restrictions on leaded gasoline.

A very important new source of information is the Toxic Release Inventory (TRI) database. The first report for 1987 is available. This source covers over 300 specific substances and provides data for all kinds of disposition of the wastes. Unlike the previous database, however, for which the government takes measurements, the TRI data are reported by those industrial and commercial establishments required to submit data to the EPA. For 1987, 19,278 manufacturing facilities submitted 74,152 individual chemical reports. The OTA provided an analysis of the 1987 TRI data in testimony before the U.S. Senate in May 1989. The OTA concluded that

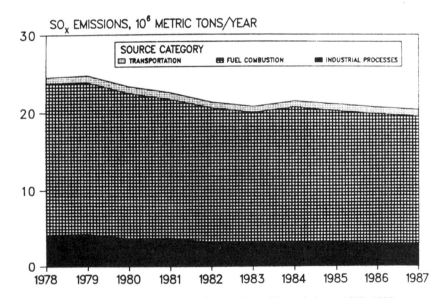

Figure 4-1(a).   National trend in sulfur oxide emissions, 1978–1987.

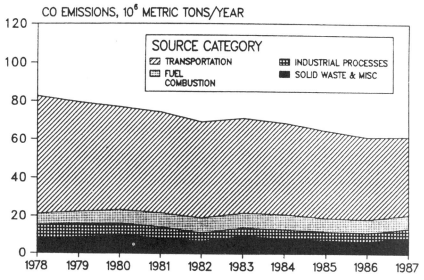

Figure 4-1(b).   National trend in emissions of carbon monoxide, 1978–1987.

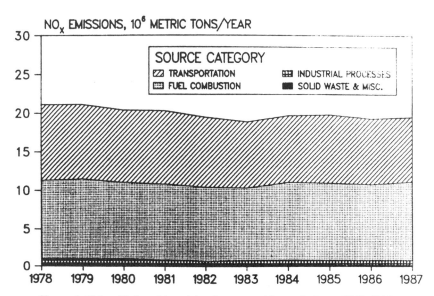

Figure 4-1(c).   National trend in nitrogen oxides emissions, 1978–1987.

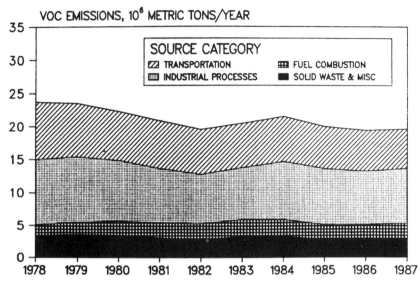

Figure 4-1(d).   National trend in emissions of volatile organic compounds, 1978–1987.

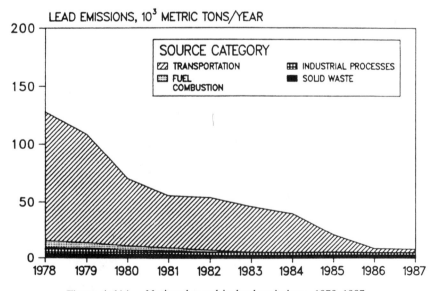

Figure 4-1(e).   National trend in lead emissions, 1978–1987.

the reported 20 billion pounds of toxic chemical wastes was "just the tip of the toxic tower." The OTA estimated that as many as 400 billion pounds of wastes were probably generated by American industry. It should be noted that the TRI data do not count environmental wastes produced by automobiles, households, and many other nonpoint sources.

Aside from this factor, the OTA pointed out that the EPA's TRI data underestimated environmental waste generation for several other reasons: There was substantial nonreporting and underreporting (including estimates rather than measurements) that probably reduced actual waste generation by 50 percent; thousands of chemicals are not covered and facilities with fewer than 10 workers are not required to report, and these factors might mean that only 10 percent of the actual amount of industrial waste was reported. For these reasons, the OTA multiplied the EPA sum by 20 to make a general correction.

Note that the monitored air pollutant database shows that about 300 billion pounds of only six wastes were produced in 1987, which strongly supports the OTA's contention that the TRI data greatly underestimate total U.S. waste production. The air data accounts for all sources, unlike the TRI data. However, one important discrepancy is that the TRI data for total lead emissions from industry is about 40,000 tons, while the air pollution data indicates only about 3,000 tons from industry. Also, the EPA's data for hazardous waste indicates a total of close to 600 million tons (1.2 trillion pounds) produced in 1987, but the majority of that total is water. We estimate that about 300 billion pounds of nonaqueous material is hazardous material which is definitely not covered by the air pollutant data. In other words, these two other sources of information alone add up to 600 billion pounds.

As another indicator of the underestimates of the TRI data, consider that the EPA reported in late 1989, when it issued final rules on benzene regulation, that total emissions were 686 million pounds. The TRI data reports a total of 33 million pounds. The EPA says that 80 percent of benzene emissions are from motor vehicles, which are not covered by the TRI. Even so, the TRI data is only 5 percent of the national total, not 20 percent.

All things considered, the total generation of many different kinds of pure toxic substances as environmental wastes may be about 1 trillion pounds (500 million tons), and not the initial reported value of about 20 billion pounds that received so much public attention when the first annual TRI data was reported. Also note that carbon dioxide emissions, of importance for global warming, are not covered in the databases discussed above. In setting national pollution goals and in measuring success over time, it will be necessary to pay much more attention to the collection of accurate data.

Another traditional way of looking at the environmental waste universe is to consider only solid (landed) types of wastes, only a small fraction of which is regulated by the government. Table 4-1 provides a summary of solid waste generation for the United States and for the European Community, which has roughly the same population. We estimate that a total of about 6.5 billion tons of solid wastes are being produced in the United States, which is about 2.5 times the amount being produced in the European Community on a per capita basis. Most of the U.S. total is accounted for by agriculture, mining, and power plant (especially fly ash from coal burning) wastes, which are generally acknowledged to be high-volume, low-hazard wastes. Nevertheless, they can pose environmental problems, depending on how they are managed, and some, but by no means most, could be reduced through preventive measures.

The problem, in general, with very large amounts of solid waste is that a large portion are not toxic substances. However, when some kinds of large-volume, low-hazard wastes are disposed of on land, they can still pose environmental problems. For example, the leaching of toxic heavy metals from large deposits of power plant fly ash can be a problem, and the burning of

TABLE 4-1.   APPROXIMATE TOTAL SOLID WASTE GENERATION IN THE LATE 1980s (IN MILLIONS OF METRIC TONS)[a]

Waste Type	European Community	United States
Agricultural	1,100	3,000
Extraction industry/power stations	400	2,200
Sewage sludge[b]	300	8
Industrial	160	1,000
Demolition rubble	160	50
Domestic	150	285
Used oils	2	5
TOTAL	2,272	6,548
Approximate per capita waste (metric tons per person)	10	25

Source: Adapted in part from United Nations (1988) and Haznews (1990) (for the European Community: Belgium, Denmark, West Germany, France, Ireland, Italy, the Netherlands, The United Kingdom) and Murarka (1987) (for some U.S. data).

[a]Data among nations is only approximate because of wide differences in definitions and methods of obtaining data.

[b]The large difference between the United States and the European Community is due to the dry weight basis for U.S. data.

ordinary wastes can lead to air pollution. Huge volumes of waste cost money and their management uses up available land, both of which could be applied to more productive purposes.

INDUSTRIAL HAZARDOUS WASTE

Industrial hazardous (toxic) waste consistently receives a lot of attention. This category of solid waste was the last to be regulated by government in the United States and nearly everywhere else, after air and water pollutants. The public's concern about industrial hazardous waste is clearly linked to toxic waste (Superfund) sites which have resulted in contamination of land and water. Starting with Love Canal in the early 1970s, people have come to fear the effects of indiscriminate dumping and even controlled land disposal of industrial toxic waste. Since passage of the first federal law governing hazardous waste in 1976, more and more types of wastes have been included in this regulatory category. Some states define an even broader universe of hazardous waste.

Table 4-2 presents a summary of hazardous waste generation in the United States, including a projection from 1988 to 1993 and 2000. The basic kinds of waste and their forms of management are also shown. The chemicals manufacturing industry accounts for over half of hazardous waste generation (and for 54 percent of the TRI releases for 1987). As explained in the table, one of the great complications in understanding U.S. data for hazardous waste is that more than half of it is usually not accounted for when data is presented. Of the total of 583 million tons produced, more than half (311 million tons) are hazardous waste by definition but are not managed in facilities that are regulated under the federal hazardous waste law, but rather in facilities regulated under the Clean Water Act.

In spite of considerable attention to hazardous waste reduction by industry, the company that assembled the data in Table 4-2 concluded that there would be a net increase in industrial hazardous waste generation in coming years. From 1988 to 2000, a 75 percent increase in U.S. hazardous waste generation is predicted. However, the annual rate of increase would be less than in past years, which could result from waste reduction as well as from less industrial activity in some sectors. It is also generally anticipated that more wastes will fall under regulation, which could offset successful waste reduction efforts. As corroboration of the projected national increase in hazardous waste, we cite the 1989 estimate made by the State of Texas, the largest state generator of hazardous waste and the second highest in the TRI database, according to the EPA. From 60 million tons in 1987, Texas industry is expected to generate 95 million tons by 2009 in spite of a planned, comprehensive waste reduction program. That is a 58 percent increase over 22 years, or just over 2 percent annually, which shows how even a small annual increase adds up.

TABLE 4-2.   GROWING HAZARDOUS WASTE GENERATION IN THE UNITED STATES DESPITE REDUCTION

	Actual	Projected			Average annual growth	
	1977	1988	1993	2000	1977–88	1988–93
Millions of Tons						
Total:[a]	131	291	380	510	7.5%	5.5%
By Type of Waste						
Heavy metals	51	114	149	196	7.6	5.5
Organic chemicals	42	100	132	180	8.2	5.7
Petroleum derived	16	33	44	60	6.8	5.9
Inorganic chemicals	17	35	43	55	6.8	4.2
Other	5	9	13	19	5.5	7.6
By Method of Disposal						
Landfill/impoundment	12	200	225	165	29.1	2.4
Deep-well injection	5	14	15	10	9.8	1.4
Illegal dumping	110	35	20	5	−9.9	−10.6
Percent						
Land disposal	97	86	68	35		
Incineration	Neg	15	35	95	—	18.5
Treatment/fixation	2	13	50	150	18.5	30.9
Resource recovery	2	12	30	75	17.7	20.1
Other	Neg	2	5	10	—	20.1

Source: Freedonia Group (1989)

[a]The total for 1988 is in good agreement with the EPA's totals of 267 and 272 million tons for 1985 from separate surveys covering federally defined hazardous waste managed in facilities requiring a permit under the federal hazardous waste regulatory program.

Data on hazardous waste generation over time in foreign nations is not readily available. Hungary is an exception. In 1982, officials reported that 1.3 million tons were generated. This increased to 1.6 million tons in 1986, for a 24 percent increase over the 5 years. This rate of annual increase is a little less than that of the United States (Table 4-2). Over that period, U.S. generation increased 34 percent, from 188 to 251 million tons, according to the data in Table 4-2.

Although industrial hazardous waste is also highly regulated in other nations, there is absolutely no consistency as to the definition of this form of waste. Another problem in comparing data from different nations is that different methods are used to obtain information. But nearly always, information comes from self-reporting by industry and sometimes from surveys of relatively small samples of industrial companies. With these caveats in mind, the data presented in Table 4–3 still offers some interesting insights. Clearly, industrial hazardous waste generation will depend on the exact level and types of industrial activity in a country. We have presented per capita generation data as well as information on GNP on a per capita basis. There is a distinct difference for three broad categories of nations: high levels of both industrialization and GNP, medium levels of both industrialization and GNP, and low levels of each.

A general observation is that increasing industrialization and economic growth without pollution prevention inevitably lead to more hazardous waste generation. For example, in looking at the difference between the most and least affluent nations, an increase of 25 times in per capita GNP is accompanied by an increase of about 150 times in per capita hazardous waste generation. An increase in per capita GNP of 8 times, in going from the low to the medium category, causes a 25 times increase in per capita hazardous waste generation. In going from the medium to the high category, an increase in per capita GNP of 3 times is accompanied by a 6 times increase in per capita hazardous waste generation. In other words, hazardous waste generation increases many times faster than increases in GNP, and the really big jump in hazardous waste generation occurs in the early stages of industrialization and then tapers off. Unfortunately, just when developing nations are experiencing enormous early increases in hazardous waste generation, they are least likely to have effective regulatory programs that stimulate safe management practices and waste reduction. By contrast, advanced industrialized, affluent nations may pursue clean technologies for themselves only.

Indeed, by closely examining the data in Table 4–3, one can see clearly that some nations stand out in their category. For example, the United States is a very high producer of hazardous waste compared to similar industrialized, affluent nations. Conversely, Switzerland, Japan, Denmark, and the Netherlands stand out as low generators of hazardous waste. The data pretty much confirms general impressions that these four nations have relatively technologically advanced and efficient clean industries, as well as vigorous regulatory programs. It is generally recognized that the United States has a regulatory system that is more inclusive than that of most nations, but it would take a 90 percent reduction in waste generation to bring the United States down to a value more typical of similar nations. And it would take about a 98 percent reduction to reach the level of one of the four low hazardous waste–generating nations in the category with the United States. It is doubtful that the broader universe of wastes regulated in the United States alone accounts for the differences between it and the

TABLE 4-3.   HAZARDOUS WASTE GENERATION AMONG NATIONS

Nation[a]	Amount (tons)[b]	Per Capita (tons)	GNP Per Capita ($)[c]
High Industrialization and GNP			
Switzerland (1987)	300,000	0.05 (low)	21,000
United States (1985)	583,000,000	2.3 (high)	17,000
Japan (1983)	1,540,000	0.01 (low)	16,000
West Germany (1988)	7,150,000	0.1	15,000
Canada (1983)	3,500,000	0.2	14,000
France (1988)	19,800,000	0.4	13,000
Austria (1984)	2,700,000	0.4	12,500
Denmark (1985)	154,000	0.03 (low)	11,000
Italy (1988)	5,000,000	0.09	10,500
Netherlands (1987)	580,000	0.04 (low)	9,000
United Kingdom (1988)	5,500,000	0.1	8,000
Belgium (1988)	1,650,000	0.2	8,000
Hungary (1986)	2,000,000	0.2	7,500
Average		0.3	
Average without high/ low		0.2	
Medium Industrialization and GNP			
Singapore (1985)	31,300	0.01	7,000
Hong Kong (1988)	105,100	0.02	6,000
Ireland (1988)	83,000	0.02	5,000
Greece (1988)	495,000	0.05	4,000
Spain (1988)	2,000,000	0.05	4,000
Taiwan (1986)	3,020,000	0.2 (high)	3,000
Malaysia (1987)	418,000	0.03	2,000
South Korea (1985)	409,000	0.01	2,000
Average		0.05	4,000
Average without high		0.03	
Low Industrialization and GNP			
Philippines (1987)	165,000	0.003	500
Thailand (1984)	163,000	0.003	800
China (1987)	44,000	0.0004	250
Average		0.002	500

[a]Year of data collection.
[b]Data among nations is only approximate because of wide differences in definitions and methods of obtaining data. Many sources have been examined. Data varies greatly because many governments do not have formal reporting systems. Definitions and methods of obtain

four low-waste industrial nations. More efficient industries which use cleaner technology appear to be an important explanation in some other nations. The other nation that stands out as a high hazardous waste producer is Taiwan. This finding too is consistent with general observations that Taiwan has been lax about environmental regulations which would pressure industry to practice even simple waste reduction measures, although that condition is changing rapidly.

WASTES FROM HOMES AND OTHER SOURCES

While industrial sources produce massive volumes of wastes of all kinds, both toxic and nontoxic, there are other sources of wastes in every country. Ignoring those sources ignores a host of areas where pollution prevention can be practiced to great advantage. Focusing on industrial sources also leads people to believe that the problem is them rather than us.

The major group of wastes other than industrial is usually called *municipal solid waste (MSW)*. The sources of MSW and the types of wastes included can vary depending on who's counting it. MSW can cover waste from households, offices, commercial firms, stores, malls, theaters, sports arenas, school, institutions, and government offices. Construction and demolition wastes and discarded automobiles are usually not included in MSW. But they can be.

MSW is picked up from generators by trucks and transported mostly to landfills and, increasingly, to incinerators, and recycling centers. It does not include waterwastes that move through sewers to municipal treatment plants or wastes that are emitted into the air.

---

ing data also vary. Many figures in the literature are not accurate. The following are especially important sources: United Nations (1988), *Hazmat World* (1990), and *Haznews* (1989) (for European nations); Keen and Jhaveri (1989) (for Asian nations); U.S. Environmental Protection Agency 1986 (for the United States). The figure is higher than the ones typically seen because it includes 311 million tons of hazardous waste which is managed in facilities permitted under programs other than the federal hazardous waste program (RCRA), as well as 133 million tons of state-defined hazardous waste. See also Craig and Warren (1988). Essentially, the same total figure was obtained by a private firm, Apogee Research, Inc., using a different methodology in a 1987 report for The National Council on Public Works Improvement.

The values are for the same year as—or within a few years of—the hazardous waste data. Thus, GNP have been rounded. Adapted from Showers (1989).

The Underestimated National Data on MSW

It is an unquestioned and widely reported "fact" that, on average, each American generates 3.6 pounds of garbage per day and that, for the country as a whole, garbage totaled 158 million tons in 1986. What few people realize is that those numbers are not measured; they are calculated. They come partly from a materials flow model and partly from waste composition studies. One firm—Franklin and Associates—designed the combination system and has been paid by the EPA since the early 1970s to update the numbers periodically.

The EPA/Franklin model data (72.5 percent of the total) can be presented in terms of the various materials (e.g., plastics, aluminum, and so on) in the waste or as the products and packaging that caused the waste. National estimates of food, yard, and miscellaneous inorganic wastes make up the balance (27.5 percent). Thus, for 1986 the model reported 115 million tons of waste and the estimates 43 million tons. Tables 4–4 and 4–5 give the model's basic output by materials and by products and packaging.

Actual waste collected varies considerably throughout the country, depending on such factors as whether an area is rural or urban and the season

TABLE 4–4.   EPA/FRANKLIN MSW DATA FOR 1986

Materials	Thousands of Tons	Percent of Total (by Weight)
Paper and paperboard	64.7	41.0
Glass	12.9	8.2
Ferrous metals	11.0	7.0
Aluminum	2.4	1.5
Other nonferrous metals	0.3	0.2
Plastics	10.3	6.5
Rubber and leather	4.0	2.5
Textiles	2.8	1.8
Wood	5.8	3.7
Other	0.1	0.1
SUBTOTAL	114.3	72.5
Food wastes	12.5	7.9
Yard wastes	28.3	17.9
Misc. inorganic wastes	2.6	1.6
TOTAL	157.7	100.0

Source: Franklin Associates, Ltd. (1988).

TABLE 4-5. EPA/FRANKLIN MSW DATA FOR 1986

Products and Packaging	Thousands of Tons	Percent of Total (by Weight)
PRODUCTS		
Durable Goods		
Major appliances	2,825	1.8
Furniture and furnishings	6,444	4.1
Rubber tires	1,769	1.1
Misc. durables	8,527	5.4
Subtotal durables	19,565	12.4
Nondurable Goods		
Newspapers	12,606	8.0
Books and magazines	4,760	3.0
Office papers	6,090	3.9
Commercial printing	3,680	2.3
Tissue paper and towels	2,925	1.9
Paper plates and cups	364	0.2
Other nonpackaging paper	5,230	3.3
Apparel	1,844	1.2
Footwear	1,198	0.8
Other misc. nondurables	2,756	1.7
Subtotal nondurables	41,453	26.3
TOTAL PRODUCTS	61,018	38.7
CONTAINERS AND PACKAGING		
Glass		
Beer and soft drink bottles	5,543	3.5
Wine and liquor bottles	2,135	1.4
Food and other bottles and jars	4,128	2.6
Subtotal glass	11,806	7.5
Steel		
Beer and soft drink cans	118	0.1
Food cans	1,777	1.1
Other nonfood cans	747	0.5

(Continued)

TABLE 4–5. CONTINUED

Products and Packaging	Thousands of Tons	Percent of Total (by Weight)
Barrels, drums, pails	91	0.1
Other steel packaging	101	0.1
Subtotal steel	2,834	1.8
Aluminum		
Beer and soft drink cans	1,317	0.8
Other cans	50	0.0
Aluminum foil	302	0.2
Closures	3	0.0
Subtotal aluminum	1,672	1.1
Paper and paperboard		
Corrugated boxes	19,444	12.3
Other paperboard	5,440	3.4
Paper packaging	4,163	2.6
Subtotal paper and paperboard	29,047	18.4
Plastics		
Plastic containers	2,871	1.8
Other packaging	2,798	1.8
Subtotal plastics	5,669	3.6
Wood packaging	2,080	1.3
Other misc. packaging	210	0.1
TOTAL CONTAINERS AND PACKAGING	53,318	33.8
OTHER WASTES		
Food	12,500	7.9
Yard	28,300	17.9
Misc. inorganics	2,550	1.6
TOTAL OTHER WASTES	43,350	27.5
TOTAL MSW 1986	157,686	100.0

Source: Franklin Associates Ltd. (1988).

of the year. There is so much variation that the EPA/Franklin data—called *national aggregated estimates*—is meaningless for many localities. For waste management planning purposes, states and local communities collect their own data on the actual amount of MSW collected or disposed of. The striking feature of this data is that it almost always results in higher generation and per capita rates compared with the EPA/Franklin data.

If the EPA/Franklin per capita data is truly a national average of MSW generation, one would expect individual state data to lie on either side of 3.6 pounds per day. Table 4-6 has data from 20 states representing 67 percent of the U.S. population. The per capita rate is 6.8 pounds per day, which refutes the EPA/Franklin data. No state has a rate lower than 3.6

TABLE 4-6. COMPILATION OF STATE DATA

State	Amount (mty)	Population (1987, millions)	Percent of Population	Pounds per Capita per Day
California	37.00	27.66	11.4	7.3
Rhode Island	0.73	0.99	0.4	4.0
New York	18.00	17.83	7.3	5.5
Missouri	3.40	5.10	2.1	3.7
Massachusetts	6.00	5.86	2.4	5.6
DC	0.72	0.62	0.3	6.3
Iowa	2.10	2.83	1.2	4.1
Washington	19.50	4.54	1.9	5.5
Oregon	2.30	2.72	1.1	4.6
Ohio	12.95	10.78	4.4	6.6
Florida	15.30	12.02	4.9	7.0
Maryland	7.10	4.54	1.9	8.6
Montana	0.63	0.81	0.3	4.3
North Carolina	6.40	6.41	2.6	5.5
New Jersey	9.40	7.67	3.2	6.7
Texas	23.10	16.79	6.9	7.5
Michigan	12.00	9.20	3.8	7.1
Connecticut	2.60	3.20	1.3	4.5
Illinois	17.00	11.60	4.8	8.0
Pennsylvania	9.20	11.90	4.9	4.2
TOTAL	205.43	164.90	67.1	6.8
Extrapolated to all 50 states	306.15	245.8	100.0	6.8

Source: Data collected by authors from various state sources.

pounds per day, and only Missouri's per capita rate is on par with the national data. At an average rate of 6.8 pounds per day per person, the United States generates 306 million tons per year, almost *twice* the official EPA rate. Corroboration of this conclusion comes from a study by the American Public Works Association, which found an average of 7 pounds per day per person.

What explains the differences? From those who assume that the EPA/ Franklin data is correct, the usual response is that state and local data are higher because they cover more kinds of waste than does the national data. For instance, states may include construction and demolition wastes that the national data does not capture. However, when the first output from the model was presented in 1974 (on 1971 generation), the EPA claimed that the model output should be *higher* than state data. The EPA argued that sampling typically does not include large, bulky items or as many sources of wastes accounted for in the model. The model calculates waste quantities from production data, but it does not tell who buys the products that make the waste. Thus, the model cannot distinguish between various sources of waste, such as household, commercial, and institutional. (When the data for Table 4–6 was collected, states were asked for MSW data that did not include construction wastes but did include household, commercial, and institutional sources.)

EPA says that "significant categories of waste" are excluded from the model but does not identify them. Close examination of the EPA/Franklin materials flow model provides several reasons for the underestimates. The model uses known data on raw material uses and some product yields, and makes assumptions about material losses and product lifetimes to estimate current waste generation. By making assumptions about future production, the model predicts future waste generation.

Underestimates occur because of several factors:

- Liquids are missing. Liquids are not part of the raw material inputs, nor are liquids added by consumers during product usage counted. All liquids are presumed to disappear prior to disposal. Examples are inks, motor oil, paints, and household and personal car products.
- Disposable plastic products are underrepresented. Other estimates say that disposable diapers alone amount to at least 2 percent of national MSW waste (or 3.6 million tons). The model predicts that waste from *all* disposables totaled 2.8 million tons in 1986.
- Only the plastic content of consumer electronics (such as TVs, stereos, video equipment, cameras, telephones, and so on) is included. Missing are the metals and glass used in the cases, motors, wiring, and electronics.
- Seemingly trivial products, such as kitty litter and light bulbs, are not

counted by the model. About 1.5 million tons of kitty litter are produced (and thrown away) each year. The model accounts only for glass containers. Therefore, light bulbs, flourescent tubes (426 million sold in 1987), and anything else made of glass are ignored.

Probably the largest source of error in the model is foreign trade. Some products, but not all, are adjusted for exports and imports. Miscellaneous durable and nondurable (e.g., disposable) products are not adjusted. In the major appliances category, only microwave ovens are adjusted. No accounting is made for the import/export of subassembled items.

For packaging, only empty glass containers are adjusted. No adjustments are made for aluminum or steel containers. All packaging (glass, metal, paper, and plastic) that comes *with* imported products is ignored. This is probably a substantial underestimate. The U.S. manufacturing trade balance has been in deficit since 1981. The 1987 imports outpaced exports by $137 billion. For instance, 20 million VCRs were imported in 1987. Consumers, then, acquired about 40 million pounds of cardboard cartons and plastic packing material that was not counted by the model. In the same year, almost 9 million microwave ovens were imported (19 million pounds of waste), almost 11 million TVs (55 million pounds), and 1.7 million refrigerators (12 million pounds). If all packaging (both transportation and consumer) of all imported goods was accounted for, the nation's waste number would surely rise by millions of tons.

This fundamental lack of accounting for trade may be the reason for the phenomena seen in Table 4–7, which shows the percent change by decade in the major categories of municipal waste. According to the EPA/Franklin data, the 1960s were the largest growth period for garbage. Since that time, has growth really abated or leveled off? It is, of course, in the 1970s and 1980s that the United States experienced a burst in consumption and an unprecedented growth in its trade imbalance. Because the total MSW

TABLE 4–7. WASTE PERCENT CHANGE, ACCORDING TO THE EPA

MSW Category	1960–70	1970–80	1980–90
Durables	52	27	14
Nondurables	44	33	35
Packaging	60	17	13
Food	5	−7	5
Yard	16	14	11
Misc. inorganics	37	27	22

Source: Adapted from: Franklin Associates, Ltd. (1988).

amount from EPA/Franklin data is too low, it is also likely that the percent changes over time are in error.

Basic Data on Materials and Products in MSW

The lastest available EPA/Franklin MSW data is for 1986. Table 4–4 shows the usual array of materials in the waste. Table 4–5 breaks down the data into the products and packaging that make up the waste. Note that for each, the EPA/Franklin waste accounting system tacks on the same set of "other wastes" that have been derived from composition studies. In both tables, the weight percent contribution of each category is shown.

As is normal for most wastes, the EPA/Franklin data has always been presented in weight amounts (millions of tons). In 1989 the weights were converted into volume, using data from other studies of landfills. By either weight or volume, paper and paperboard are the largest fraction of MSW. The amount of plastics, however, has always been underrepresented by the weight percents. On a volume basis, it moves from 6.5 to 18 percent of the waste. This characteristic of plastics—which are relatively light in weight—has masked its growth as packaging waste, at the expense of glass and paper, in the EPA/Franklin data.

Whether in volume or weight percents, 34 percent of the combined wastes still consist of packaging. Over time, products and packaging in the EPA/Franklin model have grown from 62 percent (by weight) of total MSW wastes in 1960 to 73 percent in 1986, at the expense of other wastes (yard, food, and miscellaneous). Franklin has estimated that yard and miscellaneous (dirt, stones, brick, and the like) wastes have risen over those years in a straight line fashion, the latter very slightly. Food wastes, however, have leveled off since a decrease in the mid-1970s, due, according to Franklin analysis, to the use of more prepared packaged food and of sink garbage disposals. These conclusions and national estimates are drawn from only five studies of the waste in five different locales over the period 1974 to 1985.

Per Capita Consumption and Waste Production Are Up

Consumption of goods is up in the United States, and consumption has been growing faster than the population. Here is some per capita data for a few selected products that become waste within a year or less of purchase:

- Paper and paperboard consumption grew 26 percent between 1970 and 1987 (4 percent per year).
- Consumption of shoes grew 50 percent between 1980 and 1986 (8 percent per year).

- Soft drink consumption grew 46 percent between 1970 and 1986 (3 percent per year).
- Consumption of packaging grew 9 percent between 1978 and 1985 (1.3 percent per year).

The EPA/Franklin data is the only historical national data on *waste* production. That data for 26 years (1960 to 1986) is presented in Table 4.8. It shows that total waste has grown at an average rate of 2.4 percent per year. The rates appear to have been consistently higher or less erratic in the 1960s. Waste generation since then has undoubtedly been affected by economic events, such as the energy crisis in the 1970s and the recession of 1981–1982. The years in which waste growth was very low or negative (and per capita growth was negative) can be explained by looking at the details in the EPA/Franklin model. In 1974, 1975, and 1982, waste from nondurables and packaging was down, an immediate reaction to events. In 1980 and 1985 waste from durables and packaging was down, a delayed reaction to events because of the longer lifetime of durables.

Since 1960 the U.S. population has grown steadily at an average rate of 1.2 percent per year. Compared with waste generation growth, this indicates that half of the annual increase in waste is due to factors other than population. This phenomenon is also captured in the pounds generated per person per day growth rates (pcd), which has averaged 1.2 percent over 26 years (see Table 4–8). The years in which per capita growth rates were down are, as discussed, the ones affected by economic conditions. That effect proves the connection between consumption and waste generation. Even with a growing population, garbage does not necessarily increase. While permanent economic crises may be too dramatic a cure for excessive wastes, the connection suggests that, if products were manufactured and packaged with waste in mind, economic growth would not have to increase the amount of waste produced.

While questions may remain about the true U.S. rate of growth, there can be no dispute that waste generation is growing, and growing faster than the population. Overall, Americans are buying and discarding more things.

The Impact of Social Trends

There are a number of social trends that help to explain the per capita rates of growth. According to the Congressional Research Service, the generation of solid waste is affected by numerous factors, including increasing population, increasing affluence, and technological and social changes. In general, as population and wealth increase and as the ability to produce disposable packaging and products improves, waste volumes increase.

Several studies have shown these relationships. Per capita waste genera-

TABLE 4-8.   PERCENT CHANGE IN TOTAL WASTE AND PER CAPITA GENERATION (1960-1986)

Year	Total Waste (thousands of tons)	Percent Change	Pounds per Capita	Percent Change
1960	87,532	—	2.65	—
1961	90,542	3.4	2.70	1.7
1962	92,511	2.2	2.72	0.6
1963	94,811	2.5	2.75	1.0
1964	97,994	3.4	2.80	1.9
1965	102,305	4.4	2.89	3.1
1966	107,824	5.4	3.01	4.2
1967	109,919	1.9	3.03	0.8
1968	113,058	2.9	3.09	1.8
1969	118,528	4.8	3.20	3.8
1970	120,449	1.6	3.22	0.4
1971	121,366	0.8	3.20	−0.5
1972	127,310	4.9	3.32	3.8
1973	132,550	4.1	3.43	3.1
1974	133,488	0.7	3.42	−0.2
1975	125,361	−6.1	3.18	−7.0
1976	133,413	6.4	3.35	5.4
1977	136,436	2.3	3.39	1.2
1978	141,132	3.4	3.47	2.4
1979	144,016	2.0	3.51	0.9
1980	142,659	−0.9	3.43	−2.1
1981	144,876	1.6	3.45	0.5
1982	142,059	−1.9	3.35	−2.9
1983	148,369	4.4	3.46	3.4
1984	153,648	3.6	3.55	2.6
1985	152,505	−0.7	3.49	−1.7
1986	157,686	3.4	3.58	2.4

Total Change +80%   +35%

Average   2.4   1.2

Source: Franklin Associates, Ltd. (1988).

tion tends to rise with increasing income and urbanization. Household size and number of occupants also affect the rate. Fewer people per household means less total waste. However, as the number of persons per household *decreases,* per capita waste generation *increases.* In addition, as house size *increases,* per capita waste generation *increases.*

The following changes in living patterns have occurred in the last couple of decades:

• In 1950 just over half of the population (56 percent) lived in urban areas; by 1980, 75 percent did. The rate of shift from rural to urban living continues at the same pace; in 1987, 77 percent of the population lived in urban areas.
• The number of households increased from 53 to 81 million between 1960 and 1980 and to 90 million in 1987, while the number of persons per household decreased from 3.1 (1970) to 2.7 (1987).
• Household changes are a result of smaller families, an increase in the number of single people living alone (from 18 to 21 million between 1980 and 1986), and the growth in single-parent families (from 10 to 13 million between 1980 and 1986).

Income has an obvious direct effect on garbage generation rates. More money means more consumption, which means more garbage. (Not always, of course. Wealthy people can afford to pay for high-quality, longer-lasting products. Conversely, poor people may be drawn to low-cost, disposable products.) Disposal income (money left over after taxes) continues to grow in the United States, but not as rapidly as it once did. While severe economic stress reduces garbage rates, most of the 1980s was a period of steady, but historically slower, growth of income that did not cause an overall downturn in consumption. In fact, consumption was up. First, while the poor got poorer, the richest 20 percent of the population became 32 percent richer. So, some people had a lot more money to buy things with in the 1980s. Second, those in the middle class, whose incomes were not growing as rapidly as their parents' had, coped by saving less and buying more on credit.

Another social trend has helped to create more waste. The growth in the workforce of married women and female heads of households has reduced home and leisure time. To make better use of the time available, working women and men today seek convenience in the products they buy. As we show in Chapter 7, a consequence of convenience is increased waste. Disposable products and excessive packaging for microwavable foods are two sources of increased household waste. The extraordinary growth in fast food restaurants (from 32,600 in 1970 to 90,800 in 1988) has increased the amount of packaging waste thrown out by the commercial sector. Fax machines, copying machines, and computers are new sources of paper wastes in offices.

Should these trends continue, and all evidence implies that they will, the U.S. per capita waste generation rates should continue to increase. If less developed countries continue to emulate the United States, then it can be expected that solid waste growth rates in those countries will mirror that of the United States. When McDonald's opened its first outlet in Moscow in early 1990, the local inhabitants were astounded by the throwaway wrappings, utensils, and cups. Explained one amazed patron who had saved his disposables: "You use these once only, I tell you. You use them and then throw them away" (*Washington Post* 1990).

The Garbologists

There is another source of information on MSW. It is based on work by William Rathje and others who have dug in landfills and picked through household trash before it gets to landfills. This landfill work has been instrumental in debunking the myth that everything except plastic materials degrade in landfills.

Some conclusions drawn by William Rathje in "Rubbish!", an article on *household* waste in the December 1989 issue of *The Atlantic Monthly,* need close examination. While Rathje does not say what the per capita rate is, he thinks that 3 pounds per day per person "may be" too high for "many parts" of the country. Another conclusion is that per capita household waste has not grown over time. Increases are due solely to population growth.

The first point appears to be but is not necessarily at odds with the information presented above. Per capita rates are averages, and MSW waste generation varies for a number of reasons. That means that in some parts of the country (especially rural areas), per capita rates will be lower than in others. Also, most MSW data (from EPA/Franklin and from individual states) includes the waste that Americans generate in their households *and* while eating out, shopping, and working (except those who work in industrial plants). Those per capita rates of 3.6 to 6.8 pounds should, therefore, be higher than household per capita rates alone.

The claim that per capita *household* waste generation is stable runs counter to all other information, both from the United States and from other countries. And it is difficult to believe that increases in other components of MSW can explain all the MSW increases over time. Rathje presents only one plausible factor to explain household waste stability: increases in packaging wastes have reduced home food wastes. Other possibilities are mentioned (lack of coal ash and dead horses), but they have not been part of American household waste since at least the 1920s.

There are problems with all MSW data, as we have pointed out. One problem with Rathje's is a very small, skewed data set: 16,000 pounds (8 tons) of garbage extracted from seven landfills. On a weight basis, that amounts to a microscopic 0.000005 percent of the garbage generated each year when compared with the low EPA/Franklin data. Put another way, the 16,000 pounds represents the total annual garbage from fewer than 10 people in the United States! No real scientist would have reached such broad conclusions on such flimsy evidence. In addition, the seven landfills represent just over 0.1 percent of the estimated 5,500 currently in operation, and none are located in the eastern United States.

We agree with Rathje's observation that "we have more reliable information about Neptune than we do about this country's solid-waste stream" (Rathje 1989). But we disagree with his general view that nothing much needs to be done about reducing waste generation. His view may have something to do with the sources of financing for his research, such as the plastics industry.

Foreign Data Confirms U.S. Wastefulness

There can be no mistake: Americans are more wasteful than people in other countries. No matter how the data is collected or who correlates it, the U.S. per capita waste generation rate is higher than that of any other country. When compared with less developed countries, the United States has higher rates. When compared with Western European nations, the United States has higher rates.

Table 4-9 presents data from various cities around the world. It shows that in large cities with relatively high income levels, per capita waste generation rates can vary widely. Two U.S. cities and Sydney, Australia (a country often likened to the United States), have comparable rates. The rest of the cities with comparable living standards have rates half of those of the top three cities. Some of the former have rates comparable to those in cities in less developed nations.

Higher rates of waste generation are often excused as simply differences in definitions. As with the EPA/Franklin data versus state data, the differences can be in the kinds of sources of wastes included, such as household versus commercial. Difference can also occur when data concerns not total generated wastes but wastes left over after recycling has occurred. The Organization for Economic Cooperation and Development (OECD) regularly deals with all kinds of national data and has developed methods to adjust national differences so that the data can be fairly compared across nations.

Table 1-1 is based on 1985 MSW data from the OECD. It shows that Americans generate 65 to 174 percent more waste per capita as people in the Netherlands, Switzerland, the United Kingdom, Japan, West Germany,

TABLE 4–9. PER CAPITA MSW GENERATION, VARIOUS
CITIES (CIRCA 1980)

City	Amount (pounds per day)	Source
Large, High-Income Cities		
New York	4.0	a
Sydney	3.9	d
Washington, D.C.	3.9	c
Tokyo	3.0	a
Paris	2.4	a
Singapore	1.9	a
Hong Kong	1.9	a
Hamburg	1.9	a
Rome	1.5	a
Large, Low-Income Cities		
Caracas, Venezuela	2.1	b
Lima, Peru	2.1	b
Mexico City	1.5	b
Rio de Janeiro, Brazil	1.2	b
Calcutta, India	1.1	a
Manila, Philippines	1.1	a
(city)	0.9	b
Small, Low-Income Cities		
Asuncion, Paraguay	1.4	b
Lahore, Pakistan	1.3	a
Bandung, Indonesia	1.2	a
Medellin, Colombia	1.2	a
Tunis, Tunisia	1.2	a
Kano, Nigeria	1.0	a
Yemen, Aden	1.0	b
Olongapo, Philippines	0.7	b
Sasha Settlement, Nigeria	0.4	b

Sources:
[a]Pollack (1987).
[b]Diaz and Golueke (1985).
[c]Washington, D.C., Department of Public Works.
[d]New South Wales (1989).

Note: Data among nations is approximate because of wide differences in definitions and methods of obtaining data.

and France. When a probable rather than a conservative comparison is made, the United States is 132 to 282 percent more wasteful as these other nations.

Other countries are not immune to increasing per capita rates of waste generation year after year. Table 4–10 shows that Australia over 5 years (1983 to 1988) increased its per capita waste generation by 61 percent. Australians have increased their waste significantly faster than Americans have. The Dutch, who are usually considered to be less wasteful and are still below U.S. rates, have a growth rate of 28 percent over 14 years (1975 to 1988). All styles of living, therefore, seem to be victims of products, packaging, and technological changes that result in more garbage.

Toxics from Solid, Air, and Water Wastes

The same sources that put much more than 160 million tons of MSW into landfills and incinerators yearly also put significant amounts of waste into the air and into the sewer system. Land, air, and water wastes are not only a quantity problem, they also contain toxics.

Earlier in this chapter we showed that relatively little is known about the quantity of national air—and even less about water—releases from industrial sources. Hardly anyone bothers to collect data routinely on air and water releases from nonindustrial sources. Some attempts have been made,

TABLE 4–10. INCREASES IN PER CAPITA MSW GENERATION

Year	Pounds per Person per Day	Percent Increase
MSW in Sydney, Australia		
1983	3.8	—
1984	4.3	13
1985	4.8	12
1986	5.4	13
1987	5.7	6
1988	6.1	7
	5-year change	61
Household waste in the Netherlands		
1975	1.8	—
1981	2.2	22
1988	2.3	5
	14-year change	28

Source: New South Wales (1988) (for Australia) and Netherlands (1989).

however, to quantify and qualify those releases. There is enough information available to suggest that *individuals,* and not just industry, are contributing substantial environmental damage to the air and water, as well as the land. Pollution prevention is the viable solution to many of these problems. For instance, once stable toxics are in wastewater, they are not removed. Thus, if households do not use toxics, they will not end up as pollutants in rivers and groundwater.

Into the Air

Consumers are directly responsible for automobile emissions and indirectly for the air emissions from generating plants that supply their electricity and gas (Chapter 2). Consumers also release many different organic chemicals when they use a multitude of products. These are called *volatile organic compounds (VOCs)* and are covered in Chapter 8. Some, but not all, VOCs react in the atmosphere to produce ozone, covered in Chapter 2. Others destroy the planet's ozone layer, the subject of Chapter 5.

A California state study found over 60 different VOCs in consumer products. Estimated 1985 emissions totaled 168 tons per day. California has 11 percent of the U.S. population. Assuming that the rest of the population consumes similarly to California, an estimated 1,527 tons of VOCs enter the air each day and over 550,000 tons per year, just from everyday consumer products such as cosmetics and paints.

Into the Water

In most areas of the country, waterwaste flows through sewers into municipal treatment plants and is discharged into surface waters. Households, commercial and retail firms, institutions, and offices, along with many industrial plants, use the sewer system. The exception is in rural areas, where waterwaste goes into septic tanks. It has been estimated that 25 to 30 percent of American households (23 to 27.3 million persons) use septic tanks. Once in septic tanks, some wastes are degraded by natural biological processes. Those that are not ultimately end up in groundwater.

The more than 15,400 municipal water treatment plants in the United States handle 27.7 million gallons of raw sewage daily. Only 12 percent comes from industrial plants and commercial facilities, along with 92,000 metric tons of toxic pollutants. The industrial pollutants include corrosives, toxic metals (such as lead and cadmium), and toxic organic chemicals. Households also contribute toxic pollutants from household products (such as cleaners and detergents) and pesticides that rainwater has washed off their lawns. The largest portion of household wastewater is from toilets (35 percent), which includes, of course, fecal matter, as well as chemical cleaners.

In stages, conventional water treatment plants screen out paper and plastic solids (that go to landfills), separate sludge and water, and treat both sludge and water with bacteria. (Some VOCs are released into the air during treatment, since it includes aeration of the wastewater.) Sixty percent of the country's plants then release the water and sludge; the other 40 percent follow up with tertiary treatment, during which nitrogen and phosphorus are removed. That treatment, however, adds high concentrations of chlorine that produce nitrogen gas, chloroform, and chloramine (a carcinogen), which must be removed by activated carbon filtration. (The carbon filters are either cleaned in furnaces or put in landfills.) An aluminum process removes phosphates but contaminates the sludge.

Toxic metals and organics that originate in households and industrial plants pass through the treatment stages and end up either in the sludge or in water effluent. In fact, if treatment plants try harder to clean the water, the sludge gets more contaminated. The sludge—7.7 million metric tons of it annually—is dumped in landfills (42.3 percent), incinerated (21.4 percent), dumped at sea (5.5 percent), used as a fertilizer (15.7 percent), composted (9.1 percent), or otherwise (6 percent) disposed of.

The EPA estimated in a 1986 study, based on 1980 data, that households annually send 6,119 tons of toxic metals and 2,893 tons of certain toxic organic chemicals to wastewater plants, a total of 9,015 tons. In a study of two Washington State treatment plants, 223 pounds per day of just 13 organic chemicals were estimated to come from households. This data from two plants is 99 percent above the average implied by the EPA estimates. It suggests either that the EPA data is far too low or that there is huge variation among treatment plants, or both.

Toxics in MSW

What is known about toxics that end up in MSW landfills and incinerators comes from a number of sporadic studies that have focused either on what is found in landfills or what is picked up from residences. Other studies have tried to quantify the amount of toxics that commercial and industrial sources generate that are exempt from the federal regulatory system. All of these toxic wastes—from households and other sources—end up in MSW facilities along with the nontoxic solid wastes.

From households, toxic wastes come from the billions of dollars worth of toxic household products bought each year (see Chapter 8). Sampling studies have concluded that from 0.3 to 0.6 percent of household trash, or an average of 30 pounds of toxic waste, are generated per year. With over 90 million households, that means 1.5 million tons per year for the country (see also Chapter 2). However, some studies (primarily in highly urban areas) have concluded that this amount of toxic wastes may be as high as 1 percent of MSW wastes. One percent of the data from states (in Table 4–6) means that about 3 million tyons of household toxic wastes are in MSW.

The percentages and amounts seem small and are, relative to over 583 million tons of hazardous wastes generated by *industrial* sources. This leads analysts tied to the industries that produce the toxic products to conclude that placing household hazardous wastes in landfills and burning them in incinerators is not a problem. Several things should be kept in mind, however. Many households sending small amounts of hazardous materials to a landfill may create the same environmental problem as a large plant sending its hazardous waste. Federal regulations require that hazardous waste generated, by *any one plant*, in amounts greater than 100 kg per month (just over 1 ton per year) be placed in landfills and burned in incinerators that meet special design criteria. MSW landfills and incinerators meet no equivalent criteria; thus the impact of 1.5 to 3 million tons may be greater than the tonnage or percentage implies. While that total amount is spread among hundreds of facilities, urban areas may send 500 to 20,000 tons of toxic household wastes to their facilities in a year.

There is little information on the toxic waste from the estimated 455,000 firms or plants that generate less than 1 ton per year. A 1985 EPA study surveyed only 11 percent of the facilities and concluded that all of them generate about 197,000 tons of hazardous waste per year. If the household and commercial data is correct, then, on average, each firm generates 866 pounds per year, while each household generates 30 pounds.

Medical Wastes

Hospitals, medical offices, laboratories, and research facilities are yet another source of toxic wastes. Hospitals also appear to generate wastes at a rate considerably over the national MSW averages. However, again, the numbers are uncertain.

The EPA says that hospitals alone generate approximately 3.2 million tons of waste each year. Other estimates range from 2.1 to 4.8 million tons. Most of this waste—maybe as much as 70 to 80 percent—is cafeteria garbage, disposable gowns and drapes, packaging, and the like that is counted as regular MSW. Perhaps 10 to 15 percent of the waste is infectious. This *red bag* waste (named after the disposal bags in which it is placed) consists of swabs, cultures, operating room wastes, blood serum, and body parts (pathological waste). The balance of medical wastes is toxic metals (e.g., mercury from dental work), toxic chemotherapy substances, and low-level radioactive materials.

Like the rest of MSW, most medical wastes are landfilled. Some are illegally dumped, as evidenced by the many stories of medical wastes on beaches. The fate of infectious and toxic wastes depends on individual state laws and regulations and hospital practices. The greatest portion of these special wastes may be burned in hospital incinerators; some, due to lack of state or local (and no federal) regulations, end up in regular landfills. Groups such as the Environmental Defense Fund and the National Wildlife Federation are concerned because many hospital incinerators are old-fashioned technologically and poorly control air emissions. A host of toxic metals have been found in emissions, and there is some evidence that hospital incinerators emit dioxins and furans at higher rates (compared with the amount of waste burned) than do municipal incinerators.

As with all other kinds of wastes, simply relying on management technology is not the best solution. Health care workers and hospital administrators say that the use of disposable items has increased dramatically in recent years. While some disposables have been adopted for sanitary reasons, others have undoubtedly been chosen to cut hospital labor costs. The EPA estimate for national hospital wastes is based on an average rate of 13 pounds per bed per day. Other studies, however, have suggested rates of 16 to 23 pounds per bed per day. Even 13 pounds is almost twice the average MSW rate of 6.8 pounds per person per day and almost four times the EPA/Franklin MSW rate of 3.6 pounds. This indicates that preventive health care that keeps people out of hospitals also reduces waste generation. But in-hospital waste reduction programs are also necessary.

SUMMARY

There are many sources of a wide variety of environmental wastes in any country. Rarely is there ever a complete accounting of them all or knowledge about where they end up or the damage they do. Despite this, there is compelling evidence that Americans (in both their personal and their occupational lives) are more wasteful than people in other nations, even those with equally high (or higher) standards of living. It is also clear that as the population grows and living standards improve, waste amounts and toxicity increase.

Once pollutants get into landfills, incinerators, water bodies, and the air, further dispersal of them at some time is assured. The consequences are evident, and the remaining chapters of this book cover them. The solution is pollution prevention. Changes in production methods and products, as well as in personal habits, can reduce the generation of pollutants, halting their inevitable spread, and thereby increase the livability of the planet.

THE OZONE GROAN:
DO WE STILL HAVE TIME?

Experts may argue about whether global warming or the destruction of planet Earth's protective upper atmospheric ozone layer is today's most remarkable and unremittingly disastrous environmental problem. It is a choice not worth making. Both problems pose problems that humanity is not likely to solve with conventional thinking. As discussed below, the chemicals that destroy the ozone layer also contribute to global warming and other problems.

Ozone depletion means that human life on planet Earth is directly and undisputedly threatened. An examination of what is being done about this dangerous environmental problem should, theoretically, reveal extensive use of preventive problem solving. There has been significant but cautious use of pollution prevention tactics. Most important was a ban of one product by several nations and an international commitment, the Montreal Protocol, to phase out dangerous chemicals that quickly become deadly pollutants.

But there are also serious uncertainties about whether sufficient preventive actions are being taken rapidly enough. The chemicals that destroy the ozone layer are ubiquitous, and dependence on them is extreme. There is little time left to be safe instead of sorry. Lack of foresight in the past and demands for scientific certainty have eaten up time. It is literally a case of acting rapidly enough to save the roof over our heads. Yet some companies that make ozone depleting products want to keep making profits from them. Governments want more economic growth from more consumption of those products, even if it is unsustainable and dangerous growth. Consumers want to keep their comforts and conveniences or, in poor nations, to get them for the first time. The desires of companies, governments, and consumers are pitted against effective protection of health and the environment.

HEALTH AND ENVIRONMENTAL EFFECTS

Losing stratospheric ozone means destroying the relatively fragile layer that serves as planet Earth's protective shield against solar ultraviolet radiation.

To give an idea of how vulnerable the ozone layer is, if it existed as a single pure layer of ozone molecules, it would be 0.25 inch thick. Ozone makes up only a few parts per million of our atmosphere—a tiny fraction of 1 percent. It is a vulnerable protective layer.

Ozone depletion depends on the season and location. The average global concentration of ozone decreased by about 1.7 to 3.0 percent from 1969 to 1986, increasing to 2.3 to 6.2 percent in winter. A hole in the ozone layer over Antarctica the size of North America (or 10 percent of the Southern Hemisphere) has been confirmed, and an average loss of 50 percent of ozone has been reported, with some areas having no ozone whatsoever. In March 1990, NASA reported that it had confirmed the presence of a hole in the ozone layer over the North Pole, and said that eventually there would be a severe depletion of ozone in the Arctic. Even if a total ban of all ozone depleters were instituted today, it would still take 100 years of natural ozone formation to replace all the ozone already destroyed. Another 2 to 10 percent of the ozone blanket could be destroyed by about 2050 because of the ozone destroyers already in the atmosphere. There is no time to waste to practice prevention.

Letting more ultraviolet radiation reach the Earth's surface as a result of ozone depletion inevitably means an increasing incidence of human skin cancer; an increasing incidence of cataracts, retinal damage and blindness; attacks on human immune systems, which might cause an increase in the occurrence or severity of infectious diseases and a decrease in the effectiveness of vaccination programs; harm to aquatic systems, especially small ones at the base of the marine food chain; damage to phyoplankton, a major sink for carbon dioxide, which could increase the greenhouse effect; reduced yields of crops such as soybeans; a decrease in the quality of the lower atmosphere (troposphere), which creates urban smog-type conditions in nonurban areas; and accelerated weathering of outdoor plastics, rubber products, paints, and possibly wood, paper, and textiles. The impact on the ocean's ecology is especially significant. Research has found that high levels of ultraviolet radiation can slow the growth rate of phytoplankton by as much as 30 percent. That effect could threaten krill—shrimplike creatures that eat phytoplankton and are a key link in the marine food chain—ultimately affecting fish, whales, penguins, and birds. Talk of ultimate effects that rival the effects of global nuclear war is not without substance and possibility.

A 1988 report by scientists from NASA, the United Nations, and the World Meteorological Organization said that "a 2.5 percent ozone decrease year-round, such as that detected in some areas of the Northern Hemisphere, could lead to a 10 percent increase in skin cancer." More detailed forecasts were made by the Environmental Policy Institute and the Institute for Energy and Environmental in 1988 for a 3 percent decrease; 4,000 additional cases and 1,000 deaths from the most severe kind of skin cancer; 200,000 additional cases of less severe skin cancer; and 400,000 additional cases of cataracts, many leading to blindness.

A second dimension of ozone-depleting pollutants—from the use of what were originally described as "wonder" chemicals—is that, in the lower atmosphere, they are a key contributor to global warming, which offers its own threats to life-sustaining resources. The chemicals that deplete ozone also reflect back some of the Earth's heat in the form of infrared radiation, raising surface temperatures—the well-known greenhouse effect. Global warming will exacerbate the rising sea level problem. Major global warming, which could melt the polar ice caps, would flood and destroy coastal areas, disrupt agriculture, and extinguish some plant and animal species.

During years, decades, or centuries of upward movement through the atmosphere, ozone-depleting chemical molecules contribute to global warming long before they reach high altitudes where they destroy the upper atmospheric ozone layer. An ozone-depleting CFC molecule can cause a thousand times as much global warming as a molecule of carbon dioxide, which is the major greenhouse gas. CFCs cause about 30 to 40 percent as much global warming as carbon dioxide does, and their release is increasing at nearly 10 times the carbon dioxide rate. Experts agree that cutting CFCs is the most potent way to cut global warming in the near term.

At the molecular level, ozone-depletion molecules have the unique capability of absorbing infrared radiation at certain frequencies which carbon dioxide cannot absorb. The concentration of carbon dioxide in the upper atmosphere is already so high that increasing adsorption by additional increments of it is relatively small. But the concentration of ozone-depleting molecules is still low enough, about a thousand times less than that of carbon dioxide, that additional infrared adsorption from incremental addition of waste gases is substantial. The end result is that what may seem to be small absolute amounts of ozone-depleting chemicals (parts per trillion in the atmosphere), compared to the enormous quantities of carbon dioxide from fossil fuel burning (parts per million in the atmosphere), are really potent contributors to global warming, accounting for about one-fifth of global warming overall.

Moreover, as the ozone layer is destroyed and more ultraviolet radiation strikes the Earth, more smog will be created. The reason is increased photochemical reactions, which create smog. Reactions between VOCs and nitrogen oxides in the air produce lower levels of atmospheric ozone if sunlight occurs. It has also been predicted that the upper atmospheric ozone loss will increase acid rain. Another dimension of ozone-depleting chemicals and wastes is that most of them pose significant health risks, especially to industrial workers. Table 5–1 lists the key health effects of some common ozone depleters.

TABLE 5-1. HEALTH EFFECTS OF SOME OZONE DEPLETERS

Chemical	Health Effects
CFC-11	Exposure may cause headaches, abdominal pains, vomiting, coronary thrombosis, lung lesions, fatal asthma-like attacks, cardiac arrythmia
CFC-12	Narcosis, cardia arrythmia, potentially fatal asthma-like attacks
Carbon tetrachloride	Depression of central nervous system, destruction of liver and kidneys, liver cancer
Methyl chloroform	Heart or respiratory failure

Source: National Toxics Campaign, n.d.

The bright side of the ozone depletion problem is that there has been an unprecedented amount of international cooperation and action by governments and industries. However, they have taken place more than 15 years after the cause and nature of the ozone depletion problem was first discovered. Even now there are problems in moving ahead with the necessary speed and effectiveness. The dire health effects of ozone depletion remain a major threat even though a host of pollution prevention solutions are available. Indeed, it is now technically feasible to reduce emissions of ozonene-depleting chemicals by 90 percent. Considering the peril, the ozone layer problem provides an important opportunity for individual consumer and worker pollution prevention actions. Relying on government and industry is too risky. Complete participation by developing countries will not be easy.

Destruction of the ozone layer is self-evidently a problem for which a traditional pollution control fix seems completely infeasible. Could new ozone be continually transported in massive quantities to the upper atmosphere? Hardly. Do we want to deal with massive health and environmental effects after they are upon us? No. Is wearing hats and quality sunglasses and using lavish amounts of sun blockers the solution? Such control tactics cannot be completely effective. The only pragmatic approach is a preventive one in which the chemical culprits which cause ozone depletion are no longer used. All uses of them produce injurious environmental wastes.

CHEMICAL CULPRITS

The root of the ozone depletion problem is that a host of chemicals used widely in industry migrate slowly to the upper atmosphere, where they destroy ozone molecules. It can take years for ozone depleters to rise through the lower atmosphere and reach the stratosphere. But within only 1 year, they can disperse throughout the lower atmosphere. This quick long-range air transport means that ozone depleters from any one location can affect the entire global condition. Chlorine atoms are the heart of the problem. Chlorine is released from *chlorofluorocarbons*—now universally known as CFCs—and other chlorine-based chemicals and products. The link between CFCs and ozone layer destruction was first made in 1974. Over time, the number of chemicals found to cause ozone depletion has increased. For example, halons that contain bromine and other relatively common chemicals like chlorinated solvents that contain chlorine are also a problem. Table 5–2 lists some of the known ozone-destroying chemicals. One of the more interesting issues in this area is that a group of chemicals which have been widely accepted as replacements for known ozone depleters are themselves depleters, although to a less severe extent. They are called *hydrochlorofluorocarbons, or HCFCs,* and transport less chlorine into the atmosphere.

Molecules of ozone depleters react with ultraviolet radiation in the stratosphere and break down into component atoms, especially chlorine, bromine, and fluorine. For example, when a chlorine atom meets an ozone molecule, it tears away one of the three oxygen atoms in the ozone molecule, thereby destroying it (and converting it to a normal oxygen molecule). Thereafter, the chlorine monoxide molecule can meet another oxygen atom, allowing the two oxygen atoms to combine and freeing the chlorine atom

TABLE 5–2.   OZONE-DEPLETION COMPOUNDS

Compound	Common Name	Type
Trichlorofluoromethane	CFC-11	CFC
Dichlorodifluoromethane	CFC-12	CFC
Chlorodifluoromethane	CFC-22	CFC
Trichlorotrifluoroethane	CFC-113	CFC
Dichlorotetrafluoroethane	CFC-114	CFC
Chloropentafluoroethane	CFC-115	CFC
Chlorodifluoromethane	HCFC-22	HCFC
Carbon tetrachloride	CC14	Chlorinated solvent
Methyl chloroform	C2H3C13	Chlorinated solvent
Bromochlorodifluoroethane	Halon 1211	Halon
Bromotrifluoroethane	Halon 1301	Halon
Dibromotetrafluoroethane	Halon 2402	Halon

Sources: 53 Federal Register 30566 (1988) and National Toxics Campaign (n.d.).

to meet yet another ozone molecule. Chlorine concentrations in the upper atmosphere have increased from 0.6 to 2.7 percent parts per billion in the past 25 years. Bromine concentrations are much lower, at about 1 part per billion, but are increasing rapidly. Table 5–3 lists the ozone depletion potential (ODP) of different chemicals. The higher the ODP, the higher the ozone destruction capacity. For example, halon-1301 is 10 times more destructive than CFC-11, and what is being termed an interim CFC alternative, HCFC-22, is only 5 percent as destructive as CFC-11.

Also listed in Table 5–3 is the atmospheric lifetime in years for the chemicals. The longer the lifetime of an ozone depleter, the greater its destruction potential. However, the shorter lifetime of methyl chloroform of 6.5 years is largely offset by the very large amount of it that is produced and released into the atmosphere—about equal to the total of all other CFCs in the United States. A low lifetime of 1.5 years is cited as an advantage of HCFC-123, a potential substitute for CFC-11. The average lifetime of an ozone molecule is only 6 hours; that of a CFC-12 molecule 241,000 hours, and some CFC molecules may last for over 100 years. It is generally recognized that a single CFC molecule moving in the stratosphere may destroy 100,000 molecules of ozone. The stability of ozone depleters means that destroyer chemicals already discharged to the atmosphere will remain there for long periods, continually destroying ozone. It has been estimated that if all parts of the Montreal Protocol were completely implemented, the rate of ozone loss would still double in the next few years. Stopping the discharge of new

TABLE 5–3.   ODP FACTORS AND LIFETIMES

Chemical	ODP	Atmospheric Lifetime (Years)
CFC-11	1.00	75.0
CFC-12	1.00	111.0
CFC-113	0.80	90.0
CFC-114	1.00	185.0
CFC-115	0.60	380.0
HCFC-22	0.05	20.0
Methyl chloroform	0.10	6.5
Carbon tetrachloride	1.06	50.0
Halon 1211	3.00	25.0
Halon 1301	10.00	110.0
Halon 2402	6.0	[a]

Source: 53 Federal Register 47498 (1988).

[a] not available.

ozone-depleting chemicals *now* cannot by itself, therefore, stop ozone layer destruction. But not stopping the release of ozone depleters immediately surely spells increasingly serious ozone destruction in the near term because all losses of ozone can increase the adverse health effects.

Table 5-4 shows the rising levels of some ozone depleters in the atmosphere. Concentrations of two of the most important ozone depleters, CFC-11 and CFC-12, have increased at a 5 percent annual rate since 1978. CFC-113 has increased at a 10 percent rate since 1975. Data from NASA reveals a significant decrease in upper atmosphere ozone in the Northern Hemisphere: 1.7 to 3 percent annually and 2.3 to 6.2 percent during winter months. NASA explicitly cited the abundance of certain chemicals, primarily CFCs, as the cause of the ozone loss.

THE INVENTION, GROWTH OF PRODUCTION, AND INITIAL BAN ON AEROSOLS

In the 1920s, home refrigeration was limited because the common refrigerants at the time, such as ammonia, methyl chloride, and sulfur dioxide, were noxious or toxic. Charles F. Kettering of General Motors asked Thomas Midgley, Jr., a researcher for the company, to find a nontoxic, nonflammable substitute for these gases. Midgley invented dichlorofluoromethane (Freon 12) in 1930. He demonstrated its safety at an American Chemical Society meeting by inhaling a deep breath of the gas and then exhaling it to extinguish a lighted candle. The quick success of the Frigidaire Division of General Motors was based on Midgley's discovery. In World War II, CFCs were used for insecticide aerosols such as DDT. Midgley eventually was elected to the National Academy of Science and became president of the American Chemical Society. Crippled by poliomyelitis in 1940, he set up a pulley and harness system to get into and out of bed but strangled himself in the system in 1944. He never lived to see how his wonder chemicals, designed to help people, ultimately became a major threat to all of humanity.

TABLE 5-4. RISING LEVELS OF OZONE DEPLETERS
IN THE ATMOSPHERE (1978-100)

Chemical	1978	1982	1986
CFC-11	100	126	149
CFC-12	100	120	146
Carbon tetrachloride	100	106	115
Methyl chloroform	100	126	150

Source: National Toxics Campaign (n.d.).

Worldwide use of CFCs increased from about 400 million pounds in 1960 to 1 billion pounds in 1968 to 2 billion pounds in 1974 and to 2.5 billion pounds in 1988. In 1988, North America accounted for about one-third of world use, Europe and Africa for another one-third. Correspondingly, atmospheric concentrations of CFCs have increased over time, as shown in Table 5-4.

In 1978, the United States, Canada, Sweden, and Norway banned the use of CFCs in nonessential aerosols. This action was caused primarily by environmental groups which insisted that precautionary action was needed. Industry protested the ban. However, worldwide, aerosols remain the largest source of CFC emissions, some 33 percent of CFC-11 and CFC-12 emissions. Interestingly, today many companies are labeling consumer products as ozone friendly, even though these are products which had to stop using CFCs as aerosols a decade ago because of the U.S. ban.

At the time of the original ban, aerosol use of CFCs represented half of the total U.S. CFC consumption. By 1978, aerosols accounted for only 5 percent of U.S. use. By 1985 in the United States and Canada, the growth of the refrigeration, cleaning agents, and foam insulation markets had more than offset the decline of CFCs in aerosol markets. Before the ban, CFCs in aerosols accounted for 69 percent of worldwide use in 1974 and 19 percent in 1988, but total use of CFCs had increased by about 25 percent.

Banning aerosol use of CFCs, therefore, did not stem the production, marketing, or demand for ozone depleters in the United States and other nations. Poeple in industry who, naturally, oppose government bans on chemicals have pointed to the aerosol ban as proof that bans do not work. The real lessons are different. First, a comprehensive ban is necessary when there are so many applications of chemicals like CFCs and when the manufacturers are large global corporations that know how to market chemicals for new applications. Second, bans by a few nations are insufficient when there is global usage and global environmental damage is at stake. The aerosol ban was successful for its intended purpose and, in general, bans have been very effective for specific chemicals and pesticides in the United States. The banning of hazardous substances is, in fact, probably the most potent tool government can use to implement pollution prevention.

It should also be noted that the original banning of CFCs in aerosols did not address what industry would use for substitutes. In fact, much of the aerosol industry switched to hydrocarbons such as pentane, propane, and butane. But these hydrocarbons contribute to other environmental problems, such as formation of low-level ozone smog and the greenhouse effect. Currently, in the greater Los Angeles area, household aerosol use is the 12th largest source of hydrocarbon emissions. The South Coast Air Quality Management District proposed banning hydrocarbons as aerosol propellants as part of a regional antismog plan. Note that in Los Angeles the federal maximum allowable concentration for ozone, which is produced through chemical reactions in the lower atmosphere, is exceeded about half the time. Unsafe substitutions remain a concern whenever chemicals or products are banned as a response to a specific environmental problem. Just as the original ban on aerosol CFCs resulted in greater use of smog-producing chemicals, demand for methyl chloroform is increasing because of increasing government restrictions on the use of VOCs (hydrocarbons), such as in California.

CURRENT USES OF OZONE DEPLETERS

There are seven categories of current major uses of ozone depleters, and the following discussions will enable workers to identify many uses of CFCs. All represent opportunities for workers in countless industries and perhaps to a lesser extent, consumers to accomplish a great deal for preventive planet care.

Rigid Foams

CFCs are used as blowing agents to produce rigid foams. During manufacture, the CFCs are injected into polymers that are the building blocks of plastics. The CFCs create closed cells or internal bubbles that make products excellent insulators. The materials may be produced in sheets or slabs, or may be poured or injected in the final application. Urethane and isocyanurate foams primarily use CFC-11; nonurethane foams primarily use CFC-12. Smaller quantities of other CFCs are also used. Major uses of urethane foams include building insulation materials; insulation in refrigerators, freezers, and refrigerated trucks; packaging materials; and marine flotation products. Some CFC emissions occur after manufacture because some CFC remains in the foamed products. These are referred to as *banked* CFCs because they are released to the atmosphere over long periods of time.

One alternative for building insulation applications is cellulose fiber, which is recycled wood fiber made from newsprint plus flame retardants. But larger thicknesses are needed because its insulation properties are not as good as those of rigid foams. Other substitutes are fiberglass, fiberboard, plywood, gypsum, foil-faced laminated board, and asphalt board. For refrigerators, produced at a rate of 20,000 per day in the United States, substitutes include fiberglass, which would increase external sizes or reduce inside space, and vacuum panels, which are still being developed. Rigid foams in refrigerators have contributed substantially to increased energy efficiency. In the United States, refrigerators use 7 percent of all the electricity produced; the Energy Department has mandated an average 25 percent reduction in electricity use by refrigerators by 1993 above new January 1990 standards. Manufacturers may use CFCs through 1993.

Major uses of nonurethane foams include common packaging chips or curls; insulation; fast food service products such as cups, plates, and hinged containers or clam shells; egg cartons; trays for meat, poultry, and produce in supermarkets; ice chests; and flotation products. *Styrofoam,* a Dow Chemical company name for polystyrene, is a commercial type of nonurethane foam. Some nonurethane foams are manufactured with very little, if any, CFC blowing agents. Indeed, it is well known that *n*-pentane and other hydrocarbons such as isopentane and *n*-butane can replace CFC-12. However, unlike CFCs, hydrocarbons are flammable and their use requires modification of manufacturing processes.

Some manufacturers have adopted HCFC-22 as a substitute for CFCs, especially for polystyrene foam food packages. As noted above, HCFC-22 is only 5 percent as destructive as CFC-12, and it is not yet covered by the Montreal Protocol. At issue is whether the sanctioned interim use of HCFC-12 to replace CFC-12 is acceptable or whether it too should be rapidly phased out. Industry wants to keep using HCFCs for several decades. Some environmentalists are beginning to ask for a much shorter time period because their large-scale use still threatens the ozone layer. They also argue that the pressure of a short time limit will force R&D for superior solutions.

A better way to protect the ozone layer is to use other alternatives for many of these applications. Nearly always, the alternatives were widely used before ozone depleters took over. Chief among them are molded (recycled) pulp, plastic, and paper products. Indeed, molded pulp was the dominant food packaging material in the 1950s prior to the introduction of foamed products. Today it still has 61 percent of the 2.2 billion annual egg carton market, up from less than 50 percent in 1987. In Ontario (Canada), 90 percent of egg cartons are pulp and in Chicago, 78 percent. The rapid shift back to pulp is being driven by demands of food store managers who recognize the antiplastic sentiment among their customers. In the past year or so, Mobil Chemical shut four of its five polystyrene foam egg carton manufacturing plants and Keyes Fibre sold three of its polystyrene plants, saying that it wanted to concentrate on molded pulp. Fast food restaurants, which currently use 150 million pounds per year of polystyrene for clam shell containers alone, could change to molded pulp. Already, some paper disposable food containers are replacing foam products because of adverse publicity about the plastic products. However, this substitution, while addressing the ozone problem, ignores environmental issues associated with paper disposable products.

Internal recycling of CFC blowing agents to reduce emissions is also technically feasible. Work in Europe has demonstrated recovery rates of from 40 to 85 percent in industrial facilities that make foams.

Flexible Foams

To create open cells or bubbles in a plastic matrix, CFC-11 is used. It is emitted quickly, typically within hours or days of manufacture. Carbon dioxide, the chief contributor to global warming, is also used as a prime blowing agent. Methylene chloride is widely used instead of CFC-11; it is a low-ozone depleter but is toxic. Other low-ozone depleters are also being used. CFC-11 is used to achieve soft, low-density, flexible foams.

Major applications of flexible foams include furniture, car seats, bedding, carpet underlay, and other cushion uses. The chief alternative to using CFC-11 is to produce harder, higher-density, and higher-cost foams, which manufacturers claim would not be acceptable to consumers. A new innovation is discussed later in this chapter.

Refrigeration Agents

The dominant refrigerant in home refrigerators and freezers is CFC-12, which has the proper physical properties for a refrigerant and is nontoxic, nonflammable, and nonexplosive. Commercial trademarked names include Freon, Isotron, and Genetron. A home refrigerator that weighs 250 pounds and costs $700 may contain less than 3 pounds (less than $5 worth) of CFCs, and most of this is for the foam insulation, not the refrigerant. Combinations of various CFCs are used for commercial and retail refrigerators and freezers.

Refrigerators and freezers account for 6 to 7 percent of total CFC emissions. Most CFC-12 emissions occur when refrigerators and freezers become garbage, but there are significant emissions during manufacture and through leakage during use. Much attention has been given to recovering and recycling CFC-12 from old units prior to disposal, but there are many implementation problems.

Alternatives to using CFC-12 as a refrigerant include lower-ozone depleters such as CFC-22 and HFC-134a. The latter does not deplete ozone but does contribute to global warming, increases energy demands by as much as 10 percent, and is not compatible with the current lubricants necessary to protect compressor parts. Moreover, HFC-134a is toxic and nonbiodegradable, and every gallon manufactured produces a gallon of methyl chloroform, a serious ozone depleter. Du Pont is committed to the construction of a $25 million plant in Texas to make 1,000 tons of HFC-134a annually.

It is expected to sell for 7 times the price of CFC-12. For commercial food freezing, cryogenic techniques using liquid nitrogen are possible.

Complete replacement of ozone depleters is likely to prove very difficult technically; therefore, the industry focus will probably be on recycling CFC refrigerants rather than replacing them. There are about 85 million household refrigerators in the United States, an almost 100 percent penetration of households. Still, nearly 7 million new refrigerators are produced each year—a substantial market, it would seem, for CFC-free refrigerants. In China only 10 percent of homes have refrigerators now, but the government wants to raise this to 100 percent by 2000. The Republic of China has purchased 9 million CFC-free, helium-cooled refrigerators from Cryodynamics, a New Jersey company. The units are to be produced, however, in Shanghai, This company is adapting cooler units originally designed for use on NASA's Skylab and the space shuttle, both of which use helium or nitrogen refrigerants. Cryodynamics is also developing truck refrigeration systems and air conditioning for buses that do not use ozone-depleting chemicals.

Air Conditioning

About 60 million U.S. automobile air conditioners use CFC-12, accounting for 37 percent of total CFC-12 use in the United States. Emissions result mostly from leakage and repair servicing—usually venting to the atmosphere. Lesser emissions occur during manufacture and installation and then final disposal. The EPA estimates that automobile air conditioners account for at least 20 percent of U.S. releases of ozone depleters. American automobiles account for 75 percent of global CFC emissions from automobiles worldwide.

Consumers can buy cans of CFC refrigerant for home servicing of automobile units, and their inexpert use can release significant amounts of CFCs. Thus, one environmentally friendly option is for consumers to not buy such refills but instead go to professional service centers. Automobile service centers can buy recycling equipment costing from $1,800 to $6,000. The machines are an example of internal recycling of potential waste. They remove refrigerant gas from the air conditioning systems without leakage, filter out excess water and oil, and feed the recycled gas back into the car.

Room air conditioners contain CFC-22. Emissions result from servicing, leakage, and disposal, in that order. Commercial and industrial units, called *chillers*, use CFC-11, CFC-12, or CFC-22. Emissions result—in slightly different order—from leakage, servicing, and disposal. The chief chemical substitutes for CFCs appear to be lower-ozone depleters that require major redesign of air conditioning units. A more attractive alternative is fundamentally different systems. The Rovac Corporation has developed a low-pressure compressor system that uses hydrocarbons. Cryodynamics developed a unit that uses helium. The emphasis, however, will probably remain on recovering and recycling CFCs.

Solvents

CFC-113, methyl chloroform, and carbon tetrachloride are widely used as solvents. Chlorine emissions from solvents account for approximately 25 percent of total CFC emissions and about 43 percent of prompt emissions. CFC-113 is now the fastest-growing ozone depleter; U.S. production nearly doubled from 1978 to 1985. The large emissions contribution is an inevitable result of the way solvents are used. That is, solvents are exposed to the atmosphere, where they easily volatilize, and in some applications they are used in the vapor phase. Liquid solvents also get contaminated during use and become waste. Three-quarters of the emissions result from vapor losses; the rest are from waste disposal.

Some of the uses of these solvents are for cleaning and degreasing metal parts in many industries, dry cleaning of clothes, and cleaning of printed circuit boards and semiconductors in the electronics industry. They are also used in paint formulations. Methyl chloroform is used in typing correction fluid, shoe polishers, adhesives, pesticides and home foggers, aerosol water repellents, and shoe waterproofers. CFC-113 is used widely in electronic parts cleaning to remove soldering residues, in dry cleaning to remove grease, and as a machinery cutting fluid.

Major substitutes for chlorinated solvents continue to be developed. Some commercial alternatives are made from terpenes, derived from citrus fruit rinds. Other approaches are to develop nontoxic biological solvents. Some industries have found that they can greatly reduce the use of solvents, can internally recover and recycle solvents formally lost directly to the atmosphere, or can use completely different methods for cleaning. Internal recycling can reach very high levels, easily 90 to 95 percent, but most firms have not yet done so. United Nations work indicates that substitutes exist for 90 to 95 percent of methyl chloroform uses. Some hydrocarbons can be substitutes, but they pose problems of flammability and toxicity and, therefore, are not necessarily safe substitutes. Here too, lower-ozone depleters are also being favored by some companies.

Success at the IBM plant in San Jose, California, is an indication of the potential for reduction of ozone-depleting solvents. The plant was the largest source of CFC-113 releases in the nation in 1987 but cut its emissions from 1.5 million to 500,000 pounds in 1988. However, the IBM effort does not appear to be the norm. An analysis of the EPA's TRI database by the

National Resources Defense Council concluded that only 200 of 3,014 companies reported reductions in these solvents between 1987 and 1988. Total reported releases of the three solvents in 1987 were 205 million pounds. California industries release more of these ozone destroyers than those of any other state. Of the top 10 industrial facilities reporting CFC-113 releases in 1987, the top three were IBM plants, followed by a General Motors plant in Lockport, New York. In eighth place was U.S. Air Force Plant No. 4 in Fort Worth, Texas. For reported methyl chloroform releases in 1987, the largest emitter was Pratt & Whitney in East Hartford, Connecticut, followed by IBM in Endicott, New York. In seventh place was the U.S. Naval Weapons plant in Dallas, Texas. For reported carbon tetrachloride releases in 1987, the largest emitter was the Hercules plant in Parlin, New Jersey—which actually increased its releases in 1988—followed by Velsicol Chemical in Memphis, Tennessee. In ninth place was the U.S. Department of Energy facility in Golden, Colorado.

Hospital Sterilants

CFC-12 is widely used in gas mixtures to sterilize medical instruments. A large activity is presterilization of disposable medical and surgical equipment. Feasible alternatives include steam sterilization, carbon dioxide as a substitute for CFC-12, eliminating the use of CFC-12 as an additive to ethylene oxide (which itself is a suspected carcinogen), and using lower-ozone depleters.

It is ironic that a major part of the health care system is itself using ozone depleters that contribute to a potentially global health threat. Steam sterilization was used effectively long before the advent of CFCs. There is much concern today about escalating hospital costs, but little attention is paid to reevaluating high-tech practices in terms of their real costs and benefits, especially environmental ones.

Fire Extinguishers

Various halons are increasingly used instead of water or other fire extinguishing materials. Demand for halons quadrupled from 1973 to 1984 and is growing at an annual rate of 15 percent. Halon emissions result from actual usage, testing, and manufacture. Large commercial and office total flooding systems use Halon 1301. These include computer facilities, communication rooms, libraries, museums, airlines, and engine and boiler rooms. Halon 1211 is used in hand-held fire extinguishers for similar uses. Most home fire extinguishers do not contain halons, but some do. They can be avoided. Attention is being given to substitution with lower-ozone depleters.

SCIENTIFIC UNCERTAINTY DELAYED
MAJOR RESPONSES

The history of the ozone layer problem has revealed how scientific uncertainty can be used by industry and government to delay action. Of course, there is never complete scientific certainty about any health or environmental effect. What is proof of a problem or a cause of a problem? In criminal cases, guilt must be proved beyond a reasonable doubt. But in civil cases proof requires a preponderance of the evidence, and scientists deal with probabilities. If there is any lesson to be learned from history, it should be that environmental proof must be no more than substantial evidence.

The lesson to be learned from the debate about the effects of CFCs is that it is very difficult to practice the preventive philosophy of being safe rather than sorry if too much certainty is demanded. The demand for more scientific proof, for more certainty, is often no more than a tactical obstacle to a change that threatens a financial self-interest. Rather than take action, politicians call for more studies. Environmental activists have bitter memories of having recognized serious environmental problems long before credentialed professionals verified their conclusions.

In 1972, Du Pont had recognized the scientific data showing the CFCs were accumulating in the atmosphere and arranged a seminar on "The Ecology of Fluorocarbons" for the world's CFC producers. At that time, Du Pont said (Glas 1989), "it is prudent that we investigate any effects which the compounds may produce on plants or animals now or in the future." The first research paper on the implication of CFCs in the ozone layer appeared in 1974. It triggered considerable public concern. Meanwhile, the major manufacturers of CFCs repeatedly challenged the validity of the evidence linking CFCs to ozone depletion. Industry groups were formed in 1975 and 1980 to resist attempts to shrink the production of CFCs. In 1979, a report by the National Academy of Sciences cautioned that ozone depletion could reach 16 percent if CFC-11 and CFC-12 continued to be released at the then existing rates. Still, more research was repeatedly called for by industry. Research conducted by industry resulted in two studies in 1981 that said that ozone levels had increased and emissions had decreased. Academic and government reports refuted those findings. In the early 1980s, major CFC producers dropped their research for alternatives. It was not resumed until 1986. A 1982 report by the National Research Council said that atmospheric changes in ozone were likely to be insignificant, provided that there was no large growth in CFC production. But, of course, there was.

A scientific report in March 1988 by 100 international experts said that the ozone layer around the globe was being destroyed much faster than any model had predicted. Blame was firmly attributed to CFCs. In the same month, Du Pont, the largest manufacturer of CFCs in the United States, was still claiming that there was no certain need for dramatic reductions in CFC emissions. It wasn't until 1989, after nearly everyone had reached a consensus on the threat, that Du Pont announced its intention to phase out CFC production by 2000. A few months later, in testimony at a U.S. Senate hearing Du Pont (1989) was continuing to try to slow down action. "We urge the Congress to wait until it sees the impact of this first reduction before determining whether the United States and other countries can move faster than the [Montreal] Protocol now requires."

The wait-and-see strategy has paid off for industy because production levels and profits have remained high. Indeed, major expansions of CFC manufacturing capacity took place long after the general public and the scientific community had recognized the threats from CFCs:

- In 1983 Pennwalt, the third largest CFC producer in the United States, completed a $10 million plan to modernize and expand its Kentucky CFC plant to make HCFCs.
- In 1985 Du Pont expanded its CFC production in Japan to stem the export of CFCs from the United States to the Far East.
- In 1986 Allied-Signal, the second largest CFC producer in the United States, expanded its production of CFC-113.

A report by the RAND Corporation in 1987 (Wolf) observed: "In spite of the continuing possibility that nonaerosol emissions of CFCs might eventually be regulated . . . producers increased capacity by 25 percent over the period from 1978 to 1986. From 1978 to 1984, combined CFC production increased much more, by nearly 50 percent, as the excess capacity after the aerosol ban was more fully utilized. In fact, most CFC markets are growing rapidly and CFC-11 and CFC-12 have once again reached preaerosol ban levels of production."

U.S. CFC producers often remind people that the Montreal Protocol prohibits U.S. companies from manufacturing more CFCs than they did in 1986. This condition is monitored by the EPA. However, under the Protocol, producing nations can make extra CFCs and halons, not counted in their national quotas, for export to developing nations.

The opinion of a grass-roots alliance, the U.S. Public Interest Research Group (1989) about this history of industry opposition and resistance to strong controls on the production and use of ozone depleters is as follows:

Despite industry's formal support for the Montreal Protocol and its public acknowledgement that CFCs must be reduced, the CFC industry continues to lobby against any further regulation of these ozone-depleting chemicals. Even with 100 percent participation in the Montreal Protocol, ozone depleting chemicals in the atmosphere would double from present levels within the next 50 years. Clearly the Montreal Protocol does not do enough and yet the CFC industry continues to fight against further regulation of CFCs. . . . One must question the CFC industry's sincerity when they today make promises to quickly phase out CFCs and search for safe substitutes.

Positive Reactions by Governments and Industry

There is a brighter side to the ozone depletion story. Starting in 1982, the United Nations Environmental Programme sponsored meetings to develop an international agreement to protect the ozone layer. In March 1985, the Vienna Convention for the Protection of the Ozone Layer was signed. Increasing scientific evidence of a serious loss of atmospheric ozone resulted in 24 nations signing an agreement in September 1987 known as the Montreal Protocol. The Protocol went into effect on January 1, 1989. This agreement will limit the production of five CFCs and three halons.

Ratifying nations are to freeze production and consumption of CFC-11, -12, -113, -114, and -115 at 1986 levels, reduce those levels by 20 percent by mid-1993, and further cut production and consumption levels to 50 percent by mid-1998. In 1992, Halon 1211, 1301, and 2402 are to be frozen at 1986 consumption levels. Special provisions are included to encourage the participation of developing nations. Developing nations are to initially limit their consumption to 0.3 kg per capita, only one-third the current per capita consumption in some industrialized countries. After a 10-year grace period, developing countries are to cut their consumption of the controlled chemicals by 50 percent. The Soviet Union gets a catchup allowance under the Protocol.

EPA projections say that by 2075, even with 100 percent signatory compliance with the Protocol, chlorine concentrations in the atmosphere will triple. The EPA thinks that 45 percent of the projected chlorine growth will come from allowed uses of controlled chemicals, 40 percent would come from chemicals containing chlorine not covered by the Protocol (such as methyl chloroform and carbon tetrachloride), and 15 percent would come from emissions of nonparticipating countries.

Controversy remains because of chemicals not covered by the agreement and the timetable for reductions. At a meeting in May 1989, 80 nations reached a consensus that the original goals would not be sufficient to prevent environmental crisis. Although there was general agreement about the need to eliminate CFCs by 2000 instead of merely cutting output by 50 percent by 1998, other details are contentious. In November 1989 proposals were made to strengthen the Montreal Protocol. For example, the banning of halons, methyl chloroform, carbon tetrachloride, and even HCFC substitutes now seems likely. Problems have arisen for a number of reasons.

The European Community had not yet banned CFCs in aerosol propellants, which the United States and others did over a decade ago. The Soviet Union has taken the position that all halon uses are essential. The phasing out of HCFCs faces stiff opposition from industry in the United States, some of whom have already committed to their use.

There are major issues related to the ability and willingness of developing nations to participate in pollution prevention actions to prevent ozone depletion. North America, Europe, and Japan account for about 80 percent of the total consumption of ozone depleters covered by the Montreal Protocol. The United Nations (1989) noted that "developing countries are less able to pay the costs of reducing or phasing out CFCs and halons and may have other, more immediate concerns such as food supply and economic development." For an effective global prevention effort, a lot of technology and information transfer will be necessary among all nations. Much resistance will continue to come from industries in all nations that want to continue to use equipment and technologies in which considerable capital has already been invested.

At the end of 1989, the U.S. government took a very significant action. Both the Congress and the Bush administration took advantage of the ozone depletion problem. Faced with the need to raise money to comply with a required budget deficit reduction, a new tax, called a *user fee,* was imposed on the five CFCs and three halons covered by the Montreal Protocol. As a result, the U.S. Treasury may collect as much as $5.6 billion over 5 years. Starting in January 1990, firms that produce ozone depleters will pay a tax determined by multiplying a base tax by the ozone depletion factor for each chemical. The system uses the ozone depletion factors discussed at the beginning of this chapter, which means that CFC-11 and CFC-12 producers pay the highest tax and producers of other substances pay less. The base tax for 1990 and 1991 is $1.37 per pound, rising to $1.67 in 1992 and $2.65 in 1993 and 1994. It increases at a rate of 45 cents per pound each year thereafter. These levels of taxation are all the more significant because bulk chemicals sell for about $1 per pound, less than the tax itself.

Manufacturers immediately said that they would pass their increased costs on to consumers. Other than raising revenues to help balance the federal budget, it is hoped that the higher costs for the chemicals will help companies try to find safe substitutes for ozone depleters. Whether consumers react to the higher costs of products because of the increased cost of ozone depleters is uncertain because the prices of most consumer products are not likely to rise too much relative to sales prices. The more important issue is that consumers seem ready to pay increased prices for safe substitutes. A Gallup poll (1989) found that 79 percent of Americans favor banning ozone depleters "even if it means higher prices for air conditioners, refrigerators and other consumer products."

A perverse effect of the tax scheme, however, is that a number of ozone depleters that are not included in the Montreal Protocol will now have an economic advantage because they are not taxed. From a general pollution prevention perspective, the idea of taxing unwanted wastes and pollutants has received attention for many years. But attempts to tax broad classes of wastes have proved to be unsuccessful, such as interest in 1986 in imposing a federal tax on toxic wastes to help finance the Superfund toxic waste cleanup program. The combination of an enormous amount of public attention to the ozone depletion problem and the urgent need to raise federal revenues apparently did the trick in this case.

One type of government action has not received serious consideration. Consumer products could be labeled to inform consumers how the product contributes to ozone depletion. Only in this way will consumers be able to choose alternatives in the marketplace. Moreover, this approach lets inventors and entrepreneurs know that if they develop alternatives, they will have a fighting chance in the marketplace. There are also substantial opportunities for consumers to focus product reuse efforts on products whose manufacture requires ozone depleters. Examples include plastic foam egg cartons—already a successful changeover—foam peanut packing materials, and soft foam furniture.

All the industries associated with ozone depleters have, of course, attempted to address the problem technically and in terms of minimizing adverse public reactions, such as the increasingly popular local banning of foam fast food packages. Here are some examples:

- General Dynamics has cut its use of chlorinated solvents in its metal cleaning operations in aircraft production; this was done by using closed-loop (internal) recycling and aqueous solvents.
- General Motors has said that by 1991 it would require its 10,000 car dealers to recycle CFCs purged from automobile air conditioners being serviced.
- Nissan Motors said that it would end the use of CFCs entirely in its air conditioners by 1993, replacing the coolant with a substitute that does not destroy ozone. Nissan also said that it would replace harmful solvents with water-based ones.
- Toyota Motor Corporation announced that it would discontinue the use of ozone-depleting gases used for cleaning, cooling, and foaming during the manufacturing process by 1995.
- American Telephone & Telegraph said that it would cut its use of CFC solvents in hundreds of manufacturing operations by 50 percent by 1991 and end it completely by 1994. It has already made significant advances.
- Bull HN (formerly Honeywell) in Boston completely eliminated its use

of methyl chloroform by switching to a water-based cleaning solution for printed circuit boards.

- Northern Telecom found that it can eliminate chlorinated solvents for cleaning most of its soldered materials.
- General Electric said that it would offset its much criticized, preventable release of 300,000 pounds of CFCs from its nationwide refrigerator repair program by cutting back on unspecified releases elsewhere. The repair program fixes more than 1 million refrigerators with faulty compressors, and the repair in homes results in a release of about 4 to 5 ounces of CFCs. GE's action was initiated by the remarkable protest of Lynda Draper, a Maryland homemaker, who witnessed a GE repairman vent several ounces of CFCs. Although GE was ultimately pressured to address its nationwide practice, the solution it came up with may not stop the releases from the refrigerator repair program, even though they are preventable through the use of available portable recovery equipment.
- Union Carbide introduced a new chemistry system for flexible polyurethane foam that does not use CFCs or methylene chloride as auxiliary blowing agents. The company claimed that up to 70 percent of the CFCs used in flexible foam production could be eliminated. A tag designed to be attached to consumer products made from the company's ULTRACEL foam reads: "ULTRACEL foam cushioning is manufactured without fluorocarbons which are believed to have an adverse effect on the earth's ozone layer. This 'cushion of safety' against the sun's ultraviolet rays must be preserved. So reduction in fluorocarbon use can benefit us now and all generations to come." This is a very good example of industry capitalizing on an innovation aimed at solving an environmental problem. The marketing also recognizes and appeals to the green consumerism movement.

Finding safe substitutes for ozone depleters, including chemicals, equipment, and processes, offers an unlimited business opportunity for industry. A company, for example, that can market a home refrigerator or air conditioner that uses no ozone depleters for any function is bound to have an advantage in the marketplace.

WILL THE USE OF LESS SEVERE OZONE DEPLETERS AND RECYCLING SUFFICE?

There has been a lot of support, from manufacturers and users of ozone depleters, for substituting lower-ozone depleters like HCFC-22 for much more destructive ones with high ODP values. On the one hand, this seems reasonable and environmentally effective. Those substitutes are available for many applications, require less loss and replacement of existing equipment, and will not disrupt consumer demands. However, it is also true that this strategy will not necessarily prevent a major environmental catastrophe due to depletion of the ozone layer. Use of less destructive ozone depleters may only delay the planetary problem. Recent data from the United Nations indicates that some HCFCs have significantly higher ozone depletion potentials than the 0.05 value for HCFC-22 (see Table 5–3). HCFC-141b has a value of 0.10.

Nevertheless, several major U.S. environmental organizations seem to have endorsed the commercial substitution of low-ozone depleters for much more destructive chemicals. Consumers may be misinformed about this situation. In Table 5-1, CFC-22 and HCFC-22 are identical. HCFC-22, as noted above, destroys only 5 percent of the ODP of CFC-22. This is a consequence of the weaker chemical bonds in the HCFC-22 molecule that break down in the lower atmosphere, letting less reach the upper ozone layer. It has been estimated that HCFCs will cause 2 to 10 percent of the global warming problem out to the year 2030. In January 1988 Du Pont and other companies decided to change the name of CFC-22 to HCFC-22 because of the bad publicity about CFCs as the cause of the ozone problem. The EPA provided the plastic foam industry trade association with a letter that said that "HCFC-22 is not technically a CFC." Thus, McDonald's uses placemats that say that the plastic foam food containers which adorn their establishments and litter most of America are not made with CFCs. But of course, HCFC-22 was for decades CFC-22. Better-informed consumers will be less gullible consumers.

All signs indicate that there will be continuing strong assertions that using less severe ozone depleters to gain incremental improvement over current uses of strong ozone depleters is the only rational near-term approach. But there is a strong case to be made that using chemicals with 5 to 10 percent of the ODP of the prime CFCs might backfire. Use of the less severe ozone depleters may do little to curb the insatiable appetites for refrigeration, air conditioning, and foam products, to mention only a few examples. That was a key lesson of the U.S. ban on aerosol uses of ozone depleters. The loss of 50 percent of the market was overcome in less than 10 years.

Much of the world wants what the affluent industrialized nations now have. If global warming occurs, some of those applications, like air conditioning and refrigeration, could become more difficult to curtail. Even with the use of less severe ozone depleters, increasing demand together with worldwide population growth and economic development might, over a few years, result in enough destruction of the ozone layer to cause major health and environmental problems for humanity. And, as already emphasized, the ozone depleters also contribute substantially to global warming. The key problem is that the manufacturers of CFCs want to shift their business to HCFCs and have mounted a campaign that sends the message: if not HCFCs, then no phaseout of CFCs. But the shift to HCFCs makes it difficult for inventors and entrepreneurs with truly innovative substitutes for ozone depleters to gain a foothold. In late 1989, the following was reported by *Chemicalweek:*

> Major investment decisions by producers and users of CFC alternatives hinge on firm availability of such compounds, says Du Pont. . . . Because of what Du Pont calls "the threat of near-term regulation of HCFCs," less than 5 percent of the capacity needed to switch from CFCs is under construction. Hence, the company says, "CFC phase-out is threatened" unless policymakers "provide clear and prompt signals to industry and the consumer" that HCFCs and other compounds are acceptable.

And an advertisement by ICI, appearing in 1990, speaks of "ozone-friendly fluorocarbons," which are, of course, HCFCs that were and still are CFCs:

> Food must not be allowed to go rotten, and donated blood is needed to save lives. That's why ICI is developing a new generation of ozone-friendly fluorocarbons. Not only do they serve the same purposes as CFCs, but they also can help mankind protect the ozone layer, the invisible shield that protects us all.

In November 1989, the EPA introduced a proposal to amend the Montreal Protocol to phase out HCFCs from new products between the years 2020 and 2040 and to phase out existing equipment between 2035 and 2060. At the same time, Nordic countries proposed phasing out HCFCs between 2010 and 2020. The CFC industry, however, wants to delay the phaseout of HCFCs as long as possible. In early 1990, Allied-Signal was said to have decided to construct a $50 million facility in Geismar, Louisiana—the world's first commercial operation—to make HCFC-141b.

The strategy of recovering and recycling CFCs and similar chemicals from various manufacturing processes and products, such as refrigerators and air conditioners, is, of course, preferable to simply letting those wastes enter the atmosphere. Most of these recycling efforts fit into the pollution prevention category. Preventing atmospheric releases that will ultimately destroy the ozone layer is a necessity. However, the recycling approach reduces the pressure on industry to find completely safe substitutes for the primary uses of ozone depleters. Those who support recycling, therefore, need to be persistently reminded about the primacy of replacing ozone depleters with safe substitutes. It would be best to see recycling as a necessary substitute for indiscriminate release of ozone depleters. But recovery could be followed by safe destruction of the ozone depleters, or its use could be limited to older products which will slowly be phased out of existence.

SUMMARY

The ozone layer/CFC problem has revealed how apparently reasonable solutions to some poeple are far less acceptable to others. Finding solutions is enormously difficult because over $135 billion worth of equipment in the United States depends on CFCs and related ozone depleters. The United Nations (1989) has concluded:

> The first 50 percent reduction in the global use of CFCs will require modest new capital investment, will incur little or no net cost, will result in some business disruption, and will require very little capital abandonment. This relatively easy step will be accomplished through reduction in the use of CFCs in the manufacture of flexible foams and as aerosol propellants, the more efficient use of CFCs as solvents, and by reductions in many other applications.

As discussed in Chapter 3, however, pushing prevention of ozone depleters beyond the initial relatively easy reduction options will become increasingly difficult. How serious will the commitment to protect planet Earth be if substantial sacrifices are necessary because completely safe substitutes for all applications of ozone depleters are not readily available? Is the choice between having cool, air-conditioned houses to stay in because going outside is too dangerous and having less comfortable houses but being able to go outside safely? Are sacrifices really necessary to protect the planet?

Consider that the consumption of CFCs per $1 billion dollars of GNP in the European Community is over eight times greater than in Norway. Even Australia's consumption is about six times greater than Norway's, and Australia is sitting nearly under the ozone hole over Antarctica. The U.S. consumption is about five times greater than Norway's. The conclusion: there is a large opportunity to reduce the use of ozone depleters without destroying a high standard of living.

United Nations projections show that even if CFCs are completely eliminated by 2000 *and* there is a phaseout of methyl chloroform and carbon tetrachloride *and* there is 100 percent global participation *and* HCFCs capture half of the CFC market, the amount of destructive chlorine in the atmosphere will not return to the 1985 level by the year 2100. With this knowledge, pressure will remain strong at the grass-roots level in the United States and internationally to obtain faster reductions than those mandated by the Montreal Protocol. The City of Irvine, California, for example, passed a law in 1989 to ban the use of CFCs as of July 1, 1990. Similarly, the city of Newark, New Jersey, passed an ordinance that prohibits the manufacture and use of CFCs, halons, methyl chloroform, and carbon tetrachloride on Earth Day 1990. Rhode Island has passed regulations prohibiting the manufacture or purchase of products containing CFC-11 or CFC-12. Hawaii, Vermont, and Oregon have banned the sale of small refrigerant refill cans. The Vermont law, passed in May 1989, requires recycling of CFCs during car air conditioner repair, bans the sale of small cans of CFCs, and prohibits the sale or registration of cars that use CFC coolant beginning with 1993 models.

Nevertheless, many people in industry, who use or make ozone depleters, are likely to resist more stringent timetables for phasing out usage. A Du Pont official (Glas 1989) has warned:

> a ban on CFCs before alternative chemicals or technologies can be put into place would mean lapses in the distribution of blood, other medical supplies, and up to 75 percent of the U.S. food supply. It could also force shutdowns of many modern office buildings that require air conditioning, as well as many U.S. manufacturing operations.

On the other hand, the phasing out of ozone depleters opens up new opportunities for companies *not* in the CFC business, other than the recycling discussed above. The consulting firm Mathtech estimated in late 1989 that the replacement of CFCs used as solvents in detergents for aqueous cleaning systems will create new business totaling $1 billion per year by the end of the 1990s. Such systems will be used by the metal and electronics industries. Total sales could add up to 150,000 units at a cost of $3 to $8 billion during the 1990s.

In conclusion, what the world's nations do about ozone depleters will, more than almost any other environmental problem, test humanity's resolve to be safe rather than sorry, even if it necessitates some inconvenience, sacrifice, and disruption. Is it so much trouble to put an end to CFC aerosols and solvents and to use available substitutes for 90 percent of current applications of ozone depleters? British and Swiss manufacturers label products so that consumers will know if they are ozone friendly. That needs to be a global practice, and "ozone friendly" must mean that *no* ozone-depleting chemicals have been used. Manufacturers have made enormous profits from ozone depleters for decades. Now there are more important objectives for many more people. As Tom Wray (1989) has said:

> Martin Luther King Jr. once said, "We must either live together as brothers or perish as fools." Failure to address the global implications presented by CFCs is akin to acting as an accomplice to genocide. Economics must not take precedence over technology, or our legacy may well be global catastrophe.

6

HARM AT THE FARM
AND HOME
FROM CHEMICAL PESTICIDES

One of the shining examples of pollution prevention is taking advantage of many opportunities to eliminate or reduce the use of chemicals generally referred to as *pesticides*. Pesticides are a key raw material or input to the agricultural industry, just like petroleum for chemical manufacture or coal for a power plant. The problem is that pesticide wastes are also produced and end up contaminating the products of farms and the land, water, and air where the pesticides are used. Consumers of foods, therefore, are exposed to pesticide waste. The irony is that at a time when Americans have been changing their eating habits and shifting to more fresh vegetables and fruits, they have learned that they are ingesting more toxic pesticides unless they seek pesticide-free foods and uncontaminated drinking water. People are also exposed to many other sources of pesticide pollution because of their increasing use nearly everywhere, including in and around homes.

In some American communities, such as Whately, Massachusetts, pesticide use has declined dramatically. Unfortunately, pollution prevention is practiced after a very harsh lesson that has not yet been learned by most people. For the people of Whately, mysterious skin rashes on children were one of the first signs that the community's water was poisoned with high levels of pesticides like Temik and ethylene dibromide, which were used on nearby tobacco and potato farms. Thus, 175 families with contaminated private wells had to get water at a school for 3 years. Then they paid $3,500 to $5,000 per home to hook-up with a new deep well in 1987.

It has been estimated that about 50 percent of agricultural pesticides could be eliminated and replaced by currently known alternatives. Totally pesticide-free farming is a proven technique but is practiced only on a small scale. Pesticides are, by definition, killers of life forms and consist of classes of chemicals, such as insecticides (killers of insects), fungicides (killers of fungi), and herbicides (killers of plants). Of course, most pesticides are not consumed by the species they are supposed to kill, but instead become environmental waste. In true pollution prevention, the problem of pesticide-contaminated soil, water, air, and foods is solved by replacing pesticides with other materials or practices to provide the protective function of pesticides. There is no need to question the problems facing farmers, landscapers, homeowners, and others in minimizing the damage from insects, fungi, and weeds. There is only a need to use solutions which are not excessive and which are environmentally safe.

Recent public attention to foods contaminated with pesticides is proving the most effective incentive for the agriculture industry to adopt techniques known for years to be effective, safe substitutes. It is a prime example of how consumers can act in the marketplace and bring about change at the source of environmental problems. Of course, consumers themselves use pesticides around their houses for many purposes, from killing roaches to killing dandelions to killing fleas on pets. In addition to eating and breathing pesticides, there have been a host of environmental problems and costs from pesticide use over many years which affect people and provide motivation to practice pollution prevention. Important examples include the need to clean up Superfund sites because of pesticide-contaminated soil, contaminated groundwater because of general pesticide use, and the presence of pesticide containers and residues in municipal sewer systems and wastestreams, making municipal landfills and incinerators sources of problems. Moreover, exposure to pesticides by workers making and using them is a major occupational health hazard. This is a problem for farm workers, exterminators, and lawn care applicators.

Often overlooked is the problem that fish, oysters, and clams may also be contaminated with pesticides. Although there has been publicity about bird kills from pesticides, fish problems are also significant. The problem is that runoff of water from farms and other areas contaminated with pesticides carries toxic chemicals into lakes, streams, rivers, and estuaries. For example, fish from the Great Lakes are heavily contaminated with toxic chemicals, and one cause is pesticide runoff (another is legally permitted direct discharges of industrial chemical wastes into the lakes). Even though Great Lakes fish are contaminated enough to threaten the health of people, especially children and pregnant women, the $4 billion dollar sports fishing industry has opposed public and government attention to the fish problem.

USE OF PESTICIDES

Nearly every account of pesticides starts with the DDT story—and with good reasons. Environmental contamination and damage from DDT helped focus public and government attention on all environmental problems. Moreover, the banning of DDT in 1972 by the federal government illustrated the most powerful pollution prevention tool for government use: legally prohibiting the manufacture and use of a dangerous chemical or product that inevitably results in wastes and pollutants entering the environment and causing health and environmental damages. In 1970 the average amount of DDT in Americans' fat tissue was 8 parts per million, but by 1983 it had dropped to 2.

Like so many other substances which eventually became recognized as dangerous environmental wastes, DDT was first seen universally as a miracle chemical, a marvelous product of science, a sign of the ingenuity of humans, a historical marker of progress by technological society. Paul Muller, a Swiss chemist, received the Nobel Prize in medicine in 1948 for discovering the ability of DDT to kill insects. Before that time, DDT had been an effective killer of the organisms responsible for malaria, typhus, and other deadly diseases. DDT was recognized as the key factor in preventing massive deaths and injuries during World War II. But on the darker side, DDT was equally effective in becoming an environmental killer. Like so many other chemicals, DDT had the ability to spread quickly throughout the environment, to persist chemically, and to contaminate many kinds of living species. Contamination would be found to kill some living species and to accumulate in other species, so that the entire food chain, ending with consumption by people, was affected. The irony is that people saved from diseases because of DDT's use as a pesticide may have become exposed to the chemical itself, only to suffer different kinds of health effects.

There has been ample evidence of the effectiveness of the DDT ban. As noted above, the average DDT levels in human body fat decreased by 75 percent from 1970 to 1983 in the United States. Various bird species have returned to health. On the darker side, however, DDT was sometimes replaced by other harmful chemicals. For example, toxaphene was used to replace DDT, especially for cotton crops, and the toxaphene content of fish increased more than 20-fold between 1970 and 1980. This problem of finding truly safe substitutes is urgent when dangerous substances are banned.

The advent of DDT triggered a widespread conversion of farm practices from a number of other approaches to control insect damages. Indeed, it must be acknowledged that DDT at first appeared to offer a number of advantages, even in terms of safety, compared to previously used substances such as arsenic, nicotine, and cyanide compounds. DDT was comparatively inexpensive and cost effective because it was persistent in soil and effective against a broad spectrum of insects. The last two factors, of course, also explain DDT's nasty environmental characteristics.

The era of chemical pesticides created a large, profitable business for many American companies. The current total dollar value of the U.S. agriculture pesticide market is about $6 billion a year, and worldwide it is about $20 billion. Pesticide use in agriculture increased 170 percent between 1964 and 1982, from about 110,000 to 300,000 tons. In 1987, when about 260,000 tons of pesticides were used, it was estimated that about 70 percent of all U.S. crop land, not counting alfalfa and hay areas, pasture, and range land, received some amount of pesticide, including 95 percent of the land used for growing corn, cotton, and soybeans. But historical uses of insecticides, herbicides, and fungicides have been difficult.

National data for 1987 show the following breakdown for agricultural use: 73 percent for herbicides, 15 percent for insecticides, and 12 percent for fungicides. This distribution has changed substantially over time, with insecticides and herbicides having smaller fractions of total agricultural pesticide use. In other words, U.S. use of pesticides has steadily consisted of relatively greater use of herbicides, at least in terms of tonnages of chemicals used.

In fact, the total amount of insecticide use started to decline in 1976 and leveled off in 1986 at about 50 percent of the 1976 amount. Unfortunately, government data is not presented in the format necessary for careful evaluation of pollution prevention measures. That is, having data about the national quantities of chemicals used can be deceiving because there may have been changes in farm acreage or crop production which changed the need for chemicals. Also, insecticides could have, and in some cases did, become more potent and effective, so that smaller amounts produced the desired effects. In part, this has been necessary because of resistance to pesticides by species over time. So, not all of the observed decline in insecticide use can be attributed to pollution prevention actions.

Resistance to pesticides is an important phenomenon. One sign of the significance of pest resistance to pesticides is that crop losses each year from pests have remained at about one-third, in spite of the long-time use of chemicals. For example, over a 45-year period, the number of insect and mite species with acquired resistance to pesticides increased from 7 to 447. In nearly 20 years, 48 weed species became resistant to pesticides. In some places, a ''pesticide treadmill'' has developed: pests become resistant to chemicals and suffer less from natural enemies because they too are killed, leading to crop losses which farmers react to with still more pesticides. Costs go up and crop production goes down. This syndrome also makes groundwater contamination more likely. Companies, therefore, have to reformulate and strengthen pesticides to maintain their effectiveness against the targeted species.

In other words, the chemical nature and toxicity of pesticides and their wastes have changed over time in order to stay competitive. This means that aggregate data for pesticide use over time, in terms of pounds or tons, is not particularly informative because pesticide toxicity has no relationship to weight. Moreover, complex government policies and market conditions which affect the amount of land used for production also affect the total amount of pesticide used. From a pollution prevention viewpoint, it is necessary to look at how much pesticide is used per acre and at the concentrations of specific chemicals as waste residues in foods or groundwater, for example. Still, the most accessible information is aggregate data on national use of pesticides in terms of amounts used, and it is certainly better to see a decline in the amount used rather than an increase.

Consider the use of herbicides to combat weeds. Up to about 1968, the quantity of all insecticides used in the United States exceeded that of all herbicides. The use of herbicides, though, increased at a very rapid rate until 1982, when the national tonnage started to decrease significantly. The total for 1987 was equal to the value of 1976, about 190,000 tons. From 1982 to 1987, herbicide use decreased by 17 percent from its peak of 228,000 tons. In other words, the decline in U.S. herbicide use happened 6 years after the start of the national insecticide decline and has been much less significant. As with the decline in insecticides, the exact reasons for the herbicide decline is unclear. It may be due to less agricultural production or to the use of stronger chemicals which translate into less weight—not to waste reduction efforts. Finally, for comparison, national agricultural use of fungicides has increased at a rather slow rate from 1966 to 1987, from about 12,000 to 25,000 tons.

Internationally, pesticide use has also increased over the past several decades in many areas, especially in Third World nations in which the so-called green revolution has taken place. For example, in India pesticide use increased from about 2,000 tons annually in the 1950s to more than 80,000 tons in the mid-1980s. That enormous increase corresponded roughly to the increase in the acreage on which pesticides were used. Figures on the dollar value of pesticide imports worldwide reveal major increases between 1972 and 1984, about a 160 percent increase overall in constant dollars. The Soviet Union had the largest increase, over 300 percent. But import figures are not the whole story, because multinational corporations have constructed major pesticide plants throughout the world over the past 20 years. For example, in India there was a 13-fold increase in pesticide production between 1970 and 1980.

Finally, the above discussion of agricultural use of pesticides over the years may hide other significant factors. Pesticides are used outside of agriculture, including use by households, landscapers, and institutions for ground maintenance. Pesticide use has become ubiquitous. U.S. manufacturers of pesticides which produce wastes also export much of their product. This explains why U.S. production of pesticides is over twice the amount reported for agricultural use. The value of U.S. pesticide shipments is now close to $6 billion annually. Annual U.S. production was 590,000 tons for 1986, down from a high of 802,000 tons in 1975. Subtracting the 270,000 tons used for agricultural purposes leaves 320,000 tons in 1986. Government data reveal that close to 20 percent of U.S. pesticides were exported (net after imports, some of which are probably used in the agricultural sector). That leaves about 250,000 tons of pesticides used outside the agriculture sector.

In other words, U.S. use of chemical pesticides is about equally divided between agricultural and other applications, such as direct consumer sales, institutional users, and landscapers, lawn care companies, and exterminators. More pesticides may be used per acre around homes than on farms. Indeed, a 1983 study by Citizens for a Better Environment found that an average suburban household used 8.1 pounds of pesticides per acre, in contrast to only 2.1 pounds per acre by neighboring soybean farmers in Illinois. A similar conclusion was found for San Francisco.

As another example, in many areas, local governments carry out massive aerial spraying of pesticides to control invasions of insects, such as gypsy moths and fruit flies. Also of particular importance is the residential lawn care industry. It has been reported that this industry has grown about 20 percent annually in past years. In 1981 sales were $870 million; now they are in excess of $2 billion annually from servicing almost 10 million homes. (This is about the same amount of money that Americans spend annually on bottled water to protect themselves from contaminated municipal water supplies.)

Of the 40 pesticides commonly used by the lawn care industry, 12 are suspected human carcinogens, 21 can cause other long-term health effects, and 20 can cause short-term nervous system damage. Some of the more widely recognized dangerous pesticides used on home lawns are 2,4-D, Dursban, and Diazinon. In 1984, application of Diazinon on a golf course in New York killed 700 brant geese in 2 days, 7 percent of the state's population. The federal government does little to regulate this industry, and only a few states take significant actions to require that homeowners be informed of pesticide use or to require that pesticide applicators be trained and tested. The public's concerns about pesticide wastes, therefore, should not focus solely on farming and food products. Pesticides applied around and used in the home also pose a major threat to people. (Home pesticide use and its alternatives are covered in detail in Chapter 8.)

The EPA's TRI database provides information on the release of pesticides into the environment from industrial facilities that manufacture or process pesticides. In 1988, there were 70 companies with plants in the United States making pesticides; 6 of those companies accounted for about half of U.S. sales, with 11 more companies accounting for 35 percent. For 1987, according to the TRI, 1,550 tons of pesticide releases were reported by companies. This amount equates to about 0.2 percent of the total manufactured, but many companies may not have reported data to the EPA and, of course, industrial pesticide facilities produced many other kinds of wastes. Moreover, the TRI-reported pesticide waste from industrial facilities is very small compared to the environmental wastes from direct pesticide use by farms and other users. A very high fraction of all pesticides used becomes environmental waste. Georgia facilities accounted for the largest fraction of the TRI pesticide waste, 700 tons, followed by Texas with 450 tons and Illinois at 130 tons. Industrial facilities in these three states accounted for 83 percent of the national total.

The largest sources of the releases were manufacturing (36 percent) and industrial processing (47 percent). As to disposition of the pesticide releases, 91 percent were reportedly sent offsite, which could mean nearly anything in terms of waste management. Direct releases into air were 4 percent (62 tons); public sewers, 3 percent (47 tons); land, 2 percent (31 tons); underground injection wells, 0.8 percent (12 tons); and surface waters, 0.06 percent (1 ton). These amounts are small, but because of the toxicity of pesticides, they could pose serious health threats if they result in human exposures. For example, a few pounds of pesticide dissolved in a billion gallons of water could make the water unfit for drinking. For the top 10 chemicals, the pesticide with the largest amount of releases was chlorothalonil, at 437 tons, followed by carbaryl at 240 tons, zineb at 209 tons, captan at 199 tons, parathion at 176 tons, chlordane at 125 tons, trifluralin at 70 tons, heptachlor at 60 tons, manes at 29 tons, and fluometruron at 9 tons.

HEALTH AND ENVIRONMENTAL EFFECTS

One way to build safety into pesticides is to engineer a short lifetime so that the chemical may be effective initially and then degrade under natural environmental conditions into harmless substances. But like most industrial chemical wastes from manufacturing and the worst ozone depleters, many pesticides persist in their original chemical form for long periods of time. The large amount of information on environmental contamination by pesticides proves that many different kinds of pesticides retain their original toxic forms over long periods and in different settings. Pesticides used many years ago are just now showing up in the environment. Indeed, an often overlooked problem is that pesticides that have been banned, such as heptachlor, toxaphene, and silvex, may still be present in the environment and in foods. The reason is that, like persistent ozone depleters, some pesticides remain in the complex food chain and ultimately show up in food or water. (Another reason they continue to show up is that banning is not always total; some uses may be continued.) Toxaphene (banned by the United States in 1982) is showing up in Wisconsin's waters on Lake Superior and in fish from there because of long-range air transport from that pesticide's use in Latin America. Moreover, some of the newer, less persistent pesticides pose short-term, acute effects. Bird kills that result from fresh pesticide applications, for example, are stark evidence of the acute toxicity of many pesticides.

It is important to remember that pesticides, by definition, are *designed to kill living organisms*. For example, organophosphate compounds were originally developed to be nerve gases used in chemical warfare. They are effective because of their disruption of nerve transmission. Organophosphates account for about 80 percent of the thousands of hospitalizations related to pesticide use in the United States annually. In this broad chemical class are the insecticides parathion and malathion.

Malathion may sound familiar. In southern California, the tiny Mediterranean fruit fly (medfly) has been combated with malathion. Aerial spraying first became a major health issue in 1981. To demonstrate his faith in the insecticide's safety, the director of the California Conservation Corps swallowed a mouthful of diluted malathion. There was no immediate effect. In early 1990, the controversy erupted again because the state government ordered 8 to 12 aerial sprayings by low-flying helicopters over 270 square miles and a population of 1 million people. The Coalition Against Urban Spraying argued that various research studies show that malathion is a potential carcinogen. The State of California disagreed and maintained that the release of 140 million sterile male medflies a week as a biological control was not effective enough. *The Wall Street Journal* editorialized (1990) that public "hysteria" and "irrational fears" about the spraying of the pesticide could result in "a contagion of fear and folly."

The infamous Agent Orange, used as a defoliant in Vietnam and considered by many to be a cause of serious health problems in people who served there, was an herbicide. In fact, the federal government banned the use of 2,4,5-T as a home herbicide about 20 years ago because of reports of miscarriages, birth defects, respiratory problems, and neurological illnesses associated with exposures from aerial spraying of the chemical. Dioxin waste residues are associated with this herbicide.

For many years, professionals have warned of the dangers of pesticides. Paul E. Gray, president of the Massachusetts Institute of Technology, said in 1988 that "growing reliance on insecticides and fertilizers has contributed to widespread chemical pollution of rivers, lakes, and seas, threatening the food chain itself." In 1987, Richard N.L. Andrews and Alvis G. Turner, professors at the University of North Carolina, gave the following overview of pesticide use:

> Environmental contamination by pesticides can occur from common use, spills, accidents, disposal of excess materials, disposal of wastewater from equipment, and rinsing of empty containers. Movement of pesticides through soil and into groundwater depends on a number of factors, but groundwater contamina-

tion from the use of these substances in agricultural operations has been found in at least eighteen states. Pesticide residues in food, drinking water, and the atmosphere are not uncommon.

The number of states with identified pesticide contamination of ground-water is growing. Data from 1986 indicated that 23 states and data from 1988 showed that 38 states had groundwater contaminated by 74 pesticides. Groundwater contamination by pesticides is particularly likely in certain areas. In Florida, for example, pesticides may have to sink only 15 to 18 inches to reach the water table.

Michael Colgan, who runs a nutritional science center in California, tells a story of a runner who had hopes of competing in the Olympics. His performance started to decline, and he began to have leg cramps and an upset stomach. His hands and feet were found to be swollen, his eyelids were a bit red and puffy, his abdomen was tender, and his muscles were hyperactive. Blood tests were normal. The key clue was that he had moved to a rural community to train more effectively. His water was supposedly very pure; it was from a local deep well. The water was tested. It was slightly contaminated with arsenic, and his hair had 11.4 parts per million of arsenic in contrast to a normal range of 0 to 2 parts per million. The source of the arsenic was thought to be weed-killer chemical runoff from adjacent land. After he switched to bottled water, the runner's symptoms disappeared. Colgan said publicly (1990) that ''odds are that your water *is* polluted. Best advice, don't drink it unless you clean it yourself.''

Uncertain Toxicities

It is generally recognized that only about 10 percent of currently used pesticides have been fully tested to determine their health effects. Moreover, the range of health effects studied is narrow. For example, pesticide effects on learning ability, emotional responses, sight, hearing, activity level, and memory are not required to be identified. Sandra Postel (1987) has described the persistent lack of detailed scientific information on chemical pesticides.

> The U.S. National Research Council (NRC) estimates that no information on toxic effects is available for 79 percent of the more than 48,500 chemicals listed in EPA's inventory of toxic substances. . . . Pesticides generally have received more extensive testing, but there, too, serious gaps remain. By allowing the production and release of these compounds without understanding their damaging effects, society has set itself up for unpleasant surprises.

The lack of credible, detailed information on the health effects of pesticides was compounded when a major source of data for pesticide manufacturers, the Industrial Bio-Test Company, was discredited. In the early 1980s the company was found to have faked many of its test results. Pesticides whose safety was based on information from that laboratory have been allowed to stay in the marketplace while new data are obtained, albeit very slowly. Also, more recently, in 1988 the California Department of Food and Agriculture, in its evaluation of hundreds of health studies of laboratory animals, found that 58 of 99 pesticide chemicals had received inadequate and frequently seriously flawed chronic health studies. *Public beliefs that the pesticides sold must have been found safe by the government are completely misplaced.*

A General Accounting Office (1986a) report on nonagricultural pesticide use concluded: "It is important to minimize pesticide exposure because all pesticides pose some risk, and the degree of chronic health risks is uncertain." A 1987 study by Consumers Union of 50 common household pesticides found that the EPA lacks sufficient information to conclude confidently that the chemicals are safe. To the contrary, EPA test data that *was* available showed that 28 percent caused cancer in animals, 26 percent caused adverse reproductive effects, 18 percent are known animal mutagens, and 6 percent cause birth defects in animals. Uncertainty about safety characterizes the commonly used product Roundup, the world's first herbicide to reach $1 billion in sales. Monsanto, the manufacturer of glyphosate, the active ingredient in Roundup, maintains that the chemical is safe, but there is also some evidence that it may cause cancer. The EPA is waiting for updated studies.

Thomas H. Rawls, the editor of *Harrowsmith,* has summed up (1989) the generally unrecognized threats from household use of pesticides this way:

> It is, quite simply, a world dominated by confusion and uncertainty. . . . That the ephemera of profits and the promise of some alleged short-term benefit, like an ant-free kitchen, should not be allowed to outweigh long-term human and planetary health. . . . What we need to end is the public burden: all of us should not become unwitting laboratory specimens. While manufacturers and regulators are chasing down data to determine a chemical's effects on human health, we shouldn't be the mice in the experiment. We are talking about poisons; they are not entitled to a presumption of innocence. If a chemical has not been proven safe, it doesn't belong on supermarket shelves.

Certain Exposure

Exposures to relatively high levels of pesticides are known to lead to acute health effects. Worldwide deaths from immediate exposure to pesticides may total 10,000 to 40,000 annually. To put this in context, there were reportedly as many as 5,000 deaths and several hundred thousand injuries because of the toxic gas leak at the pesticide manufacturing plant in Bhopal, India. Deaths resulting from long-term, or chronic, exposures to relatively low concentrations of pesticides has not been estimated. Most vulnerable are farm workers, especially in developing nations. A 1985 study in a Brazilian county found that 60 percent of farmers using pesticides had suffered acute poisoning. In Sri Lanka, 13,000 patients are admitted annually to government hospitals for treatment of acute pesticide poisoning, and 8 percent die.

In the United States, about 20,000 persons are taken to emergency rooms annually because of toxic exposures to pesticides, and about 35 people die. But there may be as many as 2.5 million health incidents annually that are related to household pesticide use. Under New York City's broad mandatory poisoning reporting law, about 1,000 pesticide poisoning cases were reported in 1988, despite the fact that some of the most acutely toxic organophosphates are banned by the city. All these health impacts are, at least theoretically, preventable.

It should also be understood that the application of pesticides on farms and for other uses often results in *pesticide drift*. Traditional spraying techniques may result in only 1 percent of pesticides reaching the target organism. This means that human and animal exposure can result from inhaling gaseous pesticide chemicals which remain airborne for significant periods. Depending on atmospheric conditions, pesticide drift can be a problem over a great distance. People living near farms and other places where pesticides are sprayed, as well as farm workers, are at risk from pesticide drift. Concentrations of pesticides in air can approach the same levels in water that has run over a treated farm field. In this regard, the following cases illustrate both pesticide drift and another widespread problem that physicians are not necessarily able to deal with effectively: pesticide poisoning.

Professor John Schedel has reported that on the morning of July 7, 1989, while on his way to his office in Buffalo, New York, he was overcome after inhaling some insecticides being sprayed on a Medaille College campus lawn. He became disoriented and dizzy, had difficulty walking and breathing, and experienced muscle spasms and cramps, diarrhea, severe scrotal pain, and the loss of fine motor control. When his symptoms did not disappear, he went to a hospital emergency room at around 5:30 P.M. He requested the standard blood test for cholinesterase activity, which is used to confirm poisoning from carbamate and organophosphate insecticides. Despite having the classic symptoms of pesticide poisoning, he was not given an antidote or any other treatment before being sent home nearly 12 hours after his arrival.

The Rochester (New York) *Democrat and Chronicle* reported on November 9, 1987, the unexplained health problems of workers in a large greenhouse nursery. For example, a planter who was the mother of six said that her hands, face, and neck had swelled profusely four times in the past 2.5 years after she had walked into a greenhouse where pesticides lingered in the air. Once her hands swelled almost to the size of boxing gloves; then her skin split open and oozed a sticky, watery substance. Although she and other workers attributed their problems to pesticides, physicians could make no diagnosis. A 1988 study reviewed 20 cases of infants and children whose pesticide exposure made them sick enough to be transferred to a major hospital from another medical facility. It was found that 16 of the 20 cases were not diagnosed as pesticide-related problems, even though the symptoms were caused by carbamate or organophosphate insecticides.

For farm workers, a 1986 study found a sixfold increase in the risk of cancer of the lymphatic system for Kansas farmers using certain herbicides, especially 2,4-D, for 20 or more days a year. More recent studies in Italy, Denmark, and Sweden have confirmed this finding for workers exposed to 2,4-D. And unpublished results from the National Cancer Institute have found that Nebraska farmers who use 2,4-D develop lymphoma three times as often as those who do not. Threats to poor migrant workers on American farms from pesticide use continue to be a major concern. A study (Moses 1988) found that the United Nations Food and Agricultural Organization's minimum standards on pesticide use "are not met in the United States and Canada, two highly industrialized countries. . . . Agricultural workers are not being protected from potential acute and chronic adverse health effects from their exposure to pesticides."

Exterminators and fumigators, another class of workers subject to high exposures to pesticides, have been found to have an increased risk of lung cancer and a slightly increased risk of leukemia and brain cancer. High exposures can also result from household use of pesticides, posing a serious threat to adults who use such products carelessly and, particularly, to children. In 1985, swallowed pesticides was the second most frequent cause of accidental poisoning of children in the United States.

One of the most tragic facts is that DDT and other pesticides that were banned in the United States and most European nations years ago are still being used in developing countries. For example, a 1987 report said that about 75 percent of total pesticide use in India was accounted for by DDT and benzene hexachloride (BHC), both suspected carcinogens. Moreover, an analysis of Indian foods found that 30 percent had pesticide residues that exceeded tolerance limits set by the World Health Organization. In India's Punjab region, every one of 75 samples of breast milk contained DDT and BHC. Babies who ingested such milk received a dose of the two toxins some 21 times the acceptable limit. Similarly, breast milk in Nicaraguan women has been found to have DDT levels 45 times greater than acceptable limits. The use of all pesticides in foreign countries is a major health concern to Americans and to people in other nations who eat foods imported from places where there is poorly controlled or monitored pesticide use. There have been many findings of pesticide contamination of imported foods.

The federal government's program to measure pesticide levels in all foods (domestic or imported) has been widely criticized as being ineffective. For example, a report by the General Accounting Office (1986b) found that the Food and Drug Administration's (FDA's) practices could not detect 60 percent of the pesticides likely to appear in food and that the undetectable pesticides were among the most dangerous ones, including Alachlor, the most widely used herbicide in the United States. In 1987, the FDA's sampling and testing found that 43 percent of foods had some pesticides but only 1 percent had concentrations that exceeded the legally allowable—but controversial—EPA levels.

Moreover, there are over 1,000 so-called inert ingredients in pesticides which have not been tested or regulated. Inert ingredients are used as solvents, propellants, surfactants, preservatives, emulsifiers, stabilizers, and enhancers. Some environmental groups have helped educate the public that inert ingredients themselves may be dangerous substances. These include chemicals such as carbon tetrachloride (a suspected human carcinogen), xylene, 1,1,1-trichlroethane, and methylene chloride. Some are chemicals that often justify the spending of millions of dollars to clean up Superfund toxic waste sites. In a strange twist, active ingredients in some pesticides are listed as inert ingredients in other pesticides. Even banned DDT has been listed as an inert ingredient. The EPA has labeled over 100 inert ingredients as being "of known or potential toxicological concern" and has described another 800 to 900 as inadequately tested for the risk they pose.

An interesting example of the inert ingredient problem is that the surfactant in the common herbicide Roundup, POEA, was found to be over three times more acutely toxic than the product's key active chemical, glyphosate. The surfactant belongs to a class of chemicals that is known to cause the gastrointestinal and central nervous system symptoms reported by Roundup poisoning victims. Victims of pesticide exposure are often at risk because of unspecified inert ingredients. The final fact about inert ingredients: they make up most of many commercial pesticides, especially those for household use, typically 97 to 99 percent.

Pesticides are often used to keep foods safe, but a 1987 study by the National Research Council found that pesticide residue wastes in foods could cause significant additional cases of cancer, perhaps 20,000 per year in the United States. About 80 percent of the estimated risk resulted from 15 foods, with tomatoes, beef, potatoes, oranges, and lettuce leading the list. A summary of some commonly detected pesticides in foods with their health effects is given in Table 6–1. Of the over 300 pesticides legally allowed on American foods, over 20 percent have been classified as carcinogens by the federal government. A study by the U.S. Public Research Interest Group (1989) found that 16 commonly eaten foods are legally allowed to contain 54 different carcinogens. The foods, with the number of carcinogens in parentheses, are apples (24), beef (26), broccoli (11), corn (25), eggs (18), grapes (17), lettuce (14), milk (20), oranges (17), peaches (21), pork (24), potatoes (17), poultry (17), rice (7), tomatoes (23), and wheat (16). The study said.

Until *the EPA* can prove risk assessment to be more of an exact science than it is today, it *should move to ban the sale and phase out the use of carcinogens immediately*. These chemicals pose more than a risk to the food supply, they endanger people all along the stream of commerce, from production to application. It is unacceptable for the government to defend the use of a cancer causing agent given the paucity of scientific knowledge about the chemical's use, other chronic health effects, the formula's other constituents, synergism with other

TABLE 6–1. FREQUENTLY DETECTED PESTICIDES IN FOODS

Captan
  Potential hazard: cancer
  Major crops affected: grapes, peaches, strawberries, apples

Carbaryl
  Potential hazard: kidney damage
  Major crops affected: corn, bananas, peaches, grapes, oranges

Dimethoate
  Potential hazard: cancer, birth defects
  Major crops affected: green beans, grapes, watermelon, cabbage,
    broccoli

Endosulfan
  Potential hazard: liver and kidney damage
  Major crops affected: spinach, lettuce, celery, strawberries, cauliflower,
    pears

Methamidophos
  Potential hazard: nervous system effects
  Major crops affected: tomatoes, cauliflower, cabbage, cantaloupe,
    bell peppers

Source: Natural Resources Defense Council, as reported in *Time,* March 27, 1989.

companion pesticides and fertilizers and greater threat to sensitive groups such as infants and children.

The situation has become so threatening that, in some places, supermarkets have hired private firms to inspect and test foods for pesticides. And, of course, increasing numbers of food stores tell consumers whether foods are imported and offer food purchased from farms that do not use pesticides. The response of consumers to such information has been dramatic in recent years. Periodic scares because of well-publicized cases of pesticide contamination, such as by Alar in apples, have driven this phenomenon.

One of the most fundamental problems with pesticides is that cumulative exposures and risks resulting from exposures to many different pesticides in foods, water, and possibly air are not accounted for in government estimates of safe risks and exposures. Logically, what might be a safe exposure for one chemical from one source could be dangerous when combined with exposure to many other chemicals from multiple sources. Moreover, health effects are different for different types of people, with infants, children, and pregnant women being particularly at risk from even low doses of pesticides in food and water. Farm workers could be exposed at work and in their food and water. It has also become accepted that there is no safe level for carcinogens, so that even minute amounts of pesticides may play a subtle but significant role over years of exposure.

People are exposed to pesticides because of their use for lawn care and indoors. A General Accounting Office (GAO) study of nonhome exposures to pesticides in Boston found that a person could encounter pesticides throughout a normal day in many places. A person using Raid Home Insect Killer, which contains chlorpyrifos, or Dursban, can also be exposed to that chemical in hospitals, private office buildings, retail food stores, industrial workplaces, hotels, department stores, a federal office building, public housing projects, the transit system, a mental health facility, and four state office buildings. This chemical is also widely used in agriculture, and residues in foods are known to be significant. Another study found that measurable amounts of chlorpyrifos were found in food-handling equipment at least 6 months after their last use. On October 16, 1987, a termite control company contaminated the public water supply of Clifton Park, New York, with Dursban by connecting a hose between a container with pesticide solution to a homeowner's water pipe.

It is realistic to think that many people receiving doses of this pesticide chemical from many different sources may develop health problems over relatively long periods of time. These problems may not be associated with pesticide exposure, however. Even though Dursban has been sold for over two decades, the EPA still cannot say with certainty whether it causes cancer, mutations, or nerve damage. But this organophosphate is known to be extremely toxic to fish, birds, and aquatic invertebrates; it is also a common source of household poisoning from high exposures.

As to its health effects from household use, in 1987 the National Cancer Institute found that children in households where home or garden pesticides are used have a much higher incidence of childhood leukemia. Some key findings were these: regular use of indoor pesticides by parents was linked to a 3.8 times greater likelihood of leukemia in their children than in homes where pesticides were seldom used; regular use of herbicides and insecticides by parents outdoors meant that children were 6.5 times more likely to contract leukemia; and when mothers regularly used herbicides or pesticides outdoors, children were 9.0 times more likely to get leukemia.

Contaminated groundwater used directly by people is also a very serious problem. In 1987 the Department of Agriculture estimated that 50 million Americans, mostly in rural areas, may be exposed to pesticides in groundwater. The EPA is still carrying out a national survey for 100 commonly used pesticides in a sample of about 1,350 drinking water wells. The survey began in 1988 and was scheduled to be completed in 1990. Preliminary results from the first 295 well samples were released in late 1989. Overall, pesticides were found in 5 percent of the samples, but the rate was 3 percent for community wells and 8 percent for private wells.

National statistics, though, can be misleading. Some parts of the United States are much more vulnerable than others. When groundwater is close to the surface, the ground is permeable to water, and farms use pesticides in traditional intensive ways, 25 percent of wells may have detectable levels of pesticides and 10 percent may have concentrations above health advisory limits. It has been estimated that more than 17 million people obtain their drinking water from private wells in high-risk regions, and 29 million people use community systems in those regions. The Department of Agriculture estimates that 1,437 counties, representing 46 percent of all U.S. counties, have groundwater that is vulnerable to contamination from pesticides or fertilizers. But it is important to note that groundwater contamination from current pesticide use may not become measurable for 30 or 40 years because of long time lags as chemicals travel into deep aquifers. In innovative legislation, the State of Minnesota passed a groundwater protection law in 1989 that places fees on agricultural chemicals of 0.1 percent of state sales in 1990 and 0.2 percent in 1991.

EPA regulations for drinking water standards include fewer than 10 pesticides. Aldicarb is the most acutely toxic pesticide registered by the EPA, and it has been found in groundwater in 16 states; 2,271,000 pounds of Aldicarb were used in 1987 for agricultural purposes. New York State banned the use of Aldicarb. A 1989 report by the National Research Council concluded: "Accumulating evidence indicates that a growing number of contaminants from agricultural production are now found in underground water supplies. . . . But changes in agricultural practices to reduce ground-water contamination are not widespread.''

It should also be emphasized that there is considerable data on the unintentional killing of animals by pesticides used in agricultural, institutional, and home settings. People have more than enough cause for concern about their own health effects and just as much reason, if they choose, to be concerned about the harsh impacts on many other living organisms. For example, the pesticide carbofuran has been found to be the cause of many bird killings, such as the death of almost two dozen bald eagles over a 3-year period in the Chesapeake Bay area. The eagles had been recovering from the ravages of banned pesticides such as DDT. One EPA study estimated that as many as 2.4 million birds are poisoned annually from the normal agricultural use of carbofuran. That figure, for just one pesticide, represents 24 percent of the estimated total number of bird deaths annually in the United States. About 7 million pounds of carbofuran are used annually.

Industry's perspective on the health and environmental effects of pesticides is, not unexpectedly, different. It is a bit more sanguine. A recent article in an industry trade magazine (Meyer 1989) speaks of the "reasonable compromise" that has resulted in pest and disease control and a healthy environment. Industry is said to have been successful in its "quest for harmony with the environment." At an industry conference in 1989, Dr. Maarten de Vries of the Roussel Bio Corporation said: "Food shortages would occur on a massive scale unless pesticides are used. Pesticides sustain the health and quality of life through vegetation growth and disease prevention." Ciba-Geigy (1988), a major international manufacturer of chemical pesticides, has said, "The ideal would be environmentally harmless products manufactured by environmentally harmless processes. This, however, is a utopian dream.''

ALTERNATIVE AGRICULTURE

Even the U.S. Department of Agriculture has recognized the environmental problems of farming. In 1989 it said that "evidence implicating agriculture as a major contributor to environmental problems is rapidly accumulating." Moreover, it acknowledged that "conventional agriculture's reliance on synthetic chemical fertilizers and pesticides has caused or aggravated many problems." In looking at the agricultural pesticide problem from a pollution prevention perspective, an important limitation to the traditional pollution control or end-of-pipe approach becomes more apparent. Agricultural use of pesticides creates what environmental professionals speak of as *nonpoint sources* of pollution. In other words, wastes and pollutants are created at so many locations that it becomes infeasible to think of using the kinds of pollution control technologies that are prescribed for industrial factories. That is, it is impractical to think of somehow capturing enough pollutants from countless sources and rendering the releases immobile or harmless. Other widespread sources include automobiles, home lawns and yards, and septic tanks. To deal with environmental problems from non-point sources, pollution prevention techniques are indispensable and the front end of the systems must be changed.

From a pollution prevention perspective, what has been going on in American agriculture for years provides further proof of how difficult it is to institute preventive measures aggressively. Major alternatives to the heavy use of chemicals have been used successfully for many years by many innovative farmers. *Low-input farming* is the term used to describe such approaches. But the agriculture industry and government programs, which have enormous influence on the industry, have resisted change in spite of hard evidence of the success of the alternatives. Farmers have also resisted change because some of the costs of using pesticides, especially health and environmental damage to workers and consumers, have not been borne by farmers. This is an example of classic externalization of the true costs of producing chemical waste. The situation is very similar to that of industrial waste reduction.

Even after proof has come in from countless successful examples, the system still resists change. Indeed, some of the arguments by those who resist are about the same, including the one that every site or farm is unique, making it difficult to transfer information and successful experiences. Studies of alternative agriculture also show that the use of pollution prevention techniques requires more intensive thought about the production process. In other words, as with industrial waste reduction, to reduce inefficient use of inputs and costly waste, people must replace their lazy historical use of chemicals with expertise and more efficient operations.

Government agencies have been recognizing the benefits of alternative agriculture and providing more support for it, although much more R&D could be done. In 1989, for example, Charles E. Hess, Assistant Secretary of Agriculture, said, "Whatever we call it—alternative, sustainable, or low-input, it makes remarkable sense for farmers and for the rest of society." Nevertheless, there has been no aggressive government support for pollution prevention options for agriculture because of lack of understanding that environmentally sound practices do not threaten production efficiency and profitability but, in fact, improve them. Moreover, some government actions actually help maintain the current market for chemical pesticides and block the use of waste reduction alternatives by not recognizing their comparative benefits. For example, government policies have not supported crop rotations or production levels that are profitable but not necessarily maximum. Pollution prevention does not threaten the agriculture production system, just as it does not threaten the industrial economy. What pollution prevention requires is widespread change which addresses the whole set of obstacles discussed in Chapter 3.

At the outset, pollution prevention options for agriculture, like those for manufacturing, encompass a variety of techniques. They include integrated pest management and the use of biological pest controls, crop rotations, and tillage and planting practices. From a pollution prevention perspective, the general objective is to eliminate or reduce the use of chemicals that have been proven to become harmful environmental wastes at the production site (farm), in the products (foods), and at the original chemical manufacturing facility. It should also be noted that alternative agriculture includes preventive health care for livestock in order to reduce the widespread, intensive use of antibiotics, which become a waste residue in foods and threaten human health.

The influential 1989 study by the National Research Council, *Alternative Agriculture,* concluded:

A small number of farmers in most sectors of U.S. agriculture currently use alternative farming systems, although components of alternative systems are used more widely. Farmers successfully adopting these systems generally derive significant sustained economic and environment benefits. Wider adoption of proven alternative systems would result in even greater economic benefits to farmers and environmental gains for the nation. . . . Successful alternative farmers often produce high per acre yields with significant reductions in cost per unit of crop harvested. . . . As a whole, federal policies work against environmentally benign practices and the adoption of alternative agricultural systems, particularly those involving crop rotations, certain soil conservation practices, reductions in pesticide use, and increased use of biological and cultural means of pest control. These policies have generally made a plentiful food supply a higher priority than protection of the resource base. . . . With modest adjustments in a number of federal agricultural policies many of these [alternative] systems could become widely adopted and successful. . . . Pesticides applied solely to meet cosmetic or insect fragment standards increase pest control costs to producers and may increase residues of pesticides in food and hazards to agricultural workers.

This historic government study received considerable criticism from industry. For example, Harold F. Reetz, Jr., of the Potash and Phosphate Institute said (1990) that the NRC report was "unscientific" and that he had "not seen any evidence that . . . reducing fertilizers and pesticides would solve either our economic or environmental problems."

It should also be noted that in addition to pesticides, a variety of manufactured nitrogen fertilizers are also primary inputs to farming, and they too can cause environmental problems. Groundwater is often contaminated with nitrates. Studies have shown that fertilizer use can often be reduced by 30 to 50 percent without reducing yields or profits.

INTEGRATED PEST MANAGEMENT

One of the basic operating principles of integrated pest management (IPM) is to determine the time when the economic threat from a pest population, in terms of crop damage, is large enough to justify taking control measures. Natural events may limit crop damage from pests, and a certain level of pest population is not necessarily harmful. In practice, it is necessary to scout fields to determine pest or disease populations or infestation levels. Scouting is a major cost in IPM. For insect control, users of IPM generally use precise application of specific pesticides. IPM use, therefore, is not necessarily the same as farming without any chemical pesticides whatsoever. (Some farmers do practice pesticide-free farming, but they remain a small minority.) Use of biological controls include pest predators or parasites, insect pheromones, sterile males, and bacterial insecticides. When pesticides are used in IPM, however, the frequency of application is normally decreased and the total volume of pesticides is typically reduced compared to conventional practices. In general, IPM use has reduced insecticide use much more than the use of fungicides.

A 1987 study of IPM programs by the Virginia Cooperative Extension Service for insects on nine crops in 15 states has provided unequivocal support for IPM. All IPM users obtained a higher average per acre yield over nonusers growing the same crop in the same state. Every IPM user obtained a higher net return per acre as a result of increased yields and prices. The 3,500 farmers using IPM obtained a $54 million increase in net return compared with nonusers. Results from this study are given in Table 6–2. If IPM was adopted for the nine commodities grown in the 15 states, an additional $579 million in annual returns to farmers would be obtained.

Other examples of positive results from IPM and biological controls include the following:

- A 75 percent reduction in pesticide use for alfalfa and increased net returns of $28 per acre were achieved.
- Insecticides and fungicides were virtually eliminated at a tomato farm in California, and a savings of $46 per acre was achieved compared to a saving of about $8 per acre for IPM users who still used chemicals to grow their tomatoes in California. The yield for the insecticide-free grower was 6.5 tons per acre *above* the county average in 1986, with essentially no insect or mold damage. Also, IPM tomato growers had

TABLE 6-2.   ESTIMATED AVERAGE ANNUAL ECONOMIC BENEFITS
FROM THE USE OF IPM

| State | Crop | Increase in Net Returns from IPM | |
| | | Actual Increase Statewide[a] | |
		($/hectare)	(thousands $)
California	Almonds	769	96,580
Georgia	Peanuts	154	62,600
Indiana	Corn	72	134,230
Kentucky	Grain	1	890
Massachusetts	Apples	222	400
Mississippi	Cotton	122	29,680
New York	Apples	528	33,000
North Carolina	Tobacco	6	780
Northwest[b]	Alfalfa	132	2,420
Texas	Cotton	282	215,830
Virginia	Soybeans	10	2,570
Total			578,980

Source: Virginia Cooperative Extension Service (1987).

[a]Increases from IPM farms extrapolated to whole state use.
[b]Idaho, Nevada, Montana, Oregon, Washington.

virtually none of their tomatoes rejected for insect damage, but non-users had a rejection rate of 5.6 percent.

- In Florida, vegetable growers using IPM had direct pest control costs of $200 to $300 per acre compared to $450 to $700 per acre for non-users. At 40 tomato farms, IPM use reduced insecticide use by about 21 percent.
- At one farm, herbicides have not been used to control weeds for 15 years because the 400 acres have been rotated in crops of corn, soybeans, small grain, and red clover hay, and because tillage, cultivation, rotary hoeing, and management of planting times techniques have been employed.
- A 500-acre farm had been spending $26,000 annually to control a weed (Johnsongrass) with herbicides when only corn and soybeans were grown. That cost has been reduced to $6,000 by converting the problem weed into a crop, forage for their own livestock and area farmers, and growing corn and soybeans on only 325 acres.
- From 1971 to 1982, intensive use of IPM reduced per area insecticide use on certain U.S. crops as follows: grain sorghum, 41 percent; cot-

ton, 75 percent; and peanuts, 81 percent. The results showed up in statistics on national use of insecticides. But insecticide use on corn and soybeans in that period increased by 8 percent and 13 percent, respectively.

- A study of U.S. corn belt farms found that, over 5 years, organic farms that used no chemical fertilizers or pesticides had an output that was reduced by 8.5 percent compared to conventional farms. However, that decrease in output was offset by the decrease in spending necessary for agricultural chemicals.

- According to one expert, IPM scouting can reduce pesticide use on sweet corn crops by up to 50 percent for insects and 25 percent for diseases.

- A major tomato grower reported that IPM scouting has cut his pesticide costs by about half.

- A group of cotton growers in a Chinese province planted sorghum between cotton plants to attract the natural enemies of cotton pests, changed tillage practices, and applied chemicals selectively. They reduced pesticide use by 90 percent and pest control costs by 84 percent and increased cotton yields.

- A large grape farming operation in California and Arizona rarely uses insecticides because a combination of biological and other practices control weeds, insects, mites, and diseases. Instead of fungicides, sulfur is applied several times a year. One study found that the California output was 653 boxes per acre, compared to 522 boxes per acre for neighboring grape operations. Production costs were roughly equal.

- About two-thirds of California almond growers have been using predacious mites instead of chemical control, saving $34 per acre, for a total net economic benefit of about $24 million annually.

- Brazilian soybean growers practicing IPM reduced insecticide use by 80 to 90 percent in 1982 relative to 1975.

- Corn growers in Minnesota reduced insecticide use by 45 percent over 10 years by rotating corn with other crops.

- Use of a European beetle to control a toxic range weed in California, started in the 1940s and costing about $750,000, has resulted in accumulated savings of more than $100 million.

- Instead pesticides, use of a tiny wasp imported from Latin America has controlled the mealybug that was destroying the cassava crop in Africa.

- A natural fungus, a bioherbicide, marked as Collego by Upjohn Company has achieved 90 percent control of a weed in Arkansas rice and soybean fields.

- At least 14 weed species are now being partly or completely controlled by plant-feeding insects or pathogens.

- Insecticides are no longer required for 73 percent of the alfalfa acreage in 11 northeastern states because of the introduction of six parasites to control alfalfa weevils. Use of such biological controls in the eastern half of the United States would save $44 million per year in reduced crop losses and spending on insecticides.

Another successful alternative to conventional pesticide use is mechanical devices that literally suck organisms off of crops in the field. The world's second largest lettuce grower in Salinas, California, is using "Salad Vac" instead of pesticides. The machine vacuums bugs off of the lettuce, and variations of it may be developed for cauliflower, broccoli, and celery.

Another promising area is the development of allelopathic chemicals to serve as pesticides. Allelopathic chemicals are natural chemicals produced by plants which often act as effective herbicides against other plants considered to be weeds. For example, leafy spurge is a weed that is a problem on rangelands; sunn hemp, semitropical plant, has been found to check the growth of spurge. Not only could certain plants be used directly, their natural pesticide chemicals could be recovered and used, or they might be synthesized. However, agrichemical companies have not been pursuing this idea aggressively. For example, Shell Agriculture Chemical developed an allelopathic herbicide from sage that could be used to control grasses. But after Du Pont bought the Shell unit in 1986, it stopped all activity because it concluded that the product's sales potential was too small.

Another important, rapidly growing area of scientific research and technological innovation is biotechnology. Many companies, such as Monsanto, are using various biotechnology techniques to modify crops genetically. Developing pest- and virus-resistant crops will ultimately eliminate or reduce the use of chemical pesticides on many crops. However, the alteration of the genetic makeup of crops is itself a controversial area, in part because of concerns about possible unanticipated and unknown harmful effects of genetic manipulation.

Various alternatives also exist for nonagricultural uses of pesticides. For example, to control gypsy moths, a biological control, the use of *Bacillus thuringiensis,* a microbe often called *Bt,* which is harmless to humans and other organisms, competes with aerial spraying of diflubenzuron. Bt is slightly toxic to humans, but diflubenzuron, which contaminates water runoff, is also extremely toxic to crabs, shrimp, and other aquatic organisms. Different forms of Bt, it should be noted, are effective for different insects, such as mosquitoes, beetles, and roundworms. Household use of pesticides also can be replaced by a variety of alternatives, as discussed in Chapter 8.

Lastly, there is a special twist on the organic farming concept, called *permaculture,* that originated in Australia. The approach is much more inclusive than IPM or simple organic farming. A major research program on permaculture is going on at Slippery Rock University in Pennsylvania. Organic farming is combined with the use of computers and solar energy to create small, independent farms. The organic farm is designed to be free of petrochemicals and pollution and to produce only natural, assimilable wastes. Only natural fertilizers are used.

CONSUMER DEMAND, INDUSTRY RESISTANCE, GOVERNMENT ACTIONS

Pollution prevention for the agricultural industry cannot be assumed to proceed successfully on the basis of documented benefits to the industry or because of what government agencies do. As with ozone layer destruction and replacement of CFCs and the products that depend on them, the role of consumers in agricultural waste reduction is critical and manifold. First, demanding and preferentially buying pesticide-free foods is critical. However, consumers should be aware that even foods produced on organic farms may have pesticide residues because of previous use of pesticides on the land or drifting of pesticides from neighboring farms.

The interest in organic foods is remarkable. A Harris poll found that 84 percent of Americans would buy organic food if it were available, and 49 percent said they would pay higher prices for it, which indeed they must do. This finding is consistent with the public's stated values concerning the trade-off between environmental protection and costs. That is, Americans have responded very positively when asked by CBS News-*The New York Times* to consider the statement "Protecting the environment is so important that requirements and standards cannot be too high, and continuing environmental improvements must be made regardless of cost." But the increased endorsement of this view over time is striking. In 1981 the percentage in agreement was 45 percent, in 1986 it was 66 percent, and in 1989 it was 74 percent.

As another measure of increasing consumer demand, the 1988 sales of organic foods in California were double those of 1987. The California Certified Organic Farmers organization, founded in 1973 by 13 farmers, has increased its membership from 370 growers in 1988 to 460 in 1989 and its acreage from 26,600 to 36,700. One problem for consumers is whether or not they can trust claims that food is pesticide free. Only two states—Texas and Washington—have state-run certification programs, and only 15 states legally define organic foods.

A second factor to consider in maximizing the demand for pesticide-free foods is that it may be necessary to change consumers' preferences from foods that look terrific to foods, particularly fruits and vegetables, that look imperfect but are, in fact, fine—especially when they cost significantly more money. It is a paradox: pay more for worse-looking food. The need, of course, is to show consumers that the hidden cost of cosmetically perfect foods has been intensive use of chemical pesticides and the incorporation of pesticide wastes in the foods themselves, as well as widespread contamination of U.S. drinking water and land. It is estimated that between 60 and 80 percent of pesticides are used on produce chiefly to enhance eye appeal and to prevent superficial defects which are insignificant.

Government policies and programs which now promote pesticide use and stand in the way of preventive alternatives still abound. Only strong public support for agricultural policy change will produce the scope and pace of pesticide reduction that are necessary. The evidence is in. Pesticide use can be dramatically curtailed quickly if the correct market incentives, government policies, and public tastes prevail.

As with CFCs, pushing agricultural pollution prevention to its limits threatens established companies. Greater efforts are underway to convert from older, nasty pesticides to newer, less threatening ones than to switch to pesticide-free farming and greater development and use of biocontrols. It is estimated that only 1 percent of the fruits and vegetables now available to Americans have been grown without the use of pesticides. The national policy goal should be to increase that fraction markedly. Like the push to use less severe ozone depleters, the interest in IPM techniques that depend on judicious use of pesticides can stand in the way of an even better solution: pesticide-free farming.

Companies in the pesticide business understand the threat and prefer to market pesticides that break down more easily into supposedly harmless substances, that are targeted to specific species, and that promise less environmental damage than previous pesticides. There is also a lot of interest in having the government allow "safe" levels of pesticide residues in foods rather than ban pesticide use altogether. But more bans and restrictions are inevitable. Recently, Iowa ordered farmers to cut the use of atrazine, a suspected human carcinogen and commonly used farm herbicide, by at least 25 percent because of its frequent contamination of groundwater. Iowa farmers used 11 million pounds of atrazine in 1988. Nebraska and Minnesota are expected to follow Iowa's lead. Meanwhile, American industry lobbies for federal laws that prohibit states from being more stringent than the federal government.

In October 1989, California environmental groups proposed a ballot initiative that would gradually discontinue 23 widely used farm chemicals by 1996. The California agriculture industry and pesticide manufacturers have opposed the measure. Albert H. Meyerhoff (1990), one of the authors of the environmental initiative, said: "What the growers have proposed is a plan to continue the use of cancer-causing chemicals. They want to manage cancer, not prevent it.'' Although the pesticide industry said that the environmentalists' initiative would result in a loss of $6 billion in California's annual economy, farmers who stopped using pesticides are among the most profitable in the state. Past pesticide bans have not resulted in economic catastrophe.

There is increasing evidence that governments worldwide are making serious commitments to reducing pesticide use. The Province of Ontario, Canada, started a program in 1987 called Food Systems 2002 that aims to reduce pesticide use by 50 percent within its 15-year life. Sweden and Denmark have instituted bold programs to cut pesticide use by half in just 5 and 12 years, respectively. In 1986, Vermont required two state agencies to develop long-term plans that include a schedule of pesticide reduction, and state agencies must demonstrate through impact statements that no reasonable nonchemical alternative is available and that the public health and environmental impacts of pesticide use are negligible. Rhode Island uses 25 percent of its pesticide registration fees to encourage and develop nonchemical methods of pest control in agriculture and elsewhere. In early 1987, Governor Mario Cuomo of New York directed three state agencies to develop a strategic plan by year's end ''to reduce dependency on and exposure to pesticides while achieving appropriate levels of pest control.'' However, as of mid-1989, no report had been issued and no action taken.

No one should underestimate the difficulty of reducing aggressive pesticide use. Still, there are many significant examples of progressive management:

- Cornell University achieved a 25 percent reduction in pesticide use on lawns and other turf, and Harvard University, the County of Milwaukee, Wisconsin, and Berkeley, California, have committed to the use of least toxic IPM alternatives to conventional pest control.
- The U.S. National Park Service has maintained the heavily trafficked National Mall in Washington, D.C., with zero pesticide use for close to a decade and has reduced total agency use of pesticides on its 79 million acres by about half since 1980.
- The Tornoto public school system has banned the use of all pesticides both inside and outside of its buildings.
- In a 1989 report on a strategy for compliance with environmental law in Antarctica, the National Science Foundation said: ''Avoiding the introduction of chemical pesticides in the first place is the best preventative measure available. NSF should consider imposition of a ban on the importation to Antarctica of substances governed by [the law] during the development of pollution control permitting systems under the [Antarctic Conservation Act].''

SUMMARY

No one can escape the negative effects of widespread pesticide wastes, and everyone can benefit from pollution prevention alternatives that eliminate or reduce these ubiquitous environmental wastes. In the face of many gaps in scientific knowledge about health effects, most people choose to minimize pesticide use. For a decade, 75 to 80 percent of Americans have thought of pesticide residues in foods as a serious hazard. About 30 miles north of Tucson, Arizona, a surrogate for planet Earth called Biosphere 2 is being developed. Pesticides have been excluded from the interior of the sealed giant terrarium. Why? William Dempster (1990), the project's director of engineering, explained: "in a closed system, they'll turn up next week in your coffee cup." Biosphere 1 is planet Earth, where it takes a little longer for pesticides to turn up—everywhere. However, global sales of pesticides for all applications in 1990 are likely to be $50 billion, up from $5 billion in 1975.

Until there are no more sources of contamination of food and water by pesticides, there will be more opportunities for eliminating and reducing pesticides and their inescapable wastes. Nonagricultural uses of pesticides should not be ignored. In a survey of households, 21 percent of respondents admitted that they applied larger than recommended amounts of pesticides and 38 percent denied that the pesticides they used were toxic. The use of pesticides by homeowners, landscapers, lawn care companies, exterminators, and many different kinds of institutions, such as schools, hospitals, local government agencies, and military bases, becomes more significant as attention to agricultural pesticide use leads to reduced pesticide use on the land. As we show in Chapter 8, there are a multitude of alternatives to home pesticide use.

All uses of pesticides threaten groundwater supplies that serve a vast number of people, threaten workers and consumers who apply pesticides, contribute to the generation of industrial wastes at pesticide manufacturing plants around the world, and add to the cumulative individual exposure to cancer-causing chemicals. It makes little sense for consumers to seek and buy pesticide-free foods, only to eat the food in houses around which pesticides have been widely used (either by the homeowner or the government) to control insects and weeds or within which the homeowner has used insecticides. Eating organic foods and then playing golf on a course recently sprayed with pesticides is also misguided.

We end this chapter by citing the wise words about pesticides of Rachel Carson (1962), whose pioneering work helped start the environmental movement but whose wisdom had not been fully heeded nearly 30 years after it was communicated:

The "control of nature" is a phrase conceived in arrogance, born of the Neanderthal age of biology and philosophy, when it was supposed that nature exists for the convenience of man. The concepts and practices of applied entomology for the most part date from that Stone Age of science. It is our alarming misfortune that so primitive a science has armed itself with the most modern and terrible weapons, and that in turning them against the insects it has also turned them against the earth.

CHANGING CONSUMPTION, REDUCING GARBAGE

This chapter is about things people bring into their homes and things that are sent there: the consumer products, the packaging, and the daily mail deliveries that become the waste flowing out. All of these things eventually get thrown away, evaporate into the air, or go down the sewer. Yes, we are a throwaway society, but keep in mind that it's all disposable and it's disposed of whether or not it's called a disposable product.

A lot of attention is given today to the problems caused by the growing volume of MSW. America has embraced recycling as the solution. It promotes good feelings about taking personal responsibility. However, although resources are conserved, the roots of the MSW problem—production and consumption—are not addressed. With recycling, neither industry nor the public must alter habits born in the booming 1950s. Pollution prevention is different. It asks more. It accomplishes more. Applying it at home is the most personal kind of prevention. The results can be more important than simply reducing household waste. Habitual practice at home instills prevention as a way of thinking about all environmental problems, some of which may be more important than MSW. In many respects, the personal pollution prevention of reducing household waste discussed in this chapter is simpler than that of toxics use reduction considered in the next chapter. For that, one needs to become familiar with a sometimes bewildering array of toxic products and chemicals. Here, once a few principles are learned, the process of overcoming the obstacles to prevention can begin immediately.

CONVENIENCE AND CONSUMPTION

Americans, it seems, are obsessed with saving time. With two-income families and single households becoming the norms, there is no one left at home to do the chores. The "time crunch," *Business Week* said in 1987, has become a "time famine."

There is a code word for saving time: *convenience.* Ask people questions like, Why do you

- buy a frozen microwavable meal for dinner?
- use paper towels?
- buy children's snack raisins in little boxes?
- put disposable diapers on your baby?

and most will answer: "because it's convenient." The consequence of these and many other convenience products is waste. Lots of waste. When consumers feel compelled to choose convenience and waste, they are also choosing to pay the capital and environmental costs of new recycling centers, incineration facilities, and landfills. What seems a good deal to manufacturers—being able to make and sell more and more of a product—also means more industrial waste and more landfills and incinerators, which consumers don't want in their backyards.

Somehow, Americans seem to find time to shop. Each year they spend over $1.5 trillion on goods for their personal use. Most of it is spent on things that will become waste in less than 3 years. Americans buy 28 billion foam cups, 16 billion disposable diapers, and billions of small items encased in cardboard and plastic. They also buy some 49 million major appliances such as refrigerators and washing machines and over 30 million TV sets and microwave ovens. Americans buy a lot. Americans produce a lot of garbage. More and more each year.

Some purchases are fast trash. Most packaging is instant waste; the so-called disposables are discarded after being used once. Other things are used over and over again. Appliances and furniture may not become waste for a decade or more. Still, Americans buy so many that the annual waste they create is high. Nearly 8 million TV sets are thrown away each year. The bottom line is: if it comes in the front door, it eventually goes out the back. Each average household generates up to 4 tons of waste per year, each average person as much as 7 pounds per day.

JUST ABOUT EVERYTHING IS REDUCEABLE

Today a few consumer products or their characteristics have been tagged as the evil offenders in household waste. You hear a lot about excessive packaging, Styrofoam products, plastics, and disposable products. The message from these discussions is that if you could get rid of one or a few types of waste, the problem will go away.

We make no such simple assumptions and have no special targets. Our byword is: *anything that becomes waste has reduction potential.* Waste generation is too big a problem; the impacts are too serious to confine personal pollution prevention to a few easy pickings, some unpopular or socially unacceptable items. Pollution prevention does not stop with refusal to buy Styrofoam cups or paper plates. Pollution prevention is not an elitist backlash to cheap goods made of plastic.

Pollution prevention is new consumerism. Practicing it requires a developed sensitivity to and concern for the life cycle of the things bought and brought into your home (and office and factory as well). That life cycle does not begin when something is purchased or end when it is thrown away. The cycle begins when raw materials are extracted or harvested from the earth and ends with ultimate disposition back into the environment. The goal of this book is to lessen the impact of that disposition. Selecting a less wasteful product probably means selecting one that has minimized the use of raw materials. Pollution prevention, therefore, helps meet other goals, like energy and materials conservation.

Reducing household waste means making prevention one of the criteria by which products are evaluated. Normal decision making and choosing among alternative products require assessments and comparisons about product quality, convenience, attractiveness, and price. Does it meet your needs? Is the price what you expected? Is it worth the price? The extent to which people ponder these questions varies. People don't make this effort for everything they buy. For things bought over and over again through a process of trial and error, good experiences promote consumer loyalty. For replacements, consumers simply seek out that brand or type of product again. Consumption is habit-forming.

To practice personal pollution prevention, consumers have to add new questions and form new habits. Initially, to everything bought. Even old, familiar things. How much waste will it make? How soon is it going to become waste? Will another brand or model last longer and/or generate less waste? Sometimes the answers are easy. Sometimes they are not, and only experience will tell.

This chapter is full of examples to help you gain experience. For the items covered, the examples give the facts about specific alternatives. But the examples are intended to do more than that. They are meant to help develop or hone skills in identifying wasteful products and in becoming a less wasteful consumer. The examples show that there are always trade-offs between waste, costs, and time. Difficult as pollution prevention calculations can be sometimes, they are often easier than calculating the cost and time impacts.

Reducing waste may reduce costs, but not always. A person will save money immediately if a less wasteful product is cheaper than the alternative. Sometimes, paying more initially will save in the long run because the less wasteful product lasts longer. But consumers may not have the money to buy a more expensive product and, if they do, they often have difficulty in understanding the long-term savings or caring about them. Consumers often prefer immediate cash savings from buying inexpensive products rather than waiting for savings to accrue tomorrow. Except for people who live in areas that charge for garbage pickup by the pound or bag, any lowering of costs at the waste end from reduction will not be realized. Pollution prevention directly benefits the community and only indirectly the waste preventor.

A lack of clear cost savings is an obstacle individual consumers have that companies do not. When companies reduce their internal waste, they do so chiefly because of the money they can save. Depending on the circumstances, costs like the $500 million each spent by the auto industry and by McDonalds for packaging used internally can be important incentives for reduction. A consumer may have to struggle between what socioeconomist Amitai Etzioni calls peoples' *dual nature* in making decisions, the conflict between self-serving desires and moral and social values. When people choose pollution prevention even though it does not clearly reduce their costs, values have triumphed over desires. That value may be personal health or concern for society.

Even when the true monetary costs of wasteful buying are clearly known, they are high enough to matter to only some consumers. For others, the incentive has to be the more certain health and environmental costs and lower waste management costs for society—that is, the moral and social values. When 79 percent of respondents to a poll in 1989 agreed that environmental improvements must be made regardless of the cost, they were probably thinking of costs to industry. Personal pollution prevention does not mean recognizing problems and demanding that others solve them. It is the ultimate practice of acting locally while thinking globally. It is action under the direct control of individuals.

A major waste-driving force—convenience—may have to be sacrificed to reduce waste. But not always. While packaging wastes can be reduced by avoiding microwavable foods, you still reduce when you choose a microwavable meal like the Budget Gourmet brand that comes in a simple single-layer box. Consumers may match or increase convenience by choosing a service instead of buying a product, but they have to accept the uncertaties of depending on a service. Careful analysis may reveal that the convenience of the wasteful product is an illusion.

PREVENT IT TODAY!

Another message contained in most of today's debates about household waste is that, before pollution prevention can occur, industry has to change the way it makes products and packages them. And before industry will do that, the government has to pass laws forbidding current practices or at least making them too costly to continue. In other words, pollution prevention is for sometime in the future.

We don't agree. It can happen today, and if it doesn't, it may not happen tomorrow. There are plenty of choices available now. Some have always existed but have fallen out of favor. Others are there because industry is beginning to get the message that consumers want alternative, less wasteful products. As people change their buying habits and the messages become clear in marketing analyses and surveys, those alternatives will grow. Sometimes only a small percentage drop in sales can make a company ask why and respond.

Right now, consumers are in control. Buying habits determine the products and the amount of packaging that come home, and thus the waste generated. To reduce waste, change buying habits and let values influence decision making. To help, we have formulated nine pollution prevention principles to follow. They are listed in Table 7-1. In the rest of this chapter, we use these principles to illustrate ways to reduce many kinds of household waste. Both positive and negative trade-offs are spelled out.

The nine principles are broken down into three categories: products, packaged products, and packaging alone. The categories are independent of one another, but choosing the least wasteful product sometimes requires the application of combinations, usually in sequence. The product category covers everything from automobiles to zippers. The packaged products category deals with the ways those products are packaged. Since few products come without packaging, the evaluation of how wasteful a product is must

**TABLE 7-1. THE BASIC POLLUTION
PREVENTION PRINCIPLES**

Products
　1. Buy less.
　2. Buy things that last.
　3. Buy small.

Packaged products
　4. Buy large economy sizes.
　5. Buy concentrated products.
　6. Buy products with the fewest layers of packaging.

Packaging that comes alone
　7. Say no.
　8. Reuse.
　9. Consolidate.

often include both product and packaging. The third category—packaging alone—covers the plastic and paper sacks, bags, and wrappings that are added to products by retail stores. The chapter ends with a special section on the waste that pours through everyone's mailbox courtesy of the U.S. Postal Service and the United Parcel Service. Most of this waste is difficult to control, but there are ways to reduce it.

A POLLUTION PREVENTION ASSESSMENT

All households are different. Therefore, as in industrial situations, a simple pollution prevention assessment can identify what the opportunities are. For 1 week, separate waste into the various categories discussed in this chapter, like packaging that comes with things, single-use (disposable) products, the bags that retail stores provide, microwave packaging, and mail. Separation into material categories (such as paper, plastics, and so on) is not important here; that kind of separation is for recycling. For pollution prevention, the objective is to discover the sources, products, and packaging brought into a household. By knowing what is in waste and why, a person can figure out what and how to reduce it.

REDUCING PRODUCTS

Official statistics separate consumer products into two types: durable and nondurable goods. Durable goods are the big ticket items expected to last for more than 3 years: automobiles, furniture, consumer electronics (TV sets, stereo equipment, VCRs), and appliances (refrigerators, stoves, washing machines). Nondurables are everything else—food, paper products, clothes, chemical products, and cosmetics, to name only a few.

Our three pollution prevention principles for the products category are (1) buy less, (2) buy things that last, and (3) buy small. They apply equally to all products, whether or not the products are officially classified as durable or nondurable. All dishwashers are durable, but some models or brands are more durable than others. All plates are nondurable but paper ones are designed to be used only once, while ceramic plates are used over and over again.

Buy Less

This first principle is the simplest to understand but may be the hardest to follow. The buy less principle challenges consumers to practice restraint, which is counter to societal trends, prevalent habits, and the presumption that a high quality of life and a decent standard of living depend on consuming more. Americans are continually bombarded by messages to consume more. By contrast, this principle says: to generate less waste, consume less.

An undisputable side effect of this principle is that it saves money. Don't buy a 10-pound item today and it will not become 10 pounds of waste tomorrow. What it would have cost will be saved. Practice selective purchasing avoidance day by day, month by month, year by year, and the savings will add up and be available for something really needed or desired: emergency health care, college tuition, a vacation away from a hectic job, or an expensive luxury.

The green philosophers and practitioners argue about whether a true green is one who consumes *less* or consumes *differently.* Consuming differently—buying green products—is good, they say. But it is not enough. Ecological crises will only be averted by political and institutional change, and part of that is a radical change in consumption patterns. But it is *not* an either-or choice. Our first pollution prevention principle deals with consuming less, the other eight with consuming differently.

Disposable income is what economists call the money left over after paying taxes. It's the money available to pay for needs such as housing and food, and wants such as goods and entertainment. The term is ironic, since waste generation rises with income. The more disposable income you have, the more likely you are to increase your waste generation. Disposable income has been rising in America, although not as fast as it once did. Per capita annual disposable income doubled between 1977 and 1987 (to $13,157). When the effects of inflation are removed, it doubled between 1955 and 1987. That $13,157 doesn't seem like much, but remember that it is an average over the entire population. In reality, some people have more (a lot more) income than others. During the 1980s, the richest 20 percent of American families got richer, so they now have more than 40 percent of the nation's income.

Consumer spending accounts for about two-thirds of the nation's economic activity. "Consumers are the engine driving the U.S. economy," said the *Wall Street Journal* in 1989, showing that consumer spending as a percentage of GNP has increased along a 1947–1988 trend line from 59 to 64 percent. Americans are spenders instead of savers. In 1970, Americans saved 8 percent of their disposable income; in 1987, only 3 percent. Since 1980, the rate of saving has declined faster than incomes have risen, so that the total amount saved is also less. Personal savings totaled $137 billion in 1980 and $104 billion in 1987. Of all the industrialized nations, only the United Kingdom has a lower savings rate.

Americans not only don't save, they are very good at buying on credit.

Outstanding consumer credit, not counting housing mortgages, was $686 billion in 1987. That amount is equal to 21 percent of personal disposable income in that same year. In other words, through credit, Americans increased their purchasing power by 21 percent. While the greatest portion of this debt covers automobiles and bank loans, $111 billion is retail and other miscellaneous credit.

Would lowering America's personal consumption rate mean that factories will close and people will lose their jobs? In the short term and depending upon the rate of decrease, some dislocations would be inevitable. Where they would occur and how profound they would be are difficult to predict. Since Americans buy more imported than domestic goods, the dislocations would be spread worldwide. But in the long term, if we saved more, industry would invest more in new technology and new manufacturing plants. More savings increases capital, lowers interest rates, and promotes investment. According to this scenario, that means that America's economic future would be more competitive and jobs would be more secure.

Consumption in our society is motivated by real technological changes, contrived changes, and rapidly changing fads. In 1960, Vance Packard in *The Waste Makers* revealed an industry strategy called *planned obsolescence,* in which products were designed to fail within short periods. The purpose was to create a need for replacement goods, thereby increasing profits through increased consumption.

Today, whether or not product failure is planned or simply caused by poor quality, it does seem that new products are marketed long before the previous versions or models have had time to fail. They simply go "out of style." Clothing fashions change with the seasons. American automobile manufacturers introduce new models every year, but new technology is incorporated only every 4 to 6 years. Many "improved" products simply contain superficial changes.

"Electronic gizmos" or "technology geegaws" are the cream of the fad products. They are part of a $33 billion annual market. They aren't necessarily cheap. Their usefulness is questionable. As John J. McDonald (1990), president of Casio, told *The Washington Post* on the occasion of the annual Consumer Electronics Show: "We gratify wants, not needs." Some of the newest geegaws include a telephone that automatically lowers the volume on your stereo, stereos with "surround sound," and a chessboard that recites insults. Then there is the hot dog maker that, for $69, will broil two hot dogs and buns in a couple of minutes. Sometimes true pollution prevention products emerge, like a pocket-sized electronic dictionary and other reference books that could one day replace mammoth paper reference guides.

Not all consumers fall for gizmos, just enough to keep the industry humming. A poll in 1990 revealed that 40 to 59 percent of those surveyed thought car phones, CD players, walkmans, video cameras, and pulsating shower massagers were modern frills rather than products that make life better. The microwave oven and the smoke alarm won more than three-quarters of the respondents' votes. The microwave oven is changing Americans' perceptions of food and time while generating packaging waste. More about that later.

To convince people to buy, buy, buy, companies spent $109 billion in 1987 on advertising. The messages bombard consumers from the TV, radio, magazines, newspapers, direct mail, and outdoor billboards, as well as promotions in retail stores. The big five advertisers—Phillip Morris, Procter & Gamble, General Motors, Sears, Roebuck, and RJR Nabisco—represent most consumer products. Among the hundreds of products they make or sell are soaps and detergents, personal care products, automobiles, cigarettes, processed foods, apparel, appliances, and hardware. (Advertising itself makes waste. From one-half to three-quarters of a daily newspaper's volume is advertising rather than news. Magazines are mostly advertising. Direct mail advertising sends millions of pieces of unsolicited mail to homes each year.)

Advertising isn't always successful, of course. Witness the Edsel car, on which Ford spent $500 million in the late 1950s but failed to persuade the public to buy. Or the attempt to sell a change in the recipe of Coca-Cola in 1985. Over 2,000 products were introduced in a Boston area supermarket chain in 1986; only one-quarter of them lasted for a year. The cost of product failures, though, is added to the price of products that consumers do buy. Industry always claims, when questioned about the waste generation aspects of its products, that it is only supplying what customers demand. Well, marketing and advertising develop and shape demand. Otherwise, why would product manufacturers buy their services? In a 1989 talk, Richard D. Slocum said that his firm *created* a new market for Air Wick air fresheners by using "beautifully lithographed cans" with "elegant new products with creative names like hint of spring, light lemon." Intense media campaigns *steering* fluctuating consumer demand, said a government report in 1988, were the primary reason that the sanitary paper products industry (paper towels, toilet and facial tissue paper, napkins) had been continuously adjusting its product lines for a number of years.

Buying less confronts all of these trends: spending instead of saving, living beyond one's means, succumbing to fads and advertising messages. It requires that individuals ignore, or at least be discriminating about, the constant drumbeat of messages from Madison Avenue that improving one's standard of living and pleasure means having more things. The alternative? Focus instead on authentic *quality of life,* which excessive consumption and environmental waste generation can reduce. That focus increases the amount of money available for emergencies, decreases personal debt, decreases the time spent buying things and taking care of a household, and decreases the amount of waste generated.

Buy Things That Last

A corollary to buying less is to buy things that last (and don't throw them away until they are broken). This principle covers all durable and nondurable goods from automobiles to disposables (razors, diapers, and paper cups). When buying durables, it means finding the brand name and model that has the best lifetime rating. For disposables it means substituting a product that can be reused rather than used once and thrown away. And it means that, when replacing reusables, buy ones that will last the longest.

When buying long-lasting products, waste is reduced because the *rate* of waste generation decreases. Immediate and continuous pollution prevention occurs when one chooses reusables instead of disposables. For durables, the savings may not be realized for a number of years. A simple choice between two radios, with assumed lifetimes of 3 and 6 years, results in double the waste if the 3-year radio is chosen. The problem with this option is to know for certain what the lifetimes are.

A more mundane example is the light bulb. Philips, GE, and Osram make a compact fluorescent bulb that conserves energy and reduces waste. At 15 to 18 watts, one of these bulbs provides the intensity of an incandescent bulb in the range of 60 to 75 watts. It will last for 3 years, even if left on for 12 hours a day, outlasting an incandescent bulb by about 9,000 hours. Cost is a barrier: $15 to $20 per bulb (versus 75 cents). But after 2,000 hours the cost is paid back. Another manufacturer, Aero-Tech, sells an incandescent bulb with a claimed lifetime of 20,000 hours, five to eight times longer than normal. No energy savings accrue, and six 100-watt bulbs are priced at about $20.

Whether buying ceramic dishes instead of paper plates or a better TV set, when choosing to buy things that last, the up-front cost will be higher. Calculating the long-term savings for things like appliances can be difficult because operating costs need to be included along with projected lifetimes. For other items, like those paper plates, the initial cost differential can be so high that it is too easy to discount the cost of repeated purchases. Time savings can result from fewer trips to the store to buy replacements. A note of caution: a higher price does not necessarily equal higher quality and a longer lifetime. One valuable reference guide is *Consumer Reports* magazine, especially for durable goods. Its comparisons of brand names and models often include advice on longevity.

Household batteries are a good example of a nondurable product for which there are many choices, some of which reduce waste and save money.

In buying more expensive alkaline instead of regular (carbon-zinc) batteries, the higher cost is more than compensated for by a longer lifetime. Unfortunately, this is not easy to know, since manufacturers do not give specific information about lifetimes on battery packages. *Consumer Reports* says that alkaline batteries are the best buy; they last up to 10 times as long as regular batteries.

Consumers can do their own testing, and this is a good way to teach children pollution prevention consumption. A test is simple. It consists of recording information, which is a way of formalizing the consumer's experience. We tested D cell flashlight batteries by putting them in a flashlight and leaving the flashlight on until the batteries died. Two alkaline batteries lasted three times (32 hours) longer than two regular ones (11 hours), which matched *Consumer Reports'* more extensive testing results. The alkaline batteries cost twice as much as the regular batteries. To get 30 hours of use from either battery would require buying two alkaline batteries ($2.29) or six regular ones ($4.17). The alkaline batteries save $1.88 and reduce waste by four batteries (12 ounces). Another way to compare costs is to calculate the price per hour of use. The alkaline batteries cost $2.29 for 32 hours, or 7 cents per hour. At $1.39 per 11 hours or 13 cents per hour, the regular ones are almost twice as expensive.

This same test could be conducted with different brands of alkaline batteries to find out which manufacturer gives the longest lifetime and, thus, the least waste and best value. Another choice today is rechargeable nickel-cadmium (or lithium) batteries. GE claims that, over 4 years, one rechargeable battery could replace 150 AAA, AA, or C throwaway batteries.

Flashlight Options

Pollution prevention purchasing decisions are not always about making one of two choices. The battery test example only deals with which of two batteries to choose for a flashlight. There are better ways to reduce waste while getting the service of a flashlight. Flashlights illustrate the growing complexity of many products. They come in three basic versions: the traditional case with either replaceable or rechargeable batteries, a disposable flashlight, and a rechargeable flashlight. These options are compared in Table 7-2. A flashlight with batteries rechargeable by solar cells is also now available.

A disposable flashlight has little merit. It generates the most waste and is the most expensive. We tested one such flashlight and got only 1.5 hours of continuous service. Failure may have been caused by the batteries or the light bulb, but both are sealed inside a plastic case. When either wear out, the entire unit has no value and is thrown away. Replacement means buying a new unit and throwing away more packaging.

TABLE 7–2.   COMPARISON OF FLASHLIGHT OPTIONS

Factors	Disposable	Traditional Carbon-Zinc Batteries	Alkaline Batteries	Rechargeable Batteries[a]	Rechargeabl
Cost	$2.89				$1(
Case	—	$6.00	$6.00	$6.00	—
Batteries	—	1.39	2.29	$14–34	—
Hours of service	2	11	32	4,000–6,000	2,000–3,00(
Waste (pounds)					
Individually	0.5[b]	0.5[b]	0.5[b]	1	
For 4,000					
hours	1000	182	63	1	:

[a]Assumes the purchase of a recharger unit and two sets of batteries, which means that th service life is twice that of a rechargeable flashlight.
[b]Assumes that only batteries are thrown away at the end of the service life.

A traditional flashlight case can hold a variety of batteries: carbon-zinc, alkaline, or rechargeable. The flashlight case (about $6) can be used over and over again. Over time the light bulb might need to be replaced (although one manufacturer, Rayovac, claims that its light bulb lasts as long as "you own the light"). The cases (metal or plastic) may last forever unless it is dented or cracked or the electrical contact points rust. Dead carbon-zinc batteries can leak; if left inside a flashlight, they can corrode the points.

Use of rechargeable batteries involves the highest up-front purchase cost, but payback is swift given the substantially longer service life. Initially one needs a charger unit plus batteries, but the system can be used for all household battery needs. Charger units cost from $10 to $20; two batteries, from $2 to $7. In a *Consumer Reports* test, the batteries ran continuously for 2 to 3 hours, and only marginally longer if run intermittently. With two sets of batteries, one set can be in use in a flashlight while the other is recharging. Recharging can take from 6 to 16 hours. One problem with rechargeable batteries and chargers is that they are not as easy to find in stores as are nonrechargeable batteries, which seem to be available at the checkout counters of all grocery, drug, hardware, and variety stores. But, once purchased, a charger and a set of batteries don't need replacements for years.

Reliable rechargeable batteries and flashlights are relatively new. The charge is produced from nickel-cadmium or, more recently, lithium batteries. Like disposables, rechargeable flashlights have their batteries sealed inside a case. The difference is, of course, that when the batteries die, the system can be recharged by plugging it into an electrical outlet (or, in the case of the one with solar cells, by setting it out in sunlight). Manufacturers claim that single charges of nickel-cadmium batteries will last for 6 months to 1 year and that the batteries can be recharged up to 1,000 times. One problem with these systems is that they need to be fully discharged before they are recharged. Thus, the flashlight may be needed when it has to be recharged.

The toxicity of consumer products and their waste is covered in Chapter 8. But it is important to know that the potential toxicity of these batteries varies. Because they are sealed, the effect is primarily on waste rather than on the user. Mercury is present in both carbon-zinc and alkaline (manganese) batteries in the form of mercuric chloride. But it is lower in carbon-zinc batteries (0.01 percent) than in alkalines, which may be up to 0.8 percent mercury. However, since alkalines last 3 to 10 times longer, the effective gap between the mercury contents narrows. Rechargeable batteries contain cadmium, which is a toxic metal. Lithium batteries do not contain mercury or cadmium, but some lithium compounds can be toxic. The choice comes down to waste quantity versus waste quality.

Those Ubiquitous Throwaways

Disposable products are much discussed today as culprits that are unnecessarily increasing the national wastestream and as symbols of a convenience lifestyle. It is easy to respond that they don't seem to add much to waste because the products—razors, lighters, cameras, paper towels—tend to be small. But for millions of consumers to have daily doses of disposables, billions of each are produced and bought each year. Their lifetimes are very short. After junk mail and some packaging, they are the most rapidly produced form of waste.

The BIC Company may be the king of disposables. This firm describes itself as "a diversified corporation primarily engaged in the manufacture and sale of high quality low-cost disposable consumer products." BIC introduced the disposable razor and now produces 45 percent of the 2 billion sold each year. They also make most of the 500 million disposable lighters and the 1.6 billion disposable pens sold each year. As waste, 2 billion razors, 500 million lighters, and 1.6 billion pens add up to 160 million pounds of waste per year. And that doesn't count the 2.5 billion disposable batteries, 96,000 disposable plastic telephones, 18 billion disposable diapers, and millions of disposable contact lens, disposable cameras, disposable coffee filters, and disposable paper and plastic plates and cups bought each year. Nor does it account for the packaging waste they generate.

Are disposables really convenient? Do they save time? They certainly seem to do so. A disposable razor does not have to be cleaned between uses. A paper or plastic plate, cup, or utensil does not have to be washed and stored. A paper towel does not have to be laundered, folded, and stored. A disposable flashlight does not have to have its batteries replaced. A disposable camera does not require time to replace film, and lighters and pens do not have to be refilled.

But some of the time saved may simply be time shifted. What about the repeated trips to the store for replacements? Looking for a parking space? Standing in line at checkout counters? Carrying items home and putting them away in closets and cupboards? Then there is the bulk that throwaway products add to trash that has to be consolidated and put out for pickup. Overall time savings of some disposables may be marginal; of some, illusory.

How long does it take to reline a garbage can with a bag? If a community will pick up trash from cans, why bother with the bag? Does rinsing out the reusable trash can really take much longer than removing one filed sack, tying it shut, and putting in another bag? Moreover, at one 30-gallon bag a week, the 52 bags cost between $10 and $16 a year; at two a week, $20 to $32. Making the bags consumed energy and petrochemical materials and caused the generation of toxic industrial waste.

Reuse, Reuse, Reuse

The alternative to disposable products is reusables. Eleven products are compared in Table 7-3. The estimated annual savings from reusables is almost $500 and 120 pounds of waste. It is obvious that the largest waste savings come from using cloth diapers instead of disposable ones and cloths or sponges instead of paper towels. But no category of disposable product is cheaper either in terms of waste generated or cost to buy. For razors, it depends on the reusable option selected. A razor with a replaceable blade, assuming that it gives as many shaves per blade as a disposable razor, will cost more and make more waste. The reason is packaging. Disposable razors are packaged in simple cellophane bags; the reusable version and replacement blades come in blister packs (cardboard and plastic), with the blades packed in a plastic case. For shaving, the least waste is achieved by using an electric shaver.

The waste generated is a hidden cost of disposable products. If the cost of disposal was paid by individuals who use them instead of being spread among all users of garbage collection systems, disposable products might not seem as indispensable. The low unit cost of disposable products makes them seem valueless and encourages misuse that increases waste. Who cares if they lose a disposable pen? Another one is only $1 away. What does it matter if a foam or paper cup (at 2 cents) is replaced each time a refill is sought? Conversely, a $5 pen and a $5 coffee mug are more likely to be collected before leaving a room or meeting.

Disposable cameras are unusual because most of the waste they generate does not end up in household trash. The entire camera unit, made of cardboard and polystyrene with a plexiglass lens, is sent to the film developer, who throws it away after extracting the film. The packaging (paperboard box and internal foil wrap), however, is household waste. Theoretically, cycling of product wastes back to the producer is supposed to reduce waste because the costs of disposal become internalized. In this case, it has not prevented the development of a throwaway product because the producers—Kodak and Fujicolor—don't get the waste; an intermediate—the film developer—does. *Garbage* magazine questioned Kodak about this point. Kodak doesn't feel responsible; they claim the developer is. There is a hopeful sign that consumers are starting to make an impression on marketing. Kodak is sensitive about having its camera being labeled *disposable* or *throwaway*. They prefer the term *single use.*

Cloth or Paper Towels?

Paper towel sales grew at a rate of 4 percent a year in the 1980s; 2.6 billion rolls were made in 1988. That's almost 11 rolls for every person in America, over 2 rolls per household per month. Scott Paper Company now puts paper towels in a box (added waste) so that people can handily avoid using cloth towels in the bathroom. And Texize (a division of Dow Chemical Company) sells Spiffits, a "cleaner and towel all in one."

The 2.6 billion rolls of paper towels produced in 1988 made 2 billion pounds of waste that year. That's just one small segment of the paper products industry. Markets are increasing overall, especially for high-grade printing papers, bleached paperboard used for food packaging, and tissue papers including paper towels and facial and bathroom tissue. All of these products are fast waste.

Paper towels serve a multitude of functions. One friend, who consumes more than the average number of paper towels, gave us a list of 15 uses. They appear in Table 7-4, along with reusable alternatives. The least wasteful replacement is a cloth; it rarely wears out and can be washed when dirty. Cloths don't have to be bought; old bathroom towels, linens, or cloth diapers (see below) will do. The inconvenience of a cloth towel is, of course, that to get full use it has to be washed, dried, and folded (if one is neat). Another possibility is a sponge. Sponges eventually get dirty and greasy when used in the kitchen but will save several rolls of paper towels. There is a version of the paper towel that is reusable. Handi Wipes are made by several firms out of a nonwoven material. They are rinsable and even washable and long-lasting. One of them will save the cost of waste of several rolls of paper towels. At $2, a 4-ounce package of eight wipes, used as adjuncts to cloths and sponges, can last for years. If one package lasts for

TABLE 7-3. COMPARISON OF DISPOSABLE AND REUSABLES

		Disposables							Reusables				
Item	Unit or Type	Waste per Unit[a] (Ounces)	Cost per Unit ($)	Number Disposed of per Year	Waste per Year (Ounces)	Annual Cost ($)	Item	Unit or Type	Waste per Unit[a] (Ounces)	Cost per Unit ($)	Number Disposed of per Year[b]	Waste per Year (Ounces)	Annual Cost ($)
Razor	1	0.2	0.30	52	8.7	15.60	Electric shaver	1	32.0	50.00	0.2	6.4	10.00
		0.2	0.30	104	17.3	31.20	Razor holder	1	2.5	4.00	1	2.5	4.00
		0.2	0.30	156	26.0	46.80	Blades	Pkg of 5	1.8	2.00	20	35.0	40.00
Paper towel	roll	10.0	0.90	30	300.0	27.00	Cloth	1	3.0	1.20	2	6.0	2.40
							Sponge	Medium	1.0	0.26	6	6.0	1.56
Flashlight	1	8.0	3.00	6	48.0	18.00	Rechargeable flashlight	1	16.0	10.00	0.5	8.0	5.00
Camera	With 24 exposures	4.0	8.00	6	24.0	48.00	Camera film	1 24 exposures	16.0	25.00	0.2	3.2	5.00
									1.0	4.30	6	6.0	25.80
Plastic food wrap	Box of 200 ft	11.0	1.30	4	44.0	5.20	Plastic container	1-cup size	1.5	0.80	3	4.5	2.40

Item	Size						Reusable item	Size					
Diapers	Medium	0.4	0.18	2,600	1,040.0	472.73	Cloth diapers	Medium	2.0	0.90	(c)	(c)	36.00
Pen	1 ($1)	0.5	1.00	12	6.0	12.00	Pen refills	1	0.5	5.00	0.5	0.3	2.50
	1 ($0.20)	0.5	0.20	24	12.0	4.80		1	0.3	1.00	12	3.0	12.00
Foam cup	9 oz	0.5	0.02	182	91.0	3.20	Ceramic mug	1	8.0	5.00	0.5	4.0	2.50
Paper plate	9 in.	1.0	0.02	182	182.0	4.22	Corning		9.0	2.50	0.1	0.9	0.25
							Corelle						
Coffee filter	#6 size	0.1	0.04	365	45.6	14.78	Metal coffee filter		3.0	10.00	0.2	0.6	2.00
							Cloth coffee filter		0.5	7.00	0.5	0.25	3.50
Furnace filter	1	6.0	1.50	4	24.0	6.00	Washable filter	1	6.0	6.00	1	6.0	6.00

[a]Waste amounts include packaging waste.
[b]When a fraction, it is assumed that the lifetime is more than 1 year.
[c]May be zero. If 20 diapers were discarded in 1 year, 40 ounces.

TABLE 7-4.   PAPER TOWEL USES AND ALTERNATIVES

1. Clean up liquid spills	Sponge or cloth
2. Drain wet containers/dishes	Strainer
3. Liners for candle holders	Floral clay
4. Dry dishes	Cloth
5. Clean any kitchen surface	Sponge or cloth
6. Dust/clean mirrors and glass	Cloth
7. Substitute for facial tissue	Toilet paper
8. Napkins	Cloth
9. Plates	Use plate and wash
10. Microwave pan	Use pan and wash
11. Drain fresh vegetables	Strainer
12. Wipe hands	Cloth
13. Absorb oil from fried foods	Newspaper, junk mail
14. Wrapping/packing paper	Newspaper, junk mail
15. Line refrigerator drawers	Clean with sponge

only 1 year, 30 rolls of paper towels can be avoided and $25 saved. Waste is reduced by just over 1 pound. Cloth towels will last much longer, cost nothing to buy, and generate little if any waste.

Christmas Reuse Instead of Refuse

It is said that in the weeks after Christmas, people generate the most waste for pickup. That's understandable since some retail shops make most of their annual sales in the Christmas shopping period. Over $30 billion is spent on gifts alone. In addition, Americans spend $500 million on wrapping paper and $800 million on lights. They buy 2.2 billion cards.

A lot of what gets thrown away is reusable: the shopping bags, the boxes, the gift wrap, the bows and ribbons, and even the Christmas tree. Americans buy over 35 million cut trees for Christmas. These trees don't have to end up as trash. They can be chopped up and used as mulch in gardens. Some communities will grind them up for free. One store with outlets on the East Coast—IKEA—rents cut trees for a $10 deposit. Bring it back after Christmas and they grind it up, bag it, and give back the $10. An artificial tree is reusable each year. After 1 or 2 years, the cost (compared with that of buying a cut tree each year) is paid back. A living tree is another option, but only for those with a large enough yard for planting after indoor use. Tearing gift wrap off packages reduces it to trash; with a little care when unwrapping, gift wrap can be folded and used the next year or given to charities. Those bows that are sold in lots of dozens to a bag are almost indestructible. After their first use, the sticky tape is worn but they can be affixed to packages with a small piece of tape in successive years. However, the bows (at a few pennies each) and some wrapping paper are so cheap that the monetary savings are negligible. Reuse may take up limited household storage space, and conservation of storage space may win out over waste or monetary savings.

Obscure Disposables

There is a category of disposables that doesn't get much attention. They are not labeled disposable. They are small consumer appliances like calculators, hair dryers, and toaster ovens. These are not use-once-and-toss disposables. They are used over and over again. But they are often thrown away long before their true usefulness is over because style changes convince people to purchase a new model. They are also thrown away because a part breaks and repair is more expensive or more difficult than replacement. The fast pace of electronic product development also causes waste to be generated when existing equipment is no longer compatible with the latest technology.

The cost and convenience obstacles to pollution prevention by repair are high. First, electronics have replaced mechanical parts that were often possible to repair at home. Second, local repair shops have been replaced by corporate service centers. Texas Instruments (TI) offers a repair service for its calculators, but it seems designed to encourage resales rather than repairs. During the 1-year warranty period, TI charges only a shipping and handling fee; afterward, a repair fee is added. For a low-end model that retails for $7, TI charges $10 ($7 plus $3 for shipping and handling). For a business calculator that costs $13, the repair cost is $12 plus $4 shipping and handling. Thus, both repairs are $3 more than the price of a new calculator.

Even if the repair and replacement costs are comparable, other factors discourage repair. They include the time necessary to pack and mail the item or take it to a repair shop and loss of the use of the item while it is being repaired. It can also be easier to buy a replacement than to find and deal with a local or mail repair service.

There are plenty of examples of how Americans throw away still useful things. In a survey of microwave oven owners in 1987, 12 percent said that they threw away their old oven when they purchased a new one. In one Virginia community, volunteers scavange the local dump for discarded toys. In 1989, about 1,000 toys were found, washed, repaired, and given away. Second Harvest, a nongovernmental food program, solicits rejected foodstuffs from manufacturers and merchants and donates them to local food banks all over the country.

Some people have suggested that waste would be reduced if the parts of short-life products were designed as replacable units. When a particular unit breaks, a replacement unit could be bought and plugged in, putting the gadget back in operation. Until that happens, though, for nonelectronic wizards, breakdowns are liable to force whole replacements. Thus, before buying, find out what the breakdown rates of alternatives are and choose those with better records or longer warranties. As discussed under "Buy Less," just say no to impulses to replace items that still serve a useful purpose. Make do.

Reuse Beyond Home

Reuse can continue long after one person has exhausted a product's usefulness. Instead of throwing away something that is still serviceable, find someone for whom it has value. Organizations like Goodwill Industries and homeless shelters accept donations of both durable and nondurable goods. Just as there is a market for used automobiles, used clothing can be sold through consignment shops in some areas. A regional carpet retail store—Furniture & Carpetland—will take an old rug as a trade-in on a new one. The old one is donated to charity. Almost anything can be marketed through want ads in local newspapers or through newsletters or sold at garage or lawn sales.

Sometimes reuse can be facilitated by using a delivery service. The nice part about a service is that it provides a link to reusability while maintaining or enhancing convenience. Diapers and bottled water are two products whose waste can be reduced by buying a service. Diapers are covered here; bottled water is discussed later under packaging.

Disposable diapers have been on the hot seat lately. No wonder, since the solution—switching to cloth diapers—is readily available, saves money, and dramatically reduces waste. In choosing a diaper service or buying diapers and washing them at home, pollution prevention is being practiced.

Diapers have been discussed so often that surely everyone today knows that:

- Disposable diapers were introduced in 1961 as a convenience item for families while traveling. Today at least 80 percent of American babies wear them regularly, generating sales of $3.5 to $4.0 billion per year.
- From 16 to 18 billion diapers are manufactured or used each year.
- As waste, the diapers (not counting the packaging) add up to 3.6 to 5.0 million tons of waste per year.
- Using disposable diapers is the most expensive way to diaper a baby. The cheapest way is to buy cloth diapers and wash them at home. The cost of a diaper service is comparable to the cost of disposable diapers and can match their convenience. Disposables can be relegated to a traveling convenience.

The drumbeat against disposable diapers has been so constant that the diaper service industry is growing exponentially. Nationally, growth was 39 percent in 1989. *The New Yorker* said in January 1990 that there were only two companies left servicing New York City, and deliveries were up 40 percent. Claims and counterclaims are made for cloth diapers and new, biodegradable diapers. Procter & Gamble, one of the two largest producers of disposable diapers, is funding research into recycling of soiled disposable diapers. The prototype plant equipment cost about $500,000. At $1 per diaper, the money would buy 500,000 cloth diapers. That's enough diapers for over 6,000 babies at one time. The recycling effort covers only 1,000 babies.

What are the comparative costs? They vary, depending on local prices, diaper service availability, and what's included in the calculation. The surge of interest in diaper services might temporarily drive up the price if the demand is greater than the supply. A nurses' association in Seattle says that for a 6-month-old baby, cloth diapers cost about $8 a week, a service is $9.25, and disposables are about $12.50. The National Association of Diaper Services says that the cost of a service is significantly lower than that of disposables, both to the consumer and to society, when the disposal cost is included. Their study, "Diapers in the Waste Stream," gives comparative per piece costs of 22 cents for regular disposables, 26 to 39 cents for biodegradables, 15 cents for a service, and 13 cents for home-washed cloth diapers. The capital and operating costs of laundering diapers are included.

Our calculations, based on prices in the Washington, D.C., area for a newborn baby are these:

- Diaper service $9.56 or $12.15 per week for 80 or 90 cloth diapers.
- Small-size disposable diapers, $11.17 to $11.38 for a package of 66. If 80 diapers are needed, the cost rises to at least $13.50. Disposal costs are not included.
- Cloth diapers, $11 to $20, depending on the variety and type of closure. This works out to 80 diapers for $72.80 or 90 diapers for $81.90. If the diapers last for 2 years, their cost for 1 week is 70 to 79 cents. Washing machine water, soap, and energy costs are not included. But those costs have to total $10 per week to make cloth diapers more expensive than disposable ones.

It seems clear that on a cost basis, huge savings can accrue to those families that buy and wash their own diapers. The savings over 2 years equal the cost of a washer and dryer. The costs of using a diaper service or disposables are comparable, not counting the cost of disposal. What are the pros and cons? Other than cost, the factors are convenience and health issues, as Table 7–5 shows.

The pollution prevention benefit of using cloth diapers is enormous. Cloth diapers can last through several babies. They can be handed down to a friend or donated to a charity or daycare center. They can be used for many other purposes that will keep them out of the wastestream. Even if they are thrown away, 80 cloth diapers make 20 pounds of waste. An average of 50 disposable diapers per week for one baby for 2 years, a total of 5,200 disposable diapers, amounts to 2,600 pounds of dry waste, 130 times more than for cloth diapers. This calculation does not include the significant packaging waste that disposable diapers generate weekly or the industrial waste from their manufacture.

Society benefits further from those who use cloth diapers. Studies have shown that most users of disposable diapers do not flush the captured solid waste matter down the sewer system, which is designed to neutralize it. When soiled diapers end up in landfills, the waste and viruses that may be present can become part of the leachate that contaminates ground and surface water. More methane gas will also be produced.

Disposable diapers have been under attack for several years, and the industry is fighting back to preserve this huge market. The shopworn argument is that diapers represent only 2 percent of all municipal waste. But all threatened product segments use that reasoning. The problem is that if one adds up the many small segments, the result is a significant number. Should any segment be exempt? Not with a serious national commitment to pollution prevention, especially when the reduction possibility is so feasible. Other defensive industry strategies are recycling and the development of biodegradable disposable diapers. Neither is a prevention option. For conservation of landfill space, biodegradation is now a discredited waste solution. Landfills are simply not good reaction vessels. The mixture is too heterogeneous, has too many particle sizes, and has insufficient oxygen to degrade efficiently.

Biodegradation has been so highly promoted as a solution that even environmentalists got confused. Seventh Generation, a green products mail order firm, sells a brand of biodegradable diapers. They claim many environmental benefits, including saved trees, because less wood pulp is used to make them, and no dioxin content, because the bleaching process uses hydrogen peroxide instead of chlorine. (Of course, all of those savings occur when cloth diapers are used.) The firm offers to eliminate the inconvenience of having to continually lug home disposable diapers from the store, thereby matching the convenience of a cloth diaper service. Consumers can sign up to have a month's supply sent on a regular basis. But buying a case of diapers a month (for $44.50 plus $5.95 for shipping) means throwing away a case a month of used diapers.

TABLE 7–5.   PROS AND CONS OF VARIOUS TYPES OF DIAPERS

	Disposables	Cloth/home	Cloth/service
Convenience	Use and toss in trash	One trip to the store; one set of diapers can be used for several babies	Use and toss in bin
			No trips to store
		Diapers supply is at home instead of in the store	
Inconvenience	Frequent bringing home of supplies from the store	Washing, drying, folding	Poor service from diaper firm
		Can require more changes	Can require more changes
Health issues and claims	Fewer or more rashes because of super absorbent chemicals used	Fewer rashes	Fewer rashes
	Fecal matter, including viruses, tends to end up in landfills rather than sewer systems		
Other		Diapers can serve a multitude of uses, such as dusting, polishing, bandaging	

Procter & Gamble, a firm that is out front on some waste solutions (see the packaging examples later), emphasizes that the volume of its Pampers and Luvs disposable diapers has been reduced by 50 percent due to improved technology. The firm also has replaced cardboard carton packaging with plastic polybags and squeezes the air out of the diapers so that they fit into a smaller bag. Waste reduction? Well, sort of. It's a reduction of an increase. It's analogous to the Pentagon's argument that military spending has decreased when in fact all that has happened is that spending has not risen as fast as projected. With some pollution prevention calculations, you have to be careful of the basis for comparison. Yes, switching to thinner disposable diapers that come in less packaging reduces waste. But because the reusability of cloth diapers is so high, not switching to that alternative ignores a major pollution prevention opportunity.

Making and Reusing Waste Outside the Home

Yard wastes are a special category of household wastes. They are the only kind that households actually manufacture. People plant lawns and trees and flowers, then cut and trim all summer. In the fall the leaves come down. In some communities, yard wastes are the largest fraction of all wastes, at least during certain periods of the year.

To reduce this waste, the options are to plant with waste in mind or to reuse by composting. Planting with waste in mind means to plant evergreen instead of deciduous trees, ground covers instead of grass, and perennials instead of annuals. Evergreen trees don't lose their leaves each year. Lawn wastes can be reduced by mowing often and leaving the cuttings in place. But ground covers not only don't produce the waste of lawns, they require far less maintenence (also see Chapters 6 and 8 regarding toxic pesticides). Ground cover eliminates space for walking and playing, but how much of the surface of a lawn is used anyway? Isn't much of it for looking at rather than sitting on? Perennials may contribute as much waste during the growing season as annuals, but the entire plant doesn't have to be disposed of in the fall.

Yard wastes that are inevitably generated can be turned into nutrients for soil by composting. Composting apparatus can be as cheap as a plastic yard waste bag or as expensive as an initial investment of $100 for a specially built container. One firm—Ringer—offers composting kits that include a bag and a special compost booster at prices ranging from $7 to $36. Some communities have started to collect yard wastes for composting on a large scale. That recycling option is appropriate for those without the backyard space for composting. However, large-scale composting with many sources can result in accumulations of toxic metals. With household composting, the metals that plants take up from the soil are returned to the same soil with very little effect.

Whatever You Buy, Think Small

Everything from automobiles to microwave ovens are available in different sizes. *Buy small sizes,* and then when you throw products away (remember, everything eventually is thrown away), they will make less waste than larger versions would.

It's called *downsizing* in industry when products are deliberately reduced in size. The reasons for such reductions vary. The energy conservation reaction to high gasoline prices in the early 1970s caused the automobile industry to make lighter, less powerful cars. The 1977 standard Chevrolet had the same interior dimensions as the 1975 Chevrolet Impala but weighed 1,000 pounds less. Automobiles have crept up in size and power ever since the pressure on energy prices abated in the late 1970s. Lightness is still gained by using plastics and other new materials in place of steel. New concerns about the greenhouse effect caused by the pollutants released from automobile energy consumption may bring smaller, lighter cars back.

One phenomenon of the electronics industry has been a continual reduction in the size of consumer (and industrial) products. This trend has been driven by technological advances. Today one tiny electronic chip can accomplish the same tasks as 1 million transistors 20 years ago. The size of a pocket calculator is now defined by the capability of the human hand to manipulate it rather than by electronic technology. Cellular phones are small enough to fit in a shirt pocket, and video recorders weigh as little as 28 ounces and can be held in one hand. Similarly, personal computers have been dramatically reduced in size in the last few years. The so-called footprint (volume of the box containing the electronic components) of personal computers (PCs) has been reduced 30 to 50 percent. Comparable laptop PCs weigh less than 20 pounds and are the size of a briefcase. Hand-held PCs that weigh 1 pound are now available. The laptop and hand-held PCs have benefited from technological advances in batteries, data storage devices, and display screens, as well as from the chips themselves.

Many appliances are available in large, medium, and small models. The small (or compact) models are not popular, judging by the relatively few that are produced each year. Government statistics lump annual shipments of all compact refrigerators, freezers, washers, and dryers together. In 1987 they totaled 1.5 million units, only 7 percent of all shipments of those appliances. Pollution prevention would occur if that percentage were raised. The trends toward decreasing family sizes and increasing single-person households suggest that there are growing marketing opportunities for smaller appliances.

When purchasing anything, then—whether it be a washing machine, clothes dryer, refrigerator, microwave oven, or stereo—consumers should consider the small model first. A smaller model may cost less or, for the same cost as a larger model, may offer more features. But will it satisfy one's needs? Obviously, for a large family, a small washing machine could be a nuisance. A washing machine may last for 10 or more years; if children are nearing the age when they will leave home, double loads for a few years may make sense. Buying small also means gaining some space in a home. Microwave ovens come in compact, mid-sized, and full-sized versions. Compact versions take up the least space and eventually create the least waste. But with lower power, they cook slower than larger ones, and their interiors may not be large enough to cook everything. Don't compromise on quality, however. Buying small *and* cheap will cause more replacements, and waste will increase instead of decrease.

REDUCING PACKAGED PRODUCT WASTE

Excessive packaging is a trait of advanced industrial societies. In 1984 the Western industrialized nations consumed 87 million tonnes (96 million tons) of packaging material. Americans consumed 30 percent more than all of Europe. The consequence is that more than three-quarters of all the things Americans now buy are packaged. That makes avoiding packaging sometimes hard to do. But it's possible. First, some information about packaging.

What Is Packaging, and What Does It Do?

Packaging includes a large variety of containers—boxes, jars, cans, bottles, and bags. They are made from paper, glass, steel, aluminum, and plastics (polymer materials). Packaging materials are following the current path of general materials development: toward more complex, engineered systems. Not only do polymers exist in hundreds of combinations, but all packaging materials are combined in many ways. Milk cartons are paperboard coated with a waxy polymer. The newer, single-serving juice box is made from paperboard, polyethylene, and aluminum foil. Packaging is also wrappers of plastic, paper, cellophane, and paperboard and packing materials like polystyrene and corrugated cardboard. The containers and the wrappers are combined in many ways to make multilayered packaged products.

Advances in material and packaging design can work both for and against pollution prevention. They all work against recycling efforts. Complex engineered materials can be tougher and lighter than the traditional materials they replace, resulting in less material per package—and less waste. When separate, multiple layers of packaging are used to cope with the ineffectiveness of microwave ovens or to help sell a product, pollution prevention loses. Complex and laminated materials complicate recycling systems. The goal of recycling is to break product and packaging materials down into their basic materials. The more complex a material is, the more difficult and expensive that effort can be.

Packaging is functional. It contains a product. Packaging protects and preserves the product from the time it is manufactured until it is used. Then some or all layers of the packaging are peeled off and thrown away. With many products, the last layer of packaging (especially for food products) is retained until the product is used up. Then it goes into the trash, along with residuals of the product.

Some food is processed in its packaging (canned fruits and vegetables, boxed juices). For many products, especially liquids, some kind of packaging is mandatory. Some things, though, are packaged even when it doesn't seem necessary to protect them, like six nails, a lipstick, four batteries, or a pen encased in cardboard and plastic (called a *blister* or *bubble pack*). Although these items won't break, spoil, or spill in transit, they are being protected from shoplifters.

Packaging is a way to get people to buy products. It strives to attract attention among the plethora of products on the shelves. It sells convenience. When selling and other noncontainment functions take over, things happen that increase packaging:

- To take the place of the missing sales clerk in self-service stores and to discourage shoplifting, packaging is used to increase the size of a product and provide room for messages and labeling. This blister packaging can be conveniently hung from hooks, and it assists stores in inventory control.
- It may take only one layer of packaging to contain a product; most packaged products have several. A tube of toothpaste comes in a box. Tea is bagged, tagged, boxed, and wrapped in cellophane. Soda pop is bottled and wrapped in cardboard paper or plastic to form a six-pack. Some men's stick deodorants are not boxed; women's are, because manufacturers say that women won't buy them unboxed.
- The industry calls it *portion-controlled packaging* when they put one serving of tuna fish or yogurt or green beans in a package. Then they wrap paperboard or plastic around four or six of these packages so that customers must buy more than one serving.
- For microwavability, companies put the food on a tray, wrap it in plastic film, put it in a foil-lined sleeve, and then put it all in a box.
- Laundry detergent comes in boxes with recloseable lids and bottles with

handles or with lids large enough to serve as measuring cups. Future growth in the sales of liquid soaps will depend on these "packaging innovations" and marketing emphasis, according to a January 1990 industry analysis.

• Putting juice in single-serving boxes (called *aseptic packaging*), adding individually wrapped straws, and wrapping the boxes and straws in cellophane have helped to reverse a downward trend in per capita consumption of canned juices that has "plagued" the industry since the 1970s, according to the U.S. Department of Commerce.

• Attractive packaging is crucial to the success of new or repositioned [bakery goods] products on crowded retail shelves," says the U.S. Department of Commerce (1989). "Higher prices of materials and greater use of plastic trays" caused a rise in the cost of packaged bakery goods in 1988.

• To get kids to implore their parents to buy them a hamburger and fries, McDonald's puts the wrapped food in a box festooned with games, mazes, and pictures of Ronald McDonald.

Why Worry About Packaging?

The people who make packaging materials, design packaging, and put their products in packaging often say that packaging is not a growing part of the nation's waste problem and that reductions can't amount to much anyway. These people also say that packaging is necessary, not excessive. They argue that the costs of packaging and the profit motive minimize their use of packaging.

To prove the first point, the packaging people use the EPA/Franklin waste data discussed in Chapter 4. The "containers and packaging" category has remained a steady one-third of the total amount of MSW over the years. But this percentage by weight disguises trends in packaging. Lighter-weight plastic and paper have been replacing heavier metal and glass packaging. Plastic containers were 99 percent (by number of units) of U.S. shipments of plastic, glass, and metal containers in 1988. So, the *volume* of packaging can be increasing even if the *weight percentage* in waste remains the same.

The packaging industry also points out that packaging waste is widely distributed among steel, aluminum, paper, glass, and plastics. Plastics packaging, for instance, is only 4 percent (by weight) of the EPA's total of 158 million tons. Reduce half of it, they say, and you have saved *only* 3 million tons a year. For the big categories, like corrugated boxes (12 percent, 19 million tons), they argue, most is not thrown away. It is recycled, so there is no need to reduce.

A pollution prevention strategy views packaging differently. First, it is better to reduce something than to assume that it is going to be recycled. Second, saving 3 million tons a year of plastic packaging is worthwhile. It's better than continuing to produce 3 million tons of it and certainly better than increasing waste by 3 million tons. In addition, there are targets other than plastic packaging. A 25 percent reduction overall (instead of a 50 percent reduction for only plastic) would save 27 million tons of paper, metal, and plastic waste—a 10 percent reduction, 5.5 million tons. If the more realistic waste generation numbers in Chapter 4 are used, the waste saved is almost double.

Packaging advocates like to mention that since lighter plastic beverage bottles have replaced glass ones, reduction has occurred. That is true only in a limited way. Since 1973, the use of glass bottles has declined from 71 to 38 percent of all soft drink packaging. The major impact of the change is, however, that plastic ones are used only once. The glass ones used to be collected and refilled many times before they were thrown away. If the 5.4 billion one-time soft drink plastic bottles used in 1986 could have been used just four times, then only 1.4 billion bottles would have been needed and thrown away. That's a very conservative estimate; glass bottles can be refilled an average of 30 times.

As to the profit motive argument made by the packaging industry, the *cheapest* packaging is not necessarily the *least* packaging. Material costs are not based solely on material weights or densities. Using more of one material may cost more (thicker paper, say), but choosing between two different materials could lower costs while increasing packaging. It costs more for a manufacturer to buy packaging material than it does for a consumer to throw it away. That suggests that there could be a major economic incentive to reduce packaging up front. Material costs in 1988 were $6,000 per ton of aluminum foil, $3,000 for plastic wrap, $2,000 for plastic bottles, $1,000 for paper cartons, and $300 for glass containers. With disposal costs at $25 to $100 per ton, every dollar saved by a consumer would amount to $60 to $240 savings in materials costs.

Lastly, however, packaging *costs* are not always the major consideration. Industry people surveyed on packaging requirements rated durability, function, and aesthetic appearance more important than cost. A unique, eye-catching design is a key element when introducing a new product. This can take more packaging rather than less. For high-value products, like perfumes that retail for $150 per ounce, low-cost/least-packaging rules are irrelevant. For products like major appliances or computer equipment where the cost of packaging relative to the cost of the product is usually quite small, doubling the packaging to double the protection may not add much to the cost of the product. When competition is strong, the *kind* of packaging, rather than the cost, may provide a product with a competitive edge.

When a glass bottle instead of a plastic one is used for a food product to give it an imported image (and a higher selling price), packaging weight increases.

How Much of This Do You Believe?

Like an addict, America has been down this denial road before. In the 1960s and 1970s, when the nation last worried about wastes, packaging was seen as a growing problem and a viable prevention option.

A definitive study on packaging (Darnay and Franklin) was done in 1969 for the U.S. government. It concluded that "estimated national per capita consumption of packaging materials was 404 pounds in 1958, 525 pounds in 1966, and will be 661 pounds by 1976." The study said that all of the 51.7 million tons of packaging produced in 1966 was discarded, except for about 10 percent that was reused or recycled. The study identified all of the categories and problems of packaging that are discussed today.

The pressure on industry grew. At a waste reduction conference in 1975, John R. Quarles, Jr., then Deputy Administrator of the EPA, said that between 1963 and 1971, while food consumption had increased by only 2.3 percent per capita, food packaging had increased by 33 percent and the number of food packages by 39 percent. At the same conference, John David, Director of Environmental Control for a major food processor, stated:

> General Foods recognizes its responsibility to take practical steps to avoid pollution of the environment with packaging wastes. Every effort is being made to minimize waste by eliminating excessive packaging material and to ease the solid waste problem by utilizing the most disposable package consistent with practical considerations.

Then the pressure started to ease. The same analysts who did the 1969 study concluded in 1978, in a food packaging study done for the National Science Foundation, that, with only a few exceptions, from 1960 to 1975 "there has not been a significant change in overall packaging of non-fluid foods, when examined in terms of packaging per unit food, packaging per capita, or packaging per dollar food expenditure." The study listed a number of significant packaging reductions that were occurring for specific foods, primarily because of the inroads of plastic packaging.

Ten years later, the same analysts (Franklin and Associates) reviewed their data from 1970 to 1986 and concluded that while the U.S. population had increased by 18 percent, the packaging component of MSW waste had grown by only 9 percent (by weight). Thus, packaging waste, according to this group, has *decreased* on a per capita basis. Based on this data, 1986 packaging waste was 450 pounds per capita versus the 525 pounds reported for 1966. (See Chapter 4 for our analysis of why the official U.S. solid waste data is an underestimate.)

Reality Check

Has packaging waste really decreased in households since the 1970s? Does packaging waste in homes only amount to one-third of all waste, as EPA/ Franklin data says? That data does not differentiate between household, commercial, and institutional wastes. Other data shows that up to 77 percent of all packaging is used for products most likely sold to households. One study published in 1985 puts all *household* packaging waste at 41 percent (by weight) and food packaging waste at 33 percent. If a person generates waste at the low EPA rate of 3.6 pounds per day, then packaging waste totals 539 pounds per year; at the rate of 7 pounds per day, 1,046 pounds per year of waste are produced. One household's packaging waste each year could amount to over a ton!

Even without the numbers, consumers notice increased packaging waste at home because waste has increased, if not by weight, then certainly by volume. People also have more waste because they buy more things and more of them are packaged, and because they increasingly buy products packaged in small portions, ending up with more packaging per product.

Consider the following facts:

- Since the late 1960s, beverages have increasingly been packaged in one-way (plastic and glass) bottles. Only 6 percent of domestic beer in 1987 was packaged in refillable bottles; only 14 percent of soft drinks in 1986 was packaged in refillable bottles.
- Portion-controlled food and juice packaging is a growth industry. Aseptic packaging alone is growing at a rate of 20 percent annually. The U.S. Department of Commerce estimates that 6.5 to 7.0 billion aseptic package units were sold in 1988, over 90 percent of them for juices and juice drinks. The most common size is 250 milliliters, much smaller than traditional containers.
- If home food wastes have decreased because Americans buy more packaged foods or order home delivery pizzas, then packaging wastes have increased as a result. Now packaging of fresh fruits and vegetables is a growth area. What's emerging is atmosphere-controlled packaging that retards spoilage. *Business Week* reported in 1988 that "the market for controlled-package food is $1 billion per year and expected to triple in three years."
- An estimated 5 billion paperboard trays and containers were sold in 1988 for microwave packaging.

- Per capita consumption of soft drinks rose 3.5 percent annually between 1980 and 1987. Since 1970, it has been growing at the expense of milk and coffee. Almost 50 percent of soft drinks are packaged in small sizes (6–12 ounces), whereas ground coffee is packaged as a concentrate and two-thirds of milk cartons are 1 quart and larger sizes.
- Americans are upgrading their taste in wine. Demand for jug wine has declined an estimated 3 percent annually since 1980. Replacing jug wines (gallon bottles) are more costly premium wines in smaller bottles (750 milliliters).
- Consumers' desire for tamper-resistant seals and packaging for foods, as well as pharmaceutical products, is still high, although it may be declining.
- Only 10 percent of households had a VCR in 1980; now 68 percent do. Americans have bought 56 million of them and are estimated to have bought 12 million more in 1989. Those VCRs came in 68 million cardboard boxes (with Styrofoam packing) that have been thrown away or reside in attics waiting to be thrown away. Over 300 million videotapes are made in the United States annually, and millions more are imported. Each comes with a cellophane or polymer wrapper. A pack of three has an additional wrapper.
- No one seems to know how many items are encased in blister packs. But try to buy hardware, notions, cosmetics, or stationary supplies without one in a variety, grocery, or drug store.
- Between 1984 and 1988, there was a 44 percent per capita increase in shipments of plastic bottles. The switch to plastic packaging—undeniably lighter in weight than some of the materials it replaces—may be reaching a plateau, however. Current data from the Department of Commerce suggests that the replacement of paper with plastic is growing more slowly today.
- Even though paper was being overtaken by plastics throughout the 1980s, the production of paper and paperboard grew at a faster rate than the GNP. For every 1 percent per capita increase in GNP, there was a 1.2 increase in paper and paperboard.

Paying for Packaging

Packaging costs are hidden, but consumers pay twice. First, when they buy a packaged product, the price includes the cost of packaging. They pay again when packaging is thrown away. Depending on the cost of garbage service, the combined packaging costs may be over $300 per year. Heavy consumers of products with excessive packaging pay more.

From $200 to $230 of the average costs are what consumers compensate manufacturers for their packaging costs. Professor John Deighton of the

Amos Tuck School of Business Administration says that corporations spend $50 billion a year on packaging and that 7 percent of consumer spending goes to pay for packaging. A 1987 paper from the Michigan State University School of Packaging says that the value of shipments of the packaging industry was $55.8 billion in 1985, about 4 percent of the value of all finished goods sold in the United States.

You may be paying most of your costs for packaging at the grocery store. The Michican State paper says that about 53 percent of packaging materials were used for food and beverages. Food packaging may be costing proportionately more than other packaging. It is often reported today that $1 out of every $11 (9 percent) spent in the grocery store pays for packaging. Retail sales in food stores were estimated at $350.5 billion for 1988 by the Department of Commerce. If that 9 percent rate is still valid today (it has been cited since at least 1981), then $32 billion of it pays for packaging. That's $133 per person in 1988.

Consumers on Packaging

Each year *Packaging* magazine surveys American consumers on food and pharmaceutical packaging. The surveys reveal some interesting preferences and trends that seem to be at odds with what the packaging industry says:

- Consumers are apparently not interested in the marketing aspects of food packaging. Consumers want function at minimum cost. The attributes rated highest were ''preserves flavor'' and ''provides value for money.'' The lowest rated was ''microwavable.''
- Consumers do not demand changes in packaging; in fact, they initially resist changes. Over time, however, consumers develop a preference for the packaging that industry has given them. Thus, foods such as tuna fish, tomato sauce, and ground coffee that have traditionally been packaged in metal cans are preferred that way. Conversely, metal and glass for juices and vegetables are declining in favor of plastic and paperboard.
- Consumers may be more discriminating than they are given credit for. Most consumers read labels. In the 1989 survey, 93 percent said that they (somewhat, very much, or extremely) pay attention to labels. This belies statements such as: ''just by seeing packages on the shelf and holding them in the hand, shoppers make rapid-fire judgments. Then they either buy or walk away.''
- Packaging preferences are affected by other events. The *Packaging* survey shows that tamper-resistant packaging is still important but is losing favor and that a willingness to try irradiated food is gaining. Prod-

uct tampering cases and the pros and cons of irradiated food are not often in the news today.

- Consumers are becoming so concerned about the environmental impact of packaging that 50 percent of those surveyed say that recyclability either "often" or "sometimes" affects their purchasing decisions. (The survey did not ask about prevention.)

But are consumers often consulted about what kind of packaging they prefer? A survey of industry people said that they aren't. Asked who outside their company has the most influence over packaging, only 10 percent of the respondents listed consumers. Ahead of consumers were graphic designers, packaging material suppliers, ad agencies, and no one (i.e., all influences are internal). Thus, consumer demand is not driving packaging design. Who ever asked consumers if they want to buy products encased in blister packs? Whose interests does that form of packaging really serve?

The Future of Packaging

All indications are that packaging will increase in the future. The industry continues to seek new ways to package and old products to package anew. Opposing this trend is a growing public awareness of the true environmental costs of packaging.

Packaging is the largest single industry in the U.S. economy, as measured by the number of employees. It is said that more money is spent by corporations on packaging than on advertising. These powerful economic interests may resist consumers' environmental fears. Industry's basic strategy is clear, as Susan Selke (1990), a packaging academic, told *Across the Board* magazine: "The most effective way to combat the [pending legislative] restrictions [on packaging] is for the industry to promote recycling, because recycling is politically popular." That is exactly what has occurred, as evidenced by the growth in industry organizations now funding recycling efforts. Recycling of packaging has great appeal. It asks very little of consumers and costs industry only a small portion of its public relations budget.

On another front, the Institute of Packaging Professionals is preparing a white paper to explain to the public the many benefits of packaging. It is intended to destroy the "myths and misconceptions" surrounding the packaging industry. The paper is being funded by such industry groups as the Aluminum Association, the American Paper Institute, the Council on Plastics and Packaging in the Environment, the Council for Solid Waste Solutions, and the Flexible Packaging Association.

Notably, some firms have taken the higher road of prevention. Procter & Gamble has redesigned its Crisco Oil bottle, thereby reducing its material content by 28 percent, and has a new Downy fabric softener package on the market (see below). Budget Gourmet frozen microwavable foods are packaged in a simple, one-layer paperboard box. Glad garbage bags have been made thinner and lighter while maintaining durability. The redesign has also resulted in less material needed for the boxes in which they are packaged. These changes occurred despite the more traditional perspective of a marketing director of Nabisco, who said that "nobody would dare" reduce a package size unless all competitors did so simultaneously.

Meanwhile, the growth areas of food packaging include more complex microwavable packaging; barrier packaging for fresh fruit, vegetables, and prepared foods; and shelf-stable packaging. The demand for packaging materials is expected to grow from $29 billion in 1985 to $63 billion in 1995. Not all new packaging constitutes increased packaging waste. Substitutes for existing packaging may result in reductions, while adding packaging to previously unpackaged items always results in increases.

Barrier or controlled-atmosphere packaging (CAP) will allow consumers to have seasonal foods year round. CAP, though, adds nothing to the quality of food and can fail if the temperature is not carefully controlled during transport and handling. Shelf-stable food is sealed tight before being processed in a package made of a complex layer of different plastics. Like canned foods, shelf-stable food only requires reheating by the consumer, and doing so is quicker than heating a package of frozen food. It is replacing cans, according to Del Monte, one of the first processers to introduce a line of shelf-stable foods, because people now think metal cans are "old-fashioned" and because canned foods cannot be microwaved as is. The appeal to the food preparation industry of any packaging that improves the shelf life of foods is that it saves distribution costs both for them and for retailers.

Applying Three Principles

The foregoing discussion about packaging may convince you that packaging reduction is a hopeless battle. Yes, it is hard to avoid packaging. And, yes, there are powerful forces pushing the American public to accept more and more packaging. But knowing about packaging and about those forces helps people to resist. As with products, choices can be made to reduce waste. The choices can help the packaging industry to reconsider whether to reduce packaging.

The key is to choose items that give the most of what is wanted—the product—with the least of what isn't—the packaging. In general, this means seeking out and buying products (1) in large, economy-size packages, (2) in concentrated forms, and (3) with the fewest layers of packaging. Buying less packaging can mean paying less, as a number of our examples show.

In a 1988 survey of consumer attitudes about their supermarkets, over 60 percent of the respondents checked for grocery specials in newspaper ads and used price-off coupons to save on their food bills. Gaining proficiency in lowering the amount of packaging bought can take less time than it takes to read ads and clip coupons and in regular, better savings.

The three pollution prevention principles are easiest to apply to food, cosmetics, notions, hardware, and the like. It is difficult—seemingly impossible—to apply them to large items, such as appliances and electronics. Consumers often do not have the opportunity to assess the differences in packaging for these products because packaging is not always seen in the store. But nothing stops one from asking, when choosing between two refrigerators, how much packing material each one includes. In West Germany, where strict rules govern what and how much a household can throw away each week, shoppers have learned to leave the carton behind in the store. Another option is to tell the delivery service to take the packaging back when an appliance or a piece of furniture arrives. That requirement can be written on a sales receipt.

The Bottom Line

There is one way to find out definitely, on a weight basis, which of several comparable items comes with the least packaging. Calculate a ratio of the amount of the product (the content of the box, bag, or bottle) to the amount of packaging in which it is wrapped. This is done by dividing the weight of the contents by the weight of the packaging. The number that results is the amount of content (product) per unit of packaging. The item with the highest number has the least amount of packaging.

Unfortunately, making this calculation means buying an item first (unless a scale can be carried around the store). The labels of many packaged goods (e.g., processed foods, cosmetics, personal care products, and cleaners) give the *net weight,* which is the weight of the contents. At home, a food or bathroom scale will give the total weight. Subtracting the net weight (listed on the label) from the total weight gives the weight of the packaging alone. Then divide the content (or net) weight by the packaging weight.

To illustrate this process, Table 7-6 provides some sample calculations and comparisons for food and personal care products. In comparing packaged products, the one with the highest number is best because it means more product per unit of packaging. A number of 1 means that the contents and the package are equal. A number of less than 1 (a fraction) means that consumers are buying more packaging than product.

A drawback to this system is that weights are compared. Calculating volumes is much more complex but may be more relevant for pollution prevention purposes. Depending on the complexity of the packaging, it is often possible to estimate comparative volumes based on what is seen. That is what the fewest layers principle (the third principle in this category) depends on.

In Practice, Just Practice

We are not suggesting that people weigh and calculate everything. It can be educational, though, to try it a couple of times. In general, it is only necessary to follow the three pollution prevention principles for packaged products: large sizes, concentrates, and fewest layers.

Practice improves the ability to make judgments and lower the amount of packaging that comes home. It also instills the value of pollution prevention as an important consideration in consumption. It's not always a question of what is bought, but of how a product is bought. Consider pizza. There are at least four ways to get one:

1. Order one and eat it in a restaurant.
2. Buy the ingredients and make it yourself.
3. Buy a frozen pizza and cook it at home.
4. Order one by telephone and have it delivered.

Which is most wasteful? It depends on what is counted. Eating in a restaurant generates no household waste (unless leftovers are brought home). Eating in a fast food restaurant, though, will generate cooked food packaging waste that is left behind. That doesn't occur in a regular restaurant. Both kinds of restaurants generate food and packaging wastes to make a pizza. Because restaurants buy in larger quantities than a household does, their raw material packaging waste is probably lower than is true for a homemade pizza. But again, because a homemade pizza relies on bulk supplies of flour and other ingredients (unless a mix is used), it is probably less wasteful than a frozen pizza in a multilayered package. The frozen pizza is, in turn, better than a delivered pizza that arrives in a heavy cardboard box, along with a supply of paper napkins and throwaway utensils.

The last option has been undergoing major growth. Some analysts even credit home delivery pizza for the slowdown or decline in the growth of fast food restaurants. Marriott Corporation announced in late 1989 that it was selling its fast food restaurants. One explanation given by the president of Marriott was that "many people prefer to stay at home with their VCRs and pizzas."

TABLE 7-6. COMPARISON OF PRODUCT CONTENT TO PACKAGING

Product	Total weight	− Net weight =	Pkging weight	Net weight divided by pkging weight	Packaging as percent of total	Cost per ounce	Notes on packaging [a]
		Minute Maid Orange Juice					
Pack of three	29.75	28.10	1.61	17	5	0.04	Paperbd box/plastic straws/cello wrap
Frozen conc	54.30	53.30	1.00	53	2	0.03	Paperbd and metal can
Carton	73.25	70.75	2.50	28	3	0.03	Paperbd
		Tuna Fish					
Pack of three	13.50	9.75	3.75	3	28	0.26	Metal cans/paper labels/paperbd wrap
Reg can BumbleBee	8.00	6.50	1.50	4	19		Metal can/paper label
Reg can Safeway	8.25	6.50	1.75	4	21		Metal can/paper label
		Del Monte Green Beans					
Microwave Ready	13.75	11.50	2.25	5	16	0.11	Plastic tray and lid/paperbd box
Can	18.50	16.00	2.50	6	14	0.04	Metal can/label
Can	10.00	8.00	2.00	4	20	0.06	Metal can/label
Fresh	16.10	16.00	0.10	160	1		Plastic film sack
		Mueller's Old-Fashioned Egg Noodles					
Box	14.00	12.00	2.00	6	14	0.09	Paperbd box/cello window
Cello bag	16.50	16.00	0.50	32	3	0.09	Cellophane bag

Tea Bags

Lipton Tea							
Loose in box	9.25	8.00	1.25	6	14		Paper bags/paperbd box/cello wrap
12 bags in box	4.25	3.00	1.25	2	29		Paper bags/paperbd box/cello wrap
Twinnings Tea Bags							
25 bags in box	3.25	1.80	1.45	1	45		Paper bags/paperbd box/cello wrap
10 bags/cello	1.00	0.66	0.34	2	34		Paper bags/cello wrap
Tide Laundry Detergent							
Box	44.00	42.00	4.00	11	9		Paperbd box
Box	19.75	17.00	2.75	6	14	0.09	Paperbd box
Mennen Deodorant							
Men's Speed Stick	4.00	2.25	1.75	1.3	44	1.33	Plastic tube
Lady Speed Dry	3.75	1.50	2.25	0.7	60	1.73	Plastic tube/paperbd box
Secret Wide Solid	4.00	2.00	2.00	1.0	50		Plastic tube/paperbd box
Nine Live's Pet Food							
15-oz can	17.00	15.00	2.00	8	12		Steel can/paper label
6.5-oz can	8.00	6.50	1.50	4	19		Steel can/paper label

[a]Paperbd = paperboard; cello wrap = cellophane wrapper.

Buying Large Economy Sizes

Economy-size packages of products reduce waste and often save money. The price will, of course, be higher (for buying more product), but there is often a volume discount. Because of this pricing policy, a helpful guide to less packaging in many supermarkets (especially national chains) is *unit price* labels. The version of a product with the lowest unit price (price per weight or price per number) is, invariably, the one with the least packaging.

The effect of buying large sizes is illustrated in a number of the items listed in Table 7–6. In comparing product to packaging, an 18.5-ounce can of green beans gets a 6, a 10-ounce can only a 4. A 44-ounce box of detergent is 11, a 19.75-ounce box is only 6. A 15-ounce can of pet food is 8, a 6.5-ounce can is 4. The next column in the table shows that a 4 means that 20 percent of the amount purchased is packaging, an 8 is 12 percent packaging, and an 11 is only 9 percent packaging.

Buying a larger size does not always result in a reduction of packaging. This can happen when different products or products from different manufacturers are compared. For instance, two manufacturers may sell the same net content of product (say, 16 ounces of cereal), but in different-sized boxes. The unit price may be the same. The two cereals may differ on a weight-to-volume basis; 16 ounces of a puffed cereal will take up more volume than 16 ounces of a denser one. Or the cereals can be equivalent, but one manufacturer may attempt to catch consumers' attention with a bigger box.

Different kinds of packaging may also invalidate this principle. In a comparison between a package of 10 tea bags and a package of 25 tea bags (in Table 7–6), the smaller package wins because it has one less layer of packaging. In both cases, the tea bags are individually wrapped in paper and have an outside layer of cellophane, but the packaging for the 25 tea bags also includes a box. Buying two packages of 10 (to get 20 tea bags) results in half as much packaging waste as with the package of 25 tea bags. (In this case, the fewest layers principle is invalidating the large size principle.)

When buying large sizes, consumers have to be careful not to trade reduced packaging waste for increased product waste. This can occur when buying perishables, such as food products. There are several ways to avoid increased food wastes when buying large sizes. Among them are the following:

- Store the product in a more secure glass or plastic reusable container.
- Store the product in the refrigerator instead of on a cupboard shelf.
- Repackage and freeze part of the product (in a reusable container) for future use.

Economy and single-serving packaging are at opposite ends of the spectrum for prevention. Both help food processors move more product. Both have been accepted by consumers, in the first case because of a cost reduction and in the second case because of the anticipated convenience. This portion-controlled packaging was recognized in 1980 as one of seven "hot buttons" of consumer demand. Today, the food industry makes a multitude of products available in portion-controlled packages: single meals, snacks, yogurt, vegetables and side dishes, soups, juice drinks, and even pet food. Americans buy over 1 billion individual boxes of fruit juice, shrink-wrapped with a separately wrapped straw on the side, only part of the over $11 billion worth of single-serving containers used each year.

Buying these products violates the large size principle and, often, the fewest layers principle (see below) for reducing packaging waste. Single-serving products are usually grouped together in packs (of three to six) to encourage the purchase of more than one single serving at a time. The packs are sealed together by a wrapping of paperboard or cellophane. Some food processors (like the makers of Dannon yogurt) get this same benefit with a less wasteful packaging design. The individual packages are integrated into one unit, eliminating the second packaging layer. So, when a portion-controlled product is a must, some waste can be reduced by avoiding double wrappings.

A cost premium is paid for the convenience of single-serving foods. Some save very little time in food preparation. A single serving of vegetables (Green Giant One Serving) at 90 cents for 4 ounces includes a plastic tray, film, and box; a 16-ounce plastic bag of Bird's Eye frozen vegetables costs $1.76. For the latter, the food preparer has to measure out a single serving into a dish before putting it in the microwave oven. For the time this action and washing the dish afterward takes, the cost of the vegetables is cut in half.

Many single-serving packaged food products are marketed to busy parents. Children like raisins, and parents buy them in handy little boxes. A six-pack of the 1.5-ounce boxes (for a total of 9 ounces of raisins) makes as much waste (1 ounce) as does a 15-ounce package of raisins. That means that for every 15 ounces of raisins bought in the little boxes, a consumer generates almost twice as much waste. The cost premium is not as high. At $1.43 for the six-pack (16 cents per ounce) and $1.85 (12 cents per ounce) for the large package, the wasteful packaging version costs about one-third more. To almost match the convenience of being able to dole out 1.5 ounces of raisins at a time, 10 small, reusable plastic containers could be used to repackage a 15-ounce box of raisins. Even if the containers cost $1 each, the cost would be quickly paid back with savings from buying the economy-size package. Children, of course, would have to be taught to save the reusable containers. That is an excellent way to ingrain pollution prevention early in life.

Buy Concentrated Products

Compared with a concentrate, the cost of a ready-to-use product includes more packaging, water, and labor. Food, cosmetic, cleaning, and gardening products are available in concentrated forms. But today convenience is driving the trend toward ready-to-use packaged products. The list of ready-to-use products is growing.

Garden pesticides, which have traditionally been marketed in concentrated forms, are now offered as ready to use. Even the Safer Company, which sells nontoxic garden chemicals (see Chapter 8) that reduce toxic waste, now offers ready-to-use versions that increase plastic packaging waste. A 32-ounce bottle of concentrated insecticide soap will make 13 gallons (about 1,664 ounces). It costs $13, while a 24-ounce bottle of ready-to-use soap costs $6, or 25 times as much.

Juices are a good example of multiple packaging choices. As Table 7–6 shows, of the three orange juice options, frozen concentrate is the least wasteful, next is the ready-to-use carton, and in third place are the individual juice boxes. The advertisements for Minute Maid's ready-to-use orange juice always depict busy, distracted families who don't have the time to add water to concentrated juice.

At least one firm—Proctor and Gamble—is moving in the other direction. It is concentrating dry detergent formulas that will result in paper packaging reductions of 11 percent. The risk is that consumers will not believe that they will get as many wash loads as a normal box of detergent will provide. Canadian and European customers can already buy what P&G calls *enviropaks* of concentrated liquid products, including Ivory dishwashing liquid, liquid Tide detergent, Downy fabric softener, and Mr. Clean, which reduce plastic waste by about 85 percent.

In a test of the American market, P&G has introduced a Downy fabric softener concentrated refill. It is based on the same concept as the existing foreign products. First, consumers buy 64 ounces of ready-to-use Downy in a plastic bottle. Subsequently, they buy a concentrate (21.5 ounces) in a polymer-coated paperboard box, pour the contents into the original plastic bottle, and fill the bottle with 2 measures (using the box) of warm water. P&G has capitalized on reusability and concentration to reduce packaging waste. As long as one continues to buy this particular product, only the small paperboard boxes become waste. P&G says that this system, compared with normal packaging, will reduce waste by 75 percent. The refill costs less than the equivalent amount of their regular Downy and, as P&G notes, "because of its reduced size, it is easier to store and carry."

As this new P&G product demonstrates, the time saved by using ready-to-use packaged products equals the time needed to add water and mix. For some kinds of concentrates, like concentrated shampoo (Prell) or hand lotion (Neutrogena Norwegian Formula), there is no preparation time because less is used with each application.

Fewest Layers of Packaging

It is not always obvious whether the fewest layers means the least amount of packaging. Zero layers is, of course, the very fewest and best kind of packaging from a prevention perspective. But according to statistics, only one-quarter of consumer products come without packaging.

Creating personal food packaging saves both money and packaging waste, and can be done if a grocery store sells unpackaged items in bulk. The same is true when buying fresh instead of prepared fruits and vegetables. We compared fresh and packaged green beans. Fresh beans generated 10 times less waste even when they were put in a plastic bag. When the food waste generated by the fresh beans is included, substantial waste reduction still results. Not all fresh fruits and vegetables need bags. Their skins provide adequate protective packaging for transportation home and for storage. For those that need storage protection, reusable containers work nicely over and over again.

Sometimes the form in which people choose to buy something will reduce packaging waste. Most tea now comes individually wrapped in single-sized portions. This porous paper bag (and string for convenient removal from a cup or pot) is then wrapped in paper. A collection of these double-wrapped bags of tea are put in a box, which is wrapped in cellophane. The alternative that eliminates most of this packaging is loose tea. Loose tea can be brewed with a reusable single- or multiple-serving tea diffuser (cost, $2–$3) filled each time with the appropriate amount of tea leaves. This leaves only the box, its wrapper, and the tea leaves as waste. Another reason for using loose tea is to avoid the dioxin that may be present in the bleached paper from which the bags are made (see Chapter 8).

Is the waste quantity saving worth the bother? There is 0.07 ounce of tea in a tea bag and 0.2 ounce of paper for each bag. That extra waste does not add up to much—a little less than a pound a year for someone who uses one tea bag a day. But on a national basis, it is impressive. The Tea Association says that 63 percent (87 million pounds) of all the tea bought in 1987 was sold in tea bags, while only 3 percent (4 million pounds) was sold as loose tea. (The other 34 percent was instant tea and iced tea mix.) This means that 249 million pounds (125,000 tons) of extra waste are generated each year in the United States by tea bag users.

There are other ways to avoid packaging layers even the ubiquitous blister packs. The trick is to change to a different store or type of store. Some local hardware stores—a rapidly diminishing number, however—still sell items like nails by the pound instead of six to a pack. Unlike variety stores, stationary stores still sell unpackaged pens, erasers, and tapes. Tape dispensers come with tape refills instead of requiring the consumer to buy tape sealed in a dispenser. Bread purchased from a bakery comes only with the bag used to carry it out of the store.

Cosmetics and some personal care products (shampoos, deodorants, hand lotions, etc.) are often more packaging than product. The blister pack around cosmetics can be avoided by buying cosmetics in specialty stores (in which case the price goes up) or by buying the cheapest brands. The latter seem to have lowered their costs by reducing their packaging. The Body Shop, which sells a variety of lotions and shampoos, gives a discount for bottles returned to their stores.

Avoiding packaging sometimes requires searching the shelves instead of reaching for a familiar product. Most personal care products are packaged in plastic containers; some have an outer box, some do not. Women who want stick deodorant unboxed may have to select a men's deodorant. Women's versions (except Tussy brand) are boxed. Note on Table 7–6 that boxed stick deodorants can be 60 percent packaging.

Counting the number of layers is not a foolproof method when the layers are not equal in weight or volume. In some areas of the country, consumers can still choose between one-way glass and plastic bottles of soda beverages. Both may have two layers: the bottle and a label. Which one will reduce the wastestream? Plastic. A 1-liter plastic soda bottle weighs 3.2 ounces; a glass one, 16 ounces. The walls of a plastic bottle are thinner. So, by volume too, the plastic bottle contributes less to waste. If either bottle ends up in a landfill, the plastic one tends to compress, while the glass one does not. Thus, by any quantitative measure, the plastic bottle is less wasteful than the glass bottle.

Degradability? They are equal; neither will degrade in a landfill. Which one will reduce the *toxicity* of waste? Probably the glass one, although no one knows for sure. The glass bottle is made of silica, a common natural materal; sand is silica. The plastic bottle is some form of plastic, probably polyethylene. Actually, the best pollution prevention solution is to buy drinks in returnable glass bottles. Unfortunately, that is possible only in limited areas of the country. General Electric has developed a returnable plastic bottle, but it is not yet been marketed widely, if at all.

For those who buy water in bottles, buying from a delivery service will result in using returnable bottles, either plastic or glass. Bottled water, rather than soda pop, is the fastest-growing part of the beverage industry. During the 1980s, sales of bottled water grew by 500 percent. It is now a $2 billion industry. Americans annually consume 6.4 gallons of bottled water per capita because they fear that their traditional water supply systems (municipal piped and well water) are unsafe. Many communities have had their water wells plugged and are supplied with bottled water because of groundwater contamination from toxic waste sites. Bottled water has also become a fashionable alternative to other beverages. Unfortunately, one cost of this fashion and the desire for safe water is increased bottle waste. A high percentage of the water is sold in one-way bottles.

For a person who drinks 10 gallons of bottled water each month, waste goes to zero when water is delivered. The cost of delivered water is about $22 (including rental of a cooler) versus $30 for 20 half-gallon plastic jugs. The ability to use this alternative depends on having access to a bottled water delivery firm. Beside the waste reduction, consumers benefit from not having to carry jugs of water home from a store. The convenience of home delivery, however, may be offset by delivery and other problems.

There is another way to get water without all the bottles and without potential delivery problems. Water filters can be installed to remove tap water impurities. They can cost from $30 to over $600, and their effectiveness against different chemicals and metals varies. No one system is capable of removing all types of contaminants. Thus, it is important to test one's tap water first to find out what contaminants, if any, are present before investing in a filter system. Also, these systems are not necessarily waste free. Activated-carbon filtration, for instance, absorbs a range of chemicals, but the contaminated carbon filter must be changed periodically. If not, the system will back up and recontaminate the water.

Purifying drinking water is not a true preventive strategy. In fact, having to buy bottled water or purify one's own tap water are consequences of a general neglect of natural water sources. Either approach illustrates the costs, trade-offs, and marginal gains that accrue when prevention has not been practiced. Bottled water is not necessarily safer than tap water, as proved by the revelation in February 1990 that the Perrier spring, the source of Perrier bottled water, is contaminated with the carcinogen benzene (see Chapter 8). While homeowners can dump contaminated filters from their filtration system in the ordinary trash, industrial sources must handle them as hazardous waste.

Microwave Ovens and Packaging Waste

Microwave cooking technology has fostered what may be the finest example of the art of packaging in layers. Microwavable foods come with a cooking tray plus various combinations of inner and outer wrappers. Added to this can be miscellaneous pieces of plastic or foil-lined paper that make up for the deficiencies of microwave cooking. These are called *susceptors, receptor films,* or *guidance foils* and are expected to amount to 2.2 billion units by

1991. Sometimes a polystyrene label is attached for the express purpose of removing a hot package from the oven. (Whatever happened to reusable, washable pot holders?) So that bacon can be microwaved, a cellulose-based pad is added to absorb the grease. The result? Better cooking and a "more attractive appearance," according to industry people.

Business Week gave Campbell Soup kudos for one of 1989's best new products: the Souper-Combo. Conversely, *Garbage* magazine called the corporate mentality that spawned it *NIMBLE* (*N*ot *I*n *M*y *B*ottom *L*in*E*) for the ability of business executives to discount the social and environmental costs of their products. This soup and sandwich microwave meal comes in a paperboard box with a polystyrene foam cooking/eating tray plus a polypropylene bowl for the soup and a polystryrene foam tray for the sandwich, each of which is separately wrapped in polyester film.

The message of Souper-Combo and all other microwavable foods is convenience. From freezer or shelf to oven to table. No fuss, no pan, no plate, in no time at all. No cleanup; the packaging and the heating/eating tray go into the trash. As *Packaging* magazine said in a July 1987 article: "The packaging of [microwavable] foods meets a parallel demand for disposable utensils."

How do advertisements convey convenience (at the expense of the environment) for microwavable foods? Here is a sampling:

- Delicious wholesome meals in less than 2 minutes! Just heat in single-serving bowl! Easy clean-up, no mess! (Chef Boyardee Microwave Meals)
- Snack time, any time. In no time flat. (Pillsbury Microwave Fudge Brownie Mix)
- Now you can enjoy a proper hot lunch in under two minutes—at work, home, anywhere there's a microwave. (Doubletree Foods Lunch Bucket Microwaveable Meals)
- New. Minute Microwave Dishes. (General Foods Corporation)
- Enjoy Campbell's Soup in 3 Micro-Minutes . . . There's nothing to clean up. (Campbell's Cup)
- Now Hormel Chili and Dinty Moore Beef Stew come in single-service microwavable bowls. They heat up in just 2 minutes, so you can put your favorite dish on the table in record time.

Microwave ovens started making inroads into American eating habits in the 1960s and are now found in about three-quarters of all U.S households. In 1987 domestic production and imports totaled 21.5 million units. A survey by *Packaging* magazine has recorded a rise in microwave use from 65.2 percent (of respondents) in 1986 to 84.6 percent in 1989. Five percent of the respondents to a Good Housekeeping Institute survey in 1987 said that they owned two or more microwave ovens. The Microwave Institute predicts that by the turn of the century, one-half of all households will have two or more ovens.

The technology has spawned a new generation of packaging, with more variety to come, as discussed earlier. The value of frozen foods alone went from $1.9 billion in 1976 to $4.5 billion in 1986. Surprisingly, in the 1988 and 1989 *Packaging* magazine surveys on the packaging preferences of consumers, microwavability was voted the least important attribute. The magazine's explanation is that consumers demand the taste, texture, and quality of conventional-oven food products, and they are not getting them. It seems that Americans like the speed of microwave ovens, but not what comes out of them.

Microwave ovens, which cook from the inside out, don't brown, tend to dry out some foods, and cook unevenly. Food manufacturers try to solve these deficiencies with packaging. For browning and crisping, they use metalized films and coatings that increase cooking temperatures; these temperatures reach 600°F in some parts of the food. Questions have been raised about what happens to the chemicals in packaging when temperatures get that high. It has now been shown that plastic packaging at these temperatures in the microwave oven can leach chemicals, such as benzene (a known carcinogen), as well as the plastic itself, into the food. As is usual (see Chapter 8) when toxic chemicals are discovered in a new source, the debate focuses on how much is enough to harm and how to deal with uncertainty. *Consumer Reports* has advised "concerned consumers" to avoid this type of packaging until the tests are completed. The Center for Science in the Public Interest has urged *all* consumers to avoid them, and one trade publication, *Microwave News,* agrees. Food manufacturers are urging consumers to continue buying until more is known.

So far, despite the problems, Americans seem hooked on microwave ovens as a provider of fast food. People can reduce some waste and still microwave. One answer is to supply one's own cooking tray or pan. Pillsbury, for one, sells a packaged microwave cake mix and frosting that comes in two versions; one has a reusable pan, and the other is a refill without the pan. Not only do buyers of that product avoid getting another pan with the refill version, the package is smaller because it does not have to accommodate a 7-inch round pan. As mentioned before, Budget Gourmet brand frozen dinners come in a simple paperboard box in which they are cooked and from which they can be eaten.

If some analysts are right, Americans are and will increasingly be shifting fast food restaurant packaging waste to their homes. *The New York Times* (Kleiman 1989) called it the "Age of Zap." Eating has been reduced to a necessary function rather than a time for pleasure and social interaction. Even fast food isn't fast enough anymore. The alternatives are more microwavable foods and more home delivery of ready-to-eat meals. And more household packaging waste.

The food processing industry senses the speed trend. From General Foods, "It's gotten to the point where it is too burdensome to open a can of soup." So does the fast food restaurant industry. From Steven Reinemund, president of Pizza Hut, "Meeting the consumer's need for convenience is our challenge for the future. Speed is a major focus." McDonald's is testing to see if robots can cook french fries quicker than humans. "We are always looking at ways of shaving off 25 seconds here, 25 seconds there."

We will not try to convince people who view eating as maintenance that there are equally fast alternatives to intensively packaged food. If maintenance is what one is after, until little food pills arrive on the market, the consequence of that lifestyle will be increased packaging waste. For others, who consider mealtime (and meal preparation time) as an opportunity for families or friends to get together and be together, or who believe that eating well-prepared food enhances the quality of life, there are alternative ways to conserve time other than by relying on prepackaged microwave meals. They include eating out and budgeting some weekend time to plan, prepare, and freeze major components of next week's meals.

REDUCING SACKS, BAGS, AND OTHER WRAPPINGS

Packaging that does not automatically come with products is packaging over which consumers have *total* control. This packaging is primarily the bags, boxes, and assorted wrappings that are added to the things purchased—whether prepackaged or not—at the point of purchase. Acquiring the equivalent of one Kraft paper grocery bag a day means ending up with 46 pounds of waste a year.

The prevention principles are three: (1) say no, (2) consolidate, or (3) reuse. There is only a small amount of this kind of packaging that, when reduced, saves money up front. That is because, for packaging added by retail stores, the cost is included in the price of the products, whether or not it is accepted, although there may be stores in some areas that charge for bags. For the majority that don't, not taking the packaging gives the merchant a savings. Customers can save money, however, when reuse of retail packaging eliminates the purchase of something else. Two kitchen trash bag liners a week amount to about $15 a year and four or five paperboard boxes as waste, as well as the waste of the bags themselves.

Packaging is cheap, and stores have much incentive to lower these costs. Retail stores that sell luxury goods may not worry about the cost of the packaging they provide. But why have grocery stores changed from paper to plastic bags? Because they save money. Paper bags cost twice as much as plastic bags (4 cents versus 2 cents per bag for the national Safeway chain). The stores get an added savings because plastic bags take up less storage space; they occupy one-sixth of the volume occupied by paper bags. Collectively, the nation's food stores can save between $700 million and $1.4 billion per year (on the cost of the bags alone) by using plastic rather than paper bags. That's the same amount of money Americans pay grocery stores for the bags. It represents 0.4 percent of a grocery bill. If consumers and groups had not complained about the possible environmental consequences of plastic bags when they were introduced, grocery stores would long ago have given up offering paper bags completely.

Avoiding extra packaging is not always easy to do. First, there is the automatic tendency for clerks to bag everything. Clerks in grocery stores are trained to put frozen or wet items in separate "wet pack" bags before they put them in a carry-out bag. Some stores insist on bagging all purchases (and stapling them shut) as a tactic to lower theft. This is different from the practice of many European food shops, where customers are expected to bring their own bag or buy one there (and to bag their own groceries). In Denmark a plastic bag costs about 10 cents, a paper one 20 cents. Europeans, however, get as many bags in most other retail stores as Americans do. Italy has experimented with high charges for bags and has apparently seen major reductions in their use.

To refuse added packaging, people have to be quick to say, "No bag, please." This means planning ahead to have an alternative in mind. That may mean starting out on a shopping trip with an empty bag or bags. For those who normally shop by car, empty shopping bags can be stored in the trunk, ready for use whenever shopping occurs. For those on foot, an empty shopping bag can be easily folded up and stored in a purse or briefcase.

Alternatively, pockets or a bag from a previous store may work. The latter involves practicing the *consolidation* principle. When shopping in a couple of stores or making several stops within one store, a large bag taken at the first checkout counter can have subsequent purchases added. In grocery stores, consolidate by putting fresh fruits and vegetables at the same price per pound together in the same plastic sack. The store won't like it, though, because separate sacks facilitate entry into the store's electronic inventory system upon checkout. Consolidation reduces packaging waste for the same reason that buying large economy sizes of prepacked products does: more content per unit of packaging.

The ultimate pollution prevention answer to the environmental question "Should I ask for a paper or plastic bag?" is: "Neither." If one keeps a bag out of the wastestream by reusing it, it doesn't matter whether the bag is made of plastic, paper, cloth, or straw. Just keep using it over and over again until it falls apart. People can start with bags acquired in retail stores; they've already paid for them. Determined waste reducers can use something more sturdy—a cloth tote bag or a French string bag. The venerable string bag never really went out of style; it's always been available in gourmet kitchen shops. But it's coming back into more common usage now.

REDUCING MAIL BOX WASTE A LITTLE

Each year over 12 billion mail order catalogs, millions of magazines, and thousands of tons of advertising and requests for donations arrive at homes across America. The Post Office delivered 11 billion pieces of second-class mail and 62 billion pieces of third-class mail in 1988. That's over 300 pieces of mail for each person in America, each year, most of it unsolicited and unwanted.

Mail is fast trash. Except for some first-class mail, most mail is transformed into trash within minutes, hours, or days, certainly within weeks of receipt. Mail waste totaled 9 million tons in 1987. This doesn't include the other similar items—like newspapers and telephone books—that land on front porches. One telephone book can weigh 10 pounds. If each household in America gets only one such telephone book (and today it is likely that each household gets several), then 450,000 tons of telephone books end up in landfills and incinerators each year.

Like other categories of waste, mail waste is growing. In 1972 each household received an average of 13 pieces of mail per week, and only 23 percent was third class. By 1987 the rate was 19 pieces, half of which was third class.

Where Does It All Come From?

About half of household mail—the first-class portion and some second-class pieces—is a result of individual actions. The other half is generated mainly from mailing lists, most of which no one ever asks to be on. We illustrate both kinds of mail waste here with a section on mail that people ask for—magazine subscriptions—and one on mail they don't ask for—direct mail.

Aside from personal correspondence, first-class mail is mostly bills. But bills do not come alone. They come with double or triple their weight in advertising. Department stores and credit card companies want people to consider buying more while paying for items already bought. The mortgage company wants to sell insurance. The utility companies send required public notices, as well as public relations blitzes.

Of the second and third class mail, some of it is asked for—like magazine subscriptions and requested mail order catalogs. The rest is unsolicited. Mail order firms, charities, and political campaigners all want attention and money. This is called *direct mail* by the industry that sends it and *junk mail* by those who get it.

In a 1989 poll, 62 percent of those surveyed said that they usually opened all their mail, including direct mail. The poll did not reveal whether or not they actually read it all or how quickly it was thrown away. One trick used to convince people to open junk mail is to make it appear to be official mail from the government. Another is to make it look like first-class mail. Laser printers sprinkle individual names throughout a letter, mailing labels are replaced with direct envelope addressing, and stamps are used instead of metered postage. A giveaway, however, is the amount of postage. If a piece of mail has not been sent at the first-class rate, it is probably junk.

In another poll, 82 percent of respondents said that they read catalogs before throwing them away. Almost 92 million people in 1989 shopped by mail or phone. That's a 59.7 percent increase since 1983, almost five times the population growth. That growth, or the response to it by the industry, has led to a 155 percent increase in catalog mailings since 1980.

Based on the direct mail industry's account of 12.4 billion catalogs sent in 1988, the average household receives 133 catalogs a year. Averages, though, are deceptive. One household that collected all of its junk mail for 1 year got 126 pounds of it. In a more recent tally, one household got 100 mail order catalogs during the first 8 weeks of 1990; they weighed 25 pounds. If that rate continues throughout the year, the household will receive 650 catalogs weighing a total of 163 pounds. The same household received another 10 pounds of other junk mail during those 8 weeks—another projected 65 pounds of unsolicited waste for 1990.

Assisted by the Postal Service

When people complain that junk mail is subsidized by the taxpayers or first-class mail rates, the Postal Service denies it. They say that the law requires them to fix the rates of each class of mail such that each class is self-supporting. Only one type of mail—that of nonprofit organizations—is taxpayer supported. Each year the Postal Service gets an appropriation from Congress to help pay for special second-class *revenue foregone* rates. Many of nonprofit firms take advantage of that rate to send a lot of junk mail.

Direct marketing firms do get volume discount rates by sending second- and third-class mail. Instead of the 25 cents per ounce charged for a single first-class letter, sending mail order catalogs costs only 48 cents per pound (that's 3 cents per ounce, an 88 percent discount). Other advertising arrives via third-class mail rates that can be as low as 10.1 cents a piece if the sender has presorted the mail for the post office.

Another difference between first class and other classes of mail is that the rate for first class includes a return fee. This means that all first-class mail received can be returned, at no extra charge, to the sender, where it ends up as his waste problem. The post office will not return second- and third-class mail unless the sender has paid for a return endorsement such as "address correction requested" or "return postage guaranteed." Some junk mail, though, comes with postage paid return envelopes or cards that can be used to shift some of the garbage back to the sender.

Buying Through the Mails

Direct marketing is big business. The firms attract attention and prompt people to buy with display advertisements in newspapers and magazines, as well as with the promotions and catalogs sent by mail. American consumers and businesses responded in 1987 by spending over $148 billion buying things and making contributions as a result of direct marketing campaigns. Consumer purchases totaled $70 billion, charitable contributions $37 billion, and business purchases $42 billion. Among consumer purchases, $43 billion were for products and $26 billion were for services.

People buy from mail order firms because it is convenient and gives them access to novelty or specialty items not available locally or to a wider selection of merchandise than they can find in local retail stores. They also buy because of the price, although mail order merchandise usually does not cost less than retail items, especially when shipping and handling charges are added. In a promotional article on mail order buying, the *Washington Post* (1989) said: "shopping by mail-order catalogue . . . saves time and energy and offers a universe of unique and useful things."

The convenience of catalog shopping has been enhanced by the wide acceptance of credit cards and the growing use of WATTs lines (the 800 telephone numbers). Turning *mail* order into *telephone* order and making payment easy benefits the firms because, when customers telephone and use credit cards, they tend to order more and order more often. Now customers can even fax their orders.

Catalogs, Catalogs, and More Catalogs

While all types of direct mail generate waste, the firms that sell products through catalogs may generate most of it. Those 12 billion catalogs, if they average 0.25 pound each, amount to almost 2 million tons of waste. Common to all direct mail firms, however, are certain practices that increase their profitability at the expense of household waste. Among them are:

- The renting, buying, and exchanging of mailing lists.
- Repeat mailings and duplicate mailings.

- *Bounce-back* business.
- Excessive packaging of orders.

A firm's mailing lists are its most crucial assets. They are the backbone of the mail order industry. There are three basic kinds of mailing lists: inhouse, compiled, and rented. Their combined existence make it most difficult to stop the flow of direct mail into a household. An inhouse list is a firm's own list of customers with a history of mailings and sales. And before a customer can get off an inhouse list, his or her name will be rented from another list or compiled onto a new list.

There are firms whose business is to compile mailing lists. Brokers put together firms that want to rent and that have mailing lists for rent. A mail order firm may derive a sizable amount of its annual income from renting its mailing lists. The publication *Direct Mail Lists Rates and Data* is 2 inches thick, with rental offerings from thousands of firms. Brokers and other firms compile mailing lists for sale or rent. Compiled lists are put together with some common element, like zip codes that produce mail addressed to "Occupant." (Because Americans move so often, mailing lists can quickly become outdated. Thus, there is a tendency to address direct mail to a named person and "Current Resident" simultaneously.)

Obviously, with so many mailing list sources, duplicate mailings are inevitable. A firm may send someone a catalog because the person is on the inhouse list and mail another copy because the same name appears on a list the firm has rented. In fact, the probability is high that the same person may appear on both lists, since firms attempt to rent lists of customers with a history of buying products similar to theirs or products in the same price range. Mailing lists are rated on the basis of the average sales per listing. The best list may be a direct competitor's. In one reported search among 100,000 names from 13 upscale firms, 2 percent were duplications.

While duplicate mailings may be an annoyance and a source of unnecessary waste, they are not necessarily costly for direct mailers. First, it can be more costly to eliminate duplications than simply to mail them out. In 1988, it cost the mail order industry an average of 45 cents to produce and mail each catalog. The duplications on the above-mentioned list of 100,000 names would have cost $700 to mail unnecessarily. Whether or not that $700 is worth saving depends on the status of the firm (new versus established), the price range of its products, the size of any one mailing, and the cost of finding duplications.

The need to avoid duplications also depends on what response rate is expected from a mailing. Listen to the president of one women's apparel mail order firm: "When we mail a million catalogues out we hope that only 980,000 of them will get thrown in the trash can" (*Washington Post* 1990). *He doesn't care what happens to 98 percent of the catalogs;* a response rate of just 2 percent will keep him in business. Thus, a duplication rate of 2 percent may be irrelevant to the firm but results in 245,000 pounds of waste. The highest costs for mail order firms are postage, lists, and the catalog. It can be expensive to produce one catalog (the text, photographs, and layout), but the marginal costs for each additional one are minimal. Thus, the more catalogs produced, the cheaper one each becomes. One mail order industry report concluded that the fact that a person gets 50 catalogs a month and orders from only two or three does not constitute a glut for the industry. The reason? A catalog company may need only five orders per 100 mailed catalogs to make a profit.

A book on how to set up a mail order business advises: "You can never mail too often to your own customers. They will never tire of hearing from you about new products" (Simon 1981). This is the practice of remailings. Major mail order houses spend a lot of time and effort in determining what the optimum frequency of mailings should be. For consumer merchandise, the frequency for the best customers ranges from monthly to yearly. Seventy-eight percent of mail order firms surveyed in 1988 said that they mailed catalogs one to six times per year to their best customers. New mailings do not necessarily offer new merchandise, however. The old catalog can simply have a new cover and a new date. Remailings prompt customers to reconsider ordering and cause them to throw away the "old" catalog. A good customer may get more frequent mailings: "It pays to mail more catalogs, more often, to better customers" (Simon 1981, p. 318).

Not purchasing from a mail order firm may slow down the frequency of mailings, but it may take a long time (and pounds of waste) before total elimination occurs. One firm—from which we ordered once—has been threatening to stop mailing with notices over a year in four or five mailings. Other firms—from which we have never ordered—have been sending us catalogs regularly for several years.

Packing for Shipment

Buying mail order goods means more waste. Shipping causes more packaging than shopping in person, and customers have no control over the amount of packaging that a mail order firm uses, even though they pay for it. More waste is caused when, to attract bounce-back business, firms send along with an order brochures of related products or bunches of loose package-stuffer offers.

Mail order firms don't just pack for shipment, they overpack. The amount of packaging is not related to ensuring that a customer receives undamaged goods. Part of this problem may be caused by the practice of charging a shipping and handling fee based on the cost of the product. True packaging and shipping costs depend on the size, weight, and fragility of a product and on how far it must be sent. Another reason for this overpackaging is simply that costs for packing materials are low. For instance, the polystyrene (foam) pellets normally used for protection are extremely light, and doubling their volume in a package may not affect actual shipping costs, which are based on weight. Another reason is that it takes time to train workers to package with consideration for excess, and time is money. Examples of excessive packaging of mail order shipments include:

- Boxing items that are already boxed and filling up the space between the two boxes with foam pellets. The alternatives would be (1) to simply put a wrapper around the initial box if it is sturdy enough for shipment or (2) to select a smaller box for the second layer so that the space between the boxes is less and requires less filler.
- Combining in one shipment items of very dissimilar shapes and sizes, so that an abnormally large box must be used. Shipping separately might cost a little more but could result in substantial reductions in packing materials.

One mail order firm is at least sensitive to the environmental costs of using polystyrene pellets as packing materials. Smith & Hawkin, a firm that sells gardening supplies, has discontinued their use because their manufacture with CFCs contributes to depletion of upper atmospheric ozone (see Chapter 5). Smith & Hawkin instead reuses paper that results from their normal business operations. Several major mail order vitamin companies also use old newspaper for packing. Another firm has discovered that popcorn was cheaper than foam pellets.

Packaging appears to be the only environmental consequence of their business that concerns mail order firms. In a 1989 survey, the issues of greatest importance did not include the environment.

What Can Anyone Do?

Unfortunately, the ability to control incoming unsolicited mail is limited. As our discussion on mailing lists shows, even if people have their names removed from an inhouse list, they may be quickly reinstated through a list rented from another mail order firm or a compiled list that has been purchased. The more a person buys, the more unsolicited mail and packing material he or she will receive.

There are a few things people can do that may slow down junk mail. The best advice, of course, is to not get the vicious cycle started in the first place. That is almost impossible today. Almost every time a person gives out his or her name and address, it will go into someone's database and become fodder for a mailing list. Kraft General Foods maintains a list of the people who redeem food coupons (a sweepstakes offering will entice people to put their name and address on the coupon) and of those who call its toll-free customer service telephone number. Grocery stores with check cashing identification cards and checkout scanners can track customers' purchases and build profiles.

For those who are caught in the vicious cycle, stopping shopping is one solution. The more a person buys (responds to mailings), the more likely he or she is to get on an increasing number of lists because the person becomes classified as a "repeat buyer," the most valuable kind to direct mail firms. In a 1988 survey, most mail order firms said that they mailed out three to four catalogs before stopping due to a lack of response. But 10 percent of the firms mailed out more than six catalogs.

If the benefits of catalog buying are too compelling, then buying only from firms that have a "not to be sold or rented" list can help. Getting on this list has to be requested, usually in writing, with a copy of the mailing label. Firms that are members of the Direct Marketing Association (DMA) call it *Mail Preference Service*. The DMA maintains a master Mail Preference Service list for its members and will place anyone's name on it upon written request (6 East 43rd Street, New York, NY 10017). Unfortunately, not all mail order firms belong to the DMA.

People can also try to get their names removed from mailing lists by writing to or calling the firms and asking for removal. This, of course, only removes their name from inhouse lists. Many mailings received are through rented lists over which the mail order firm has no control. When responding to such a mailing from a firm by telephone, one is asked for the "key code" on the catalog. The code tells the firm which mailing is getting the response, and that helps them to assess the value of a rented list. Once a customer orders, the customer's name is placed on that firm's inhouse list, which will probably become part of yet another rented list, and so on.

Magazine Subscriptions

Subscriptions are different from direct mail; people ask for them. They are the product people buy in response to one of the largest classes of consumer direct mail advertising. There were 11,229 periodicals published in the United States in 1988, almost 40 percent of them monthly. It is difficult to calculate the total waste contribution of these magazines because of the varying frequencies of publication, their size, and the number of copies. But just one subset—50 periodicals in 1986, with a total of 171 million subscribers—would make 1 billion pounds (500,000 tons) of waste per year if they are published monthly and weigh, on average, 0.5 pound each.

All magazines eventually become waste, but the rate at which this happens varies greatly from person to person. Some people save magazines for months or years, but that only delays the inevitable. Regardless of how long people keep the magazines, they all create some immediate waste. Most magazines arrive—not as they once did, with the address label glued directly to the cover—but in a plastic bag (*polybag*), along with extra sheets of advertising. Inserted in the magazines are loose return post cards (called *blowins*) begging for renewal (or some other purchase). Publishers follow these up with as many as eight separate notices a year to encourage renewals.

Magazines differ in weight; they range from 0.5 pound to 2 pounds or more. Thus, one monthly 1-pound magazine subscribed to for a year makes 12 pounds of waste. The polybag for the label (at about 0.02 pound each) and five renewal notices (at 0.5 ounce each) bring the year's total to 12.55 pounds. A sample calculation for six subscriptions is provided in Table 7-7. They add up to 116 pounds of waste each year.

Polybags

Although one household doesn't make much waste by subscribing to magazines that come in polybags, on a national basis it adds up. By one estimate, the 120 polybagging machines in the United States consume 12 to 48 million pounds of plastic each year. By that measure, the waste generated ranges from 6,000 to 24,000 tons per year.

Some will argue that polybags reduce waste because they reduce spoilage and replace the heavier Kraft paper wrapper. Those who make the first

TABLE 7-7. CALCULATING MAGAZINE SUBCRIPTION WASTE

Magazine	Weight	×	Number of Issues per Year	=	Total Waste per Year
PC Magazine	1.75		22		38.5
New Yorker	0.75		52		39.0
House and Garden	1.25		12		15.0
Vanity Fair	1.37		12		12.0
Horticulture	0.40		12		4.5
National Preservation	0.40		11		4.1
SUBTOTAL	5.9		121		113.8
Plus 110 polybags[a]			2.2		
TOTAL (pounds per year)			115.9		
Annual subscription cost = $107					

[a]The National Trust for Historic Preservation does not mail its publication in a polybag.

point cannot quantify the savings. On the second, it is true that polybags are lighter and thinner than the older paper wrapper. But it is also true that before the polybag, the paper wrapper was used on fewer—primarily heavy—magazines.

The major benefactors of polybags are publishers and, of course, the industry that makes and fills the bags. Who else needs them? Who really cares if their magazines arrive a little wrinkled or spotted? Publishers, on the other hand, get to send a magazine and some advertising flyers that can't be missed (unlike gazing over the ads printed inside the magazine) for the price of mailing the magazine.

What Can Anyone Do?

Obviously, it's the magazines themselves—rather than the polybags that many complain about—that make the largest contribution to a household's waste. The first step is to assess the household's subscriptions. Figure out how much waste they actually amount to and how much they are costing. Next, everyone in the household should question the reasons for the subscriptions. Some may be simply a habit rather than a desire to read each issue. If so, they are more waste than anything else.

Some subscriptions may be desirable, but lack of time means that they are not read each month. If so, the convenience of getting them regularly by mail is no savings. Many magazine direct mail promotions offer "half price off the newsstand price" for subscriptions. Reading only half means that no money is saved. Canceling subscriptions (or, more conveniently, just letting them lapse) and buying magazines at the local newsstand only when there is time and a real desire to read them or when an issue has an article of special interest will save money, reduce waste, and support the local economy.

Maintaining the magazines that a household now gets offers an opportunity to share them with a neighbor or co-worker. Donating them to a hospital or other organization may lead to fewer magazine subscriptions and less waste.

A more radical option is to cancel subscriptions that come in bags. Write and tell editors that polybags or any other wraps are unnecessary and unacceptable. Tell them that it doesn't matter if their bag is biodegradable. It still makes unnecessary waste, and it probably won't degrade in a landfill anyway. Conversely, people need to tell editors of the magazines without bags that their pollution prevention practice is the reason they have chosen to subscribe.

Actually, public pressure against polybags may be starting to work. At least two magazines eliminated polybags in 1990. *Horticulture* magazine now comes with the label attached to the cover. *The New Yorker,* though, has substituted a brown paper outer cover that is "recyclable," the editors say. It is truly recyclable only in areas that have paper recycling systems that accept that kind of paper. The paper wrapper constitutes as much waste as the bag did and will get recycled only by those readers who bother to rip it off the magazine and put it in a recycling bin.

SUMMARY

Waste is a product of consumption. American styles of living that determine what we buy and how we buy it fix the amount of waste each household generates. In a few cases, such as junk mail, a consumer can exert little control. For others, however, there are a multitude of ways to reduce household garbage today.

Nine simple pollution prevention principles can get anyone started, but at the heart of the matter is a change in habits. That means becoming sensitive to the waste consequences of consumption, to make waste as normal a criterion of every purchasing decision as price and benefits already are. Personal pollution prevention is also a path toward becoming sensitive to other environmental problems and understanding their solutions.

Once consumers start to show manufacturers that pollution prevention matters, this criterion will be integrated into the design of consumer goods and packaging. The variety of things that make less waste will increase. Consumers can also push government to help convince manufacturers that waste matters. Most people think that bans and taxes, especially on packaging, would be appropriate. Both can be quite limited in their effectiveness. If the tax is not high, it will discourage few people from continuing old habits. Most wasteful consumerism has a monetary cost. But often it is too small, in and of itself, to effect change. It is usually politically impossible to make a tax high enough to reduce waste, especially on ordinary consumer products. Taxes do, however, raise money to educate people about pollution prevention. By helping to pay for recycling and other forms of waste management, tax revenues internalize some of the costs of wastes. But that is not prevention. Bans on certain products or materials do not reduce waste if they are not inclusive. When communities ban polystyrene packaging, the replacement is paper packaging or other forms of plastic. To truly reduce waste, bans must include a pollution prevention goal.

Personal combat against mail waste can be frustrating. Even if people never give out their name and address, never subscribe to any magazines or periodicals, never order anything by mail, never enter any contests, never contribute to organizations, or never get a credit card, they will still get unsolicited mail. Someone will put their address on a compiled list and sell or rent it to someone else.

There are some government policies that people can work for to make avoiding direct mail easier and more effective. First, make it unlawful to sell or rent mailing lists. After all, it's an invasion of personal privacy. At the very least, the government could require that permission has to be given before a name and address are rented or sold. As it is now, the opposite is true; the person has to act to try to prevent this from happening. Second, increase second- and third-class mail rates to make it less profitable to send unsolicited mail. It is unknown how high the rates would have to go to get mail order companies and advertisers to pay more attention to their mailing lists and decrease the number of mailings. Mail order firms would add the increased cost to the prices they charge for merchandise or to the fees for shipping and handling. They also might reduce the size of their catalogs or the number of mailings, both of which would reduce mail waste. And third, make all unsolicited mail returnable at the expense of the sender. This could be done by building the return cost into second- and third-class rates, as is already done with first-class rates. Then at least the cost of disposal for unwanted mail would be borne by the generators.

8

HOUSEHOLD TOXIC PRODUCTS: THINKING MORE AND BUYING LESS

Once upon a time, the word *toxic* was a technical term that was hardly used in everyday speech. Now it is applied to everything, even parents. The increased use of the word reflects the increased use of toxic chemicals and increased public fears of toxic substances and wastes. In fact, toxic products are purchased every day by nearly everyone, even at a time when the public's fears of toxics has heightened. Production of toxic products is big business. Consumption continues despite the many warnings found on packages, such as this:

> Warning: Flammable. Do not spray near flame or while smoking. Avoid excessive inhalation. Use only as directed. Intentional misuse by deliberately concentrating and inhaling the contents can be harmful or fatal. Do not apply to broken, irritated or sensitive skin. If rash or irritation develops, discontinue use. Avoid spraying into eyes. Contents under pressure. Do not puncture or incinerate container. Do not expose to heat or store at temperatures above 120°F. Keep out of reach of children.

Even with warnings like these, the problem with toxic products is that you never really know when you've got one in your hand or your household. And when you know that you've got one, you can never really be sure of the consequences. If you think that the U.S. government and industry work diligently to help you reduce the uncertainties, you are wrong.

In this chapter, we discuss the many dangers and uncertainties of toxic household products. Our premise: that prevention is the best way to reduce the uncertainties and increase personal protection. This chapter is more than a collection of facts. Each of us needs to learn how to make decisions every day that will reduce our exposure to toxic products and that of the general environment (and thus to other people). Pollution prevention requires new thinking. Buying less toxic products, however, can eventually become as easy as buying toxic ones is today.

Both government and industry assume that current product labeling solves most of the potential problems of using and disposing of toxic products. But reading labels can be as uninformative as it is informative, and it certainly is tedious. In the above example, from a seemingly innocuous skin care product without a list of ingredients, the reader has to deduce the potential harm of the product from the warning message. It clearly says that the contents are flammable, and will ignite and burn. It doesn't tell why the vapors shouldn't be inhaled (immediate lung damage or long-term cancer risk?) or how much inhalation is excessive. (The "intentional misuse" clause is there because teenagers have discovered that inhalation produces the hallucinogenic effect of a drug.) Some kind of rash can develop on skin, and something will happen if the product is sprayed into the eyes. Maybe these effects are immediate and will disappear soon; maybe the rash is an indication that something is entering the body through skin contact. And, finally, it is inferred that an explosion will occur if the aerosol can is punctured, incinerated, or otherwise heated. Should the consumer presume to know how the can may ultimately be disposed of?

Limited and uncertain as the information from this message is, is it likely to affect the decision to use this product? Do consumers anticipate inhaling the aerosol spray that this product creates (what, after all, is *excessive* inhalation?) and rubbing it into their skin? Do consumers want it sitting around in their bathroom cupboard? Is the benefit worth the risk? And, once the product is bought and used, who will remember to worry about whether the can might end up in an incinerator? Perhaps the most significant message is that manufacturers have not yet seen an adverse effect of current product labeling requirements on consumer demand. Toxic products remain a dominant part of daily consumption.

INDOOR AIR POLLUTION

If only one or two toxics were periodically released in a household, there probably would be nothing to worry about. The problem is that in most houses there are multiple sources of toxics that provide an almost continuous barrage. By one estimate, each person uses about 25 gallons per year of toxic products in the home. The cumulative effect of many sources causes indoor air pollution that may be as dangerous to health as releases from a nearby factory or toxic waste dump. This is especially true since adults spend 75 to 90 percent of their time inside their house or place of work, where the same pollutants exist.

The effects of these pollutants can be acute or chronic, reversible or permanent, and range from dizziness to cancer. As is true of all pollutants, no one really knows whether and which ones are additive, synergistic, or antagonistic in their health effects. Indoor polluted air may contribute to chemical sensitivity, a consequence of exposure to toxics that is receiving growing attention. The National Academy of Sciences in 1987 estimated that 15 percent of Americans may react to certain chemical exposures with a variety of symptoms that range from headache to irregular heartbeat. Even exposure to extremely low concentrations—in parts per billion—may trigger these reactions.

The sources of indoor air pollution are consumer products (including tobacco products), as well as biological contaminants, materials used in the construction of houses and furnishings, and the combustion of fuels. Many of the latter sources are more difficult to avoid than consumer products, since they can be major components of houses or operating defects that may be costly to repair. Building and furnishing materials are sources of asbestos fibers and formaldehyde vapors. Both can be emitted from insulating materials. Formaldehyde is also emitted from products such as pressed woods, tobacco smoke, durable press drapes and other textiles, and glues—to mention only a few. New carpeting has been suspected of producing fumes that cause adverse reactions. Radon gas is released from construction materials and from the soil or rock upon which houses are built. Unvented kerosene and gas space heaters, wood stoves, fireplaces, and gas stoves emit carbon monoxide, nitrogen dioxide, particles, polycyclic aromatic hydrocarbons, and acid aerosols. Biological contaminants include bacteria, molds and mildew, animal dander and cat saliva, house dust, mites, and pollen.

Most household products (such as cleaners, paints, disinfectants, and cosmetics, as well as degreasing, repair, and hobby products) contribute to indoor air pollution through the release of organic gases (VOCs from solvents). Studies by the EPA have shown that the levels of VOCs (such as benzene, perchloroethylene, paradichlorobenzene, and methylene chloride) can be two to five times greater inside a house than outside. (Perchloroethylene is emitted in homes from newly dry-cleaned clothing.) When products containing organic compounds are used, exposures to very high pollutant levels, which can persist in the air long after use, can be serious.

Both the active and inert ingredients in pesticides contribute to indoor air pollution. It is suspected that pollutants are released not only during use indoors but afterward from surfaces that have absorbed them and from products stored inside. Pesticide-contaminated soil and dust may also float or be tracked in from outside. A number of studies have concluded that people are exposed to more and substantially higher levels of pesticides indoors rather than outdoors. Exposures are not necessarily related to individual use. Thus, pesticides used by one household or dispersed across a community can cause other persons to be involuntarily exposed. Limited evidence suggests that, for some pesticides, higher exposures result from home use than from intake as residues on food.

CHOOSING PREVENTION

Indoor air pollution experts suggest three ways to combat the problem: (1) improve ventilation, (2) install air cleaners, and (3) eliminate the sources. Choosing the last means choosing prevention. Eliminating the sources prevents not only indoor air pollution problems but problems caused by other routes of exposure, such as absorption and ingestion, and by the wastes they create. Prevention not only lowers the user's exposure (and that of all the other people in a household) to toxics but also lessens their impact on the environment and, ultimately, on everyone else's health. Ignoring toxic products is simply not healthful.

This book is concerned with wastes, and Chapter 4 presents data on the contribution of toxic household products to waste problems. In this chapter, though, the focus is on the *use* of toxic products. Using products means being exposed to the ingredients. They get on or into our bodies, intentionally or otherwise. Like dangers from passive cigarette smoke, toxic product pollution poses threats to those who did not choose to buy the product. Considering only disposal problems, then, ignores what may be the greater harm of direct exposure. It also allows people to think that if they turn in toxic product residuals on a household hazardous waste collection day, they have safely disposed of them. Not true.

Toxic product disposal does not occur when consumers put residuals in the trash. It starts at the moment of use (and even while a toxic product is being stored in a household). For products that volatilize (like solvents or the propellants in aerosol cans), dispersal into the air occurs as they are used. The cleaner used for the toilet bowl and the detergents in washing machines go down the drain and into the sewer system. They end up in some nearby body of water, either directly or after having gone through a water treatment plant. There is a good chance that a community's water treatment plant will not destroy the toxics. For an estimated 30 percent of households, sewer wastes go into septic systems which rely totally on natural biological processes. But the toxic chemicals may kill the beneficial bacteria that degrade wastes.

Lawn pesticides wash off with rain water into the sewer system or move with the lawn cuttings to the local landfill, incinerator, or compost pile. New York's Nassau County has estimated that 83,000 gallons of organic chemicals from household products end up in their groundwater each year. Thus, when cans, bottles, and boxes of toxic products are used up to the last drop or grain, toxic chemicals have been disposed of everywhere *but* in the local landfill. Taking the containers to a special household hazardous waste collection place—which few people do anyway—may be almost irrelevant at that point.

Another consideration when choosing toxic products is that their manufacture undoubtedly caused the generation of hazardous wastes and air and water pollutants as well as worker exposure to toxic materials. Choosing prevention, then, sometimes means deciding not to support waste-intensive industries. As later discussions in this chapter show, some products, like plastics and dyes, the use of which may mean minimal exposure to toxicity because the toxic substances are tightly bound within the material structure, come from plants that may produce some of the highest relative amounts of industrial pollutants and routinely expose workers to known carcinogens.

When people choose prevention, they have not necessarily taken the simple route. The simplest thing to do is just to accept that government and industry policies regarding toxic substances protect public health and the environment. Choosing prevention acknowledges that government actions apply to limited kinds of products and in inconsistent ways, and that industry does not always have the consumer's best interests as its top priority. Like preventive health care, practicing prevention requires gaining personal knowledge about products. That is not easy. In fact, anyone who does take control will find that, because of what industry and government do or don't do, practicing pollution prevention means overcoming obstacles. It requires more attention and effort than accepting the status quo.

A lot of the information about toxic substances and their presence in products seems meant to obfuscate rather than enlighten. Part of the reason is that there are tens of thousands of chemicals, and their uses—even in similar products—can vary greatly. The chemicals in products also change over time as new products are introduced and the formulations of existing products are altered. The numbers alone make it impossible for anyone—government, industry, or individuals—to understand the true consequences of all of them.

Society has become dependent on toxic chemicals to keep houses sparkling clean, to assure that paper products are lilly white, and to kill pests and weeds in gardens. The ability to make synthetic chemicals, primarily from petroleum and natural gas, and the relatively low cost of doing so, has overcome any limits imposed by the use of natural chemicals. The changes since elbow grease was the principal ingredient in home cleaning are widespread. The result is pollution from chemicals and pollution from increased electricity generation. As the Enterprise for Education (Lord 1986) has said:

> Today, vacuum cleaners have replaced brooms and consumers use a host of specialized, powerful cleaners: scouring powders with bleach, no-wax floor cleaners, detergent for washing walls, oven cleaners, toilet bowl cleaners, furniture polish. Chemicals are doing jobs formerly done by people.

Another part of the problem of understanding is that government only tries to help consumers in selected product areas. When it does, the affected industry fights it every step of the way. There is no consistency in the government on products; no one agency, institution, or law is responsible. Rather, there are several agencies and many laws. Even when government intervenes in a whole class of products, such as to protect the public and the environment from pesticides, the outcome is quite ineffectual in many ways.

REDUCING UNCERTAINTY

The key word for toxic substances and toxic products is *uncertainty*. No one knows exactly:

- What is and what isn't toxic.
- How much exposure or how long a period of exposure to a toxic substance is required before getting sick.
- What it means to be exposed to a combination of toxic substances at one or sequentially.
- Where suspected and known toxic materials are used and in which products they end up.
- The impact on the environment of toxic products and residuals.

There are a few facts amid the uncertainties. Toxicity is a *known* problem for human health and the environment. Consumer products *do* contain toxic substances. And, while consumers have little control over industrial use and misuse of toxics, *consumers have total control over the amount of toxic products they bring into their homes.*

Choosing not to use toxic products means choosing not to accept the unknown acute and chronic risks of using them. Quality of life improves, and anxiety about toxic residuals and how they might affect the air, landfills, incinerators, sewer systems, and water treatment plants in which they end up is reduced. Conversely, when consumers decide to use a toxic product, they accept personal risks and put all others at involuntary risk when the residuals end up in the air everyone breathes and the water everyone drinks.

Some risks of use are known, although the acute effects are more easy to quantify than the chronic ones. An estimated 469,000 children under 5 years of age were treated in hospital emergency rooms in 1985 for accidental ingestions, chemical burns, and other acute injuries associated with household chemical products. Accidental ingestions accounted for 57 percent of the total. But it's not just children who are at risk. There are 2.5 million poison incidents each year from home pesticides. In June 1986 in Los Angeles, 55 people (children *and* adults) were poisoned by ant and roach baits, 46 by ant and roach sprays, 41 by foggers, 62 by rat poisons, 21 by herbicides, 96 by organophosphate insecticides, and 9 by insect repellents. Cosmetics caused a reported 47,000 injuries requiring treatment in hospital emergency rooms during 1987. Between 1985 and 1987, over 151,000 injuries resulted in 2,300 hospitalizations. A range of products—soaps, nail preparations, hair permanents, hair straighteners, shampoos and dyes, creams, lotions, and facial masks—were responsible.

Avoiding toxic products means accepting trade-offs. To reach the same apparent level of cleanliness, a nontoxic cleaning alternative may require more elbow grease. A nontoxic pesticide may have to be applied more often in a garden than a toxic pesticide. Some alternatives require consumers to mix the ingredients. All of these consequences of using nontoxics take more time. Time is also required to learn which products to avoid and what the alternatives are, if any. But time is saved by not having to take residuals of toxic products to a household hazardous waste collection site and by not having to take someone to a hospital.

Another trade-off—between reducing waste quantity and toxicity—may occur but can sometimes be avoided by careful selection. Chapter 7 contains an evaluation of battery usage in which toxicity and waste quantity reduction conflict. A more easily solvable problem is that of dioxin in bleached paper products. Alternative nonbleached paper products (such as coffee filters) are now being produced and marketed. But a person can avoid the dioxin problem *and* reduce the waste quantity by selecting a reusable coffee filter made from cloth or gold-plated mesh. Even at $20, the latter is less expensive than paper in the long run.

There are also some monetary cost savings to be gained. When consumers make their own cleaners out of, say, vinegar and water, they save twice. They don't pay for someone else (the manufacturer) to do it, and the ingredients are relatively cheap. A solution of vinegar and water to wash windows costs 4 cents per ounce. A window cleaner costs between $1.96 and $2.30 per ounce. Waste quantity is cut, too. The home mix can be prepared in a reusable pail, avoiding the throwaway plastic bottle that comes with a window cleaner.

Probably the best news about toxics in products is that not all alternatives are old-fashioned remedies. Some enterprising firms are marketing new, nontoxic (or less toxic) green products. Green products provide the same convenience of use, but often cost more than the toxic products they replace. The higher prices today are due primarily to the lower demand, a situation that hopefully will change. Note, however, that a term now commonly used to describe some of these products—*environmentally friendly*—can have a limited meaning. A product may be developed or changed to be friendly to the environment but remain unfriendly to people during use.

Being environmentally friendly may simply mean that the chemical compounds in the products degrade once released into the sewer system, but not necessarily before some harmful exposure has occurred.

New, nontoxic products are emerging because, as people learn more about the potential risks of using toxic products, they are beginning to want the benefits of prevention. A few manufacturers are listening. In a survey conducted in Massachusetts in 1986, people were asked about three household hazardous waste "reduction" options: (1) passing unused toxic products on to friends to use up, (2) being able to buy toxic products in the amount needed, or (3) having manufacturers make nontoxic products. The third option—the only true prevention one—came out on top twice. Asked whether or not they were in favor of the options, 95 percent said "yes" to new nontoxic products (versus 70 and 78 percent for options 1 and 2). The respondents also overwhelmingly preferred option 3 (77 percent) over option 1 (4 percent) and option 2 (18 percent).

WHAT IS TOXIC?

The problem with asking whether or not something is toxic is that the answer you get is: it depends. It depends on whether or not you are asking from a scientific, regulatory, or industry perspective. Some substances are considered toxic by all groups. Some may be considered toxic by many scientists, but not by all the government regulators or by the government but not industry. Usually, the last to concede is industry. To make matters more uncertain for consumers, members within each group can also disagree as to whether a substance is toxic or not.

Scientists, regulators, and industry often also disagree on how to respond to uncertainty about toxicity because each can have a different concept of acceptable risk. There are those who belittle people's fear of toxicity by saying that *any* substance can be toxic. True. Get too much pure spring water in your lungs and you will drown—an acute, fatal reaction. But that begs the issue and lacks any criteria for decision making. Technocrats tend to think of toxicity as part of an equation for risk wherein toxicity times exposure equals risk. Risk, then, can be judged minimal when exposure is given a low value even when toxicity itself is high.

Our viewpoint is risk averse and moves decision making about toxics to individuals. Barbara S. Shane, a toxicologist, in a review (1989) of the relationship between chemicals and birth defects, concluded that uncertainty is high because little is known. Still, she cautioned, "progress that has been made so far has indicated the importance of limiting exposure to chemicals as much as is practical and feasible." Prevention limits exposure and reduces toxic uncertainty.

The Uncertain Science

From a scientific viewpoint, toxicity is a range. It depends on how much of a substance (the dose) someone ingests, inhales, or absorbs through the skin *and* how often that exposure occurs. It depends on who is exposed (adult or child, male or female) and on conditions such as body temperature. Conclusions about toxicity also depend on how thoroughly a particular chemical has been studied.

The effects from exposure to toxic substances vary. An *acute effect* is an immediate reaction to an exposure. This kind of toxic reaction is commonly referred to as *poisoning* and can range from dizziness to death. For example, swallowing lye (the ingredient in household drain cleaners) corrodes tissue. Severe damage can occur; the exposure can be fatal. A *chronic effect* is a delayed reaction and can result from either a single dose or many doses over a long period of time. Cancers, birth defects, and kidney and liver damage are chronic effects. Lung cancer is a delayed reaction to smoking. Cadmium can build up in body tissue and cause kidney dysfunction.

Depending on the chemical, an acute or chronic effect can be the result of a very small or very large dose. Small amounts of the metal selenium are essential for good health, but large doses are toxic. Conversely, even a tiny dose of the pesticide parathion can be fatal. An effect can be local (affecting only the area of contact, such as a skin irritation) or systemic (in which the damage may be remote from the point of contact, such as cancer). Chemicals can cause both acute and chronic reactions. Certain solvents acutely affect the nervous system, while the chronic effect is liver damage.

Chronic toxicity is more difficult to discern than acute toxicity because of the dilemma of figuring out which chemical exposure out of many over a long period of time might be the cause. Human carcinogens are usually inferred if a series of tests in animals produce cancer. While it is not known with certainty that all animal carcinogens are also human carcinogens, most well-studied human carcinogens show some evidence of carcinogenicity in animals. More conclusive evidence of human carcinogenicity comes from epidemiologic studies.

The simplest way to express toxicity is to characterize chemicals in terms of their relative degree of acute toxicity, based on how high a dose is needed to cause the death of 50 percent of those exposed (a standard measurement in toxicity research). The higher the dose required, the lower the level of toxicity. Thus, a substance that requires a dose smaller than a teaspoonful is highly toxic, a substance that requires a dose of a pint is moderately toxic, and one that requires a dose larger than a quart has extremely low toxicity. Drawing conclusions about toxicity gets more complicated when effects other than death and when various exposure routes and chronic exposures are included. Both potassium hydroxide and PCBs are highly toxic, according to the fatal dose scale, but the reactions are different. Potassium hydroxide (lye) is a corrosive; it is a skin irritant and is highly toxic (acute) when ingested. PCBs are a suspected human carcinogen (chronic toxicity) and are moderate to highly toxic acutely when they are ingested, inhaled, or come in contact with skin.

Government Decisions Depend on Science and Politics

There are at least five government agencies and more than 12 laws under which those agencies decide whether or not a waste product is toxic enough to be regulated. To do so, they have to reconcile the often uncertain scientific toxicity findings with a legal definition of toxicity and legal criteria required for decision making. Both definitions and criteria (such as whether or not a cost/benefit analysis is called for) can vary among the laws. One major result of these variations is that the same substance can be deemed toxic enough to regulate under one law and not so under another.

There are five laws that focus on where toxics end up (in the air, in surface or ground water, or on the land). That focus and other differences among the laws means that each law covers a separate set of toxics. Even the nomenclature differs. There are 126 *priority* pollutants included under the regulations for water pollutants. Over 30 *hazardous* air pollutants have been listed as possible air toxics, but only 6 of them are actually regulated. (Changes in the Clean Air Act of 1990 may result in almost 200 air toxics being subject to future regulation.) Almost 400 substances, which have traditionally been placed in landfills and surface ponds, are regulated as *hazardous wastes*, and just over 700 substances are called *hazardous substances* under the Superfund law that covers the cleanup of toxic waste dumps. The newest set of government toxics are the 300 or more TRI chemicals.

The regulatory decision-making process differs among the various laws. This means that even when the same substance qualifies under several laws, there may be a period of time during which it is not regulated under all of them. Benzene is highly toxic, has been recognized as a potential human carcinogen since the early 1970s, and is commonly found in both indoor and outdoor air. It is released from automobile exhaust, automobile refueling operations, consumer products, cigarette smoke, and industrial operations. It was listed as an air toxic in 1977, but the development of proposed air regulations for industrial emissions did not occur until late 1989. Meanwhile, benzene has been regulated as a hazardous waste and a hazardous substance under the Superfund law since the early 1980s.

There are government standards for allowed levels of benzene in drinking water as well. But, as with all standards, that does not guarantee safety. In February 1990, Perrier recalled millions of bottles of its sparkling water from global markets (72 million from the United States alone) because benzene was found in some bottles at a concentration of 12.3 to 19.9 parts per billion—two and a half to four times the standard. Government officials quickly explained that the risk was "negligible," although drinking only a pint a day raises the risk of cancer to 1 in a million. That risk level of excess cancers triggers a Superfund toxic waste cleanup. Additionally, the usual government analysis assumed that a person drinking Perrier was not exposed to benzene from any other source. Simply pumping gasoline into one's automobile could easily elevate one's level of exposure to benzene. There are many toxic lessons in this Perrier story—truthfulness in advertising, system failures, and the chance of discovery. For years consumers have been sold Perrier as a pure source of water. The truth now appears to be that the source of gases for its carbonation is contaminated with benzene. In the normal course of collecting and bottling the water, Perrier filters out the benzene. Workers' failure to clean the filters on schedule caused benzene to be introduced into the water subsequently bottled. Lastly, the publicity revealed that bottled water is tested only on a random basis.

Who releases a chemical can be the factor determining whether or not it is a regulatory toxic. If a factory dumps toluene, it has to follow a rigorous regulatory procedure to ensure that the toluene ends up in a specially designed landfill. Consumers may be using toluene at home when they use a paint thinner. They can throw it into the trash and send it off to the local landfill, which may offer no protection against the leaching of that chemical into the groundwater. The factory was handling a *hazardous waste*. The consumer, on the other hand, legally disposed of a *nonhazardous waste*.

These inconsistencies regarding what is and what isn't considered a toxic waste were summed up in a 1989 American Chemical Society guide to chemical risk:

> Although many U.S. laws affect the production and use of chemicals, they do not cover every chemical that may be harmful, nor do they necessarily work as well as they were intended. A basic problem is that these regulations are devised and applied in a world that is part science and technology, part economics, and part politics.

When they do regulate, government agencies set many limits based on acute toxicities rather than chronic ones such as cancer. That was a finding of an OTA report in 1987 that reviewed the work of agencies responsible for regulating carcinogens. Of 144 chemicals that tested positive in animal studies by the National Toxicity Program, the highest number that any one agency regulated as carcinogens was 30.

The slowness and complexity of regulatory decision making mean that the discovery of chemical toxicity often awaits public evidence of major public health or environmental damage. Sometimes—as with the pesticide DDT, CFC propellants in aerosol cans, and asbestos—the government prevents pollution by banning a product. In other cases, as with formaldehyde, there is not strong enough information for the regulators to overcome industry pressure.

Asbestos, like many other known toxics today, was once considered a perfect material; it is nonflammable, light, and easy to cut into any shape. It was used extensively in ship building, in laboratories, and to insulate homes and public and commercial buildings. Now the consequences of inhaling the fibers are known (although still being debated by scientists). Millions of tax dollars are being spent yearly to remove it from school buildings. Prospective buyers of commercial buildings can get steep discounts if asbestos is present. Asbestos has been banned from some consumer products and from sprayed-on fireproofing and insulation. Most other uses will be banned in a gradual phaseout process to begin in August 1990.

Over 6 billion pounds (3.1 million tons) of formaldehyde are produced in the United States each year. Formaldehyde is used in hundreds of products, including pesticides, plastics, wood and paper products, cosmetics, detergents, air fresheners, and food. It is the substance that makes "permanent press" fabrics. Formaldehyde is slowly released from some products, causing irritation of the eyes, skin, nose, and throat. Inhalation can affect the nervous system. The Consumer Product Safety Commission (CPSC) banned the use of urea-formaldehyde foam insulation in 1982, but the decision was overturned by court action brought by the industry in 1983. The CPSC continues to study formaldehyde as one source of indoor pollution.

The list of once commonly used substances that are now considered too toxic is long. Often today, it is the actions of consumers or public interest groups, rather than of government regulators, that get them removed from the market. Among the pollution prevention actions have been the following:

- Almost 50 extremely toxic, environmentally persistent pesticides have been restricted by the government from use in consumer products. Many of them are still applied in and around households by commercial applicators.
- Alar, a chemical ripening agent for apples, has been suspected of being a carcinogen since 1985 by the EPA. The manufacturer, under pressure from the Natural Resources Defense Council and Consumers Union, has voluntarily removed it from market until final tests are completed in 1990. Then the government may make the voluntary ban a permanent one.
- The manufacture, processing, and distribution in commerce of PCBs were banned by law in 1976. This was long after they had been widely

used in electrical capacitors and transformers, as well as in paints, adhesives, caulking compounds, and copy paper, and for dust control during road construction.
 • Lead in household paints was reduced to a maximum of 0.06 percent, and unleaded gasoline was introduced when it became obvious that children are seriously impaired from minor exposures.

All of these withdrawals from the marketplace have been reactive. Obviously the lack of official governmental designation of toxicity does not mean that a chemical is not toxic. There are tens of thousands of known chemicals, and thousands more are being produced each year and are in commercial use. Yet, the federal regulatory lists range from a low of 6 to a high of 700 chemicals.

As a practical—but admittedly limited—guide to toxic chemicals, we have selected the TRI chemicals. We use the symbol (TRI) throughout this chapter to indicate when a product ingredient is on the TRI list. This is the only list that covers various environmental media chemicals, and it is one that the public will hear more about as annual emissions data submitted by companies to the government is made public. Some cautions, however: The chemicals on the list range broadly in relative toxicity. They were chosen through a political rather than a scientific process. They have been identified primarily because of their presence in the workplace and as emissions from plants. Thus, there are some chemicals on the list that don't appear in products, especially consumer products. On the other hand, chemicals that are part of a manufacturing process may end up as contaminants in consumer products. Because their presence is not intentional, they will not be listed as ingredients.

This list is not static. A number of chemicals on the original list have been removed because they were not justified under the criteria of the law that created the list. The EPA added nine chemicals to the list in 1989. Some public interest groups have petitioned for the addition of many more chemicals, ones already present on other lists such as those discussed above. Table 8-1 lists the TRI chemicals by name and number and indicates whether or not they appear in the most recent National Toxicity Program's *Annual Report on Carcinogens*. This report contains all known and reasonably suspected carcinogens to which people are exposed. It is viewed by some professionals as overly pessimistic because it has relied on animal test results. Also included in Table 8-1 are the results of the World Health Organization's International Agency for Research on Cancer (IARC). IARC's evaluation includes epidemiological study results.

Focusing only on carcinogens, however, can skew concerns about toxicity. It leaves out the other chronic effects of chemicals such as neurological effects, reproductive disorders, and birth defects. At least 50 compounds

TABLE 8-1. TRI CHEMICALS LIST

Chemical Name as Listed	CAS#	NTP	IARC	VOCs	Some Synonyms (and Notes)
Acetaldehyde	75–7–0		2B	V	Ethanal, acetic aldehyde
Acetamide	60–35–5		2B		
ACETONE	67–64–1			V	Dimethyl Ketone, dimethylformalde-hyde
ACETONITRILE	75–05–8		2A	V	Ethyl nitrile
2-Acetylaminofluorene	53–96–3	R			
Acrolein	107–02–8		3	V	
Acrylamide	79–06–1		2B		
Acrylic acid	79–10–7		3		Vinylformic acid
Acrylonitrile	107–13–1	R	3	V	Acrylon, carbacryl, cyanoethylene
ALDRIN	309–00–2		3		
Allyl alcohol	107–18–6				
Allyl chloride	107–05–1		3	V	1-Chloro-2-propene
Aluminum (fume/dust)	7429–90–5				
Aluminum oxide	1344–28–1				Alumina
2-Aminoanthraquinone	117–79–3	R	3		
4-Aminoazobenzene	60–09–3		2B		Aniline Yellow, Solvent Yellow 1
4-Aminobiphenyl	92–67–1	K	1		Xenylamine
1-Amino-2-methylanthraquinone	82–28–0	R	3		Disperse Orange
AMMONIA	7664–41–7				Spirit of Hartshorn
Ammonium nitrate (solution)	6484–52–2				
Ammonium sulfate (solution)	7783–20–2				
ANILINE	62–53–3		3	V	Aminobenzene
o-Anisidine	90–04–0		2B		2-Anisidine

Chemical	CAS				Synonyms/Notes
p-Anisidine	104–94–9		3		4-Anisidine
o-Anisidine hydrochloride	134–29–2	R			Fast Red BB Base (CI 37115)
Anthracene	120–12–7		3		Paranapthalene, Green Oil
ANTIMONY (and compounds)	7440–36–0				
ARSENIC (and compounds)	7440–38–2	K	1		
ASBESTOS (friable)	1332–21–4	K	1		
Barium (and compounds)	7440–39–3				
Benzal chloride	98–87–3		2B		
Benzamide	55–21–0				
BENZENE	71–43–2	K	1	V	Benzol, coal naphtha
Benzidine	92–87–5	K	1		Fast Corinth Base B
Benzoic trichloride	98–07–7	R	2B		Benzotrichloride
Benzoyl chloride	98–88–4		3	V	
Benzoyl peroxide	94–36–0		3		
Benzyl chloride	100–44–7		2B		
Beryllium (and compounds)	7440–41–7	R	2A		
Biphenyl	92–52–4				Bibenzene, diphenyl
Bis(2-chloroethyl) ether	111–44–4		3	V	Diethylene glycol dichloride
Bis(chloromethyl) ether	542–88–1	K	1		
Bis(2-chloro-1-methylethyl) ether	108–60–1		3		
Bis(2-ethylhexyl) adipate	103–23–1				DCIP (nematocide)
					Dioctyl adipate (DOA)
Bromoform	75–25–2			V	Tribromomethane
BROMOMETHANE	74–83–9		3	V	Methyl bromide, monobromomethane
1,3-BUTADIENE	106–99–0	R	2B	V	
Butyl acrylate	141–32–2		3		
N-BUTYL ALCOHOL	71–36–3				Butanol, methylolpropane
sec-Butyl alcohol	78–92–2				
tert-Butyl alcohol	75–65–0				t-Butanol

(Continued)

TABLE 8-1. (CONTINUED)

Chemical Name as Listed	CAS#	NTP	IARC	VOCs	Some Synonyms (and Notes)
BUTYL BENZYL PHTHALATE	85-68-7		3		
1,2-BUTYLENE OXIDE	106-88-7				
Butyraldehyde	123-72-8			V	
CI Acid Green 3	4680-78-8		3		Guinea Green B
CI BASIC GREEN 4	569-64-2				
CI Basic Red 1	989-38-8		3		Rhodamine 6G, monohydrochloride
CI Direct Black 38	1937-37-7	R			
CI Direct Blue 6	2602-46-2	R			
CI Direct Brown 95	16071-86-6				
CI Disperse Yellow 3	2832-40-8		3		
CI Food Red 5	3761-53-3		2B		
CI FOOD RED 15	81-88-9		3		Rhodamine B
CI Solvent Orange 7	3118-97-6				Sudan II
CI Solvent Yellow 3	97-56-3	R	2B		o-Aminoazotoluene, CI 11160
CI Solvent Yellow 14	842-07-9				Sudan I
CI Solvent Yellow 34	492-80-8		2B*		Auramine* (tech grade)
CI Vat Yellow 4	128-66-5		2B		
CADMIUM (and compounds)	7440-43-9	R	2A		
Calcium cyanamide	156-62-7				
CAPTAN	133-06-2		3		
CARBARYL	63-25-2		3		Sevin, Monsur, carbomate, Atoxan
CARBON DISULFIDE	75-15-0			V	Carbon bisulfide
CARBON TETRACHLORIDE	56-23-5	R	2B	V	Perchloromethane, Carbona
Carbonyl sulfide	463-58-1			V	
Catechol	120-80-9				CI 76500
Chloramben	133-90-4				

Chlordane	57-74-9		3		
Chlorine	7782-50-5				
Chlorine dioxide	10049-04-4				
Chloroacetic acid	79-11-8				
2-Chloroacetophenone	532-27-4				Mace
CHLOROBENZENE	108-90-7			V	Monochlorobenzene
Chlorobenzilate	510-15-6		3		
Chloroethane	75-00-3			V	Ethyl Chloride
CHLOROFORM	67-66-3	R	2B	V	Trichloroform, trichloromethane
CHLOROMETHANE	74-87-3		3	V	Methyl chloride
Chloromethyl methyl ether	107-30-2				
Chlorophenols			2B		
Chloroprene	126-99-8		3	V	
Chlorothalonil	1897-45-6				
CHROMIUM (and compounds)	7440-47-3	K	3*		*(Metal and trivalent compounds)
Cobalt (and compounds)	7440-48-4				
Cooper (and compounds)	7440-50-8				
p-Cresidine	120-71-8	R	2B		
CRESOL (mixed isomers)	1319-77-3				
M-CRESOL	108-39-4			V	3-Cresol, m-toluol
O-CRESOL	95-48-7			V	o-Toluol
P-CRESOL	106-44-5			V	p-Toluol, 4-cresol
CREOSOTE	8001-58-9		2A		
Cumene	98-82-8			V	cumol
Cumene hydroperoxide	80-15-9				
Cupferron	135-20-6	R			
Cyanide compounds*					*(Not all compounds)
Cyclohexane	110-82-7			V	

(Continued)

271

TABLE 8-1. (CONTINUED)

Chemical Name as Listed	CAS#	NTP	IARC	VOCs	Some Synonyms (and Notes)
2,4-D	94-75-7		2B		Dichlorophenoxyacetic acid, Hedonal
Decabromodiphenyl oxide	1163-19-5				
Diallate	2303-16-4		3		
2,4-Diaminoanisole	615-05-4		2B		
2,4-Diaminoanisole sulfate	39156-41-7				
4,4-Diaminodiphenyl ether	101-80-4	R	2B		4,4′-Oxydianiline
Diaminotoluene (mixed isomers)	25376-45-8				
2,4-Diaminotoluene	95-80-7	R	2B		CI Oxidation Base
Diazomethane	334-88-3		3		
Dibenzofuran	132-64-9				
1,2-Dibromo-3-chloropropane	96-12-8	R	2B	V	DBCP, Fumazone
1,2-DIBROMOETHANE	106-93-4	R		V	Ethylene dibromide (EDB)
DIBUTYL PHTHALATE	84-74-2				
DICHLOROBENZENE (mixed isomers)	25321-22-6				
1,2-DICHLOROBENZENE	95-50-1		3	V	o-Dichlorobenzene, Cloroben
1,3-Dichlorobenzene	541-73-1			V	
1,4-DICHLOROBENZENE	106-46-7	R	2B	V	p-Dichlorobenzene, paradichlorobenz
3,3′-Dichloroenzidine	91-94-1	R	2B		CI 23060
Dichlorobromomethane	75-27-4			V	
1,2-DICHLOROETHANE	107-06-2	R	2B	V	Ethylene dichloride (EDC)
1,2-Dichloroethylene	540-59-0			V	
DICHLOROMETHANE	75-09-2	R	2B	V	Methylene chloride
2,4-Dichlorophenol	120-83-2				DCP
1,2-DICHLOROPROPANE	78-87-5		3	V	Propylene dichloride
2,3-Dichloropropene	78-88-6				

Chemical	CAS No.	R	IARC	V	Synonyms / Notes
1,3-Dichloropropylene	542-75-6	R	3		
DICHLORVOS	62-73-7		3		DDVP, Vapona, vinylofos, dichlofos
Dicofol	115-32-2				
Diepoxybutane	1464-53-5	R	2B		
DIETHANOLAMINE	111-42-2				
Di-(2-ethylhexyl)phthalate	117-81-7	R	2B		DEHP
DIETHYL PHTHALATE	84-66-2				
Diethyl sulfate	64-67-5	R	2A	V	Ethyl sulfate
3,3'-Dimethoxybenzidine	119-90-4	R	2B		CI Disperse Black 6
Diethylamine*	109-89-7				*(proposed for TRI list)
4-Dimethylaminoazobenzene	60-11-7	R	2B		Brilliant Oil Yellow, DAB
3,3'-Dimethylbenzidine	119-93-7	R	2B		o-Tolidine, CI 37230
Dimethylcarbamyl chloride	79-44-7	R	2A		
1,1-Dimethyl hydrazine	57-14-7	R	2B		Dimazin
2,4-Dimethylphenol	105-67-9			V	m-Xylenol
DIMETHYL PHTHALATE	131-11-3				Solvanol
Dimethyl sulfate	77-78-1	R	2A	V	
m-Dinitrobenzene	99-65-0				
o-Dinitrobenzene	528-29-0				
p-Dinitrobenzene	100-65-0				
4,6-Dinitro-o-cresol	534-52-1				
2,4-Dinitrophenol	51-28-5				
Dinitrotoluene (mixed isomers)	25321-14-6				
2,4-Dinitrotoluene	121-14-2				
2,6-Dinitrotoluene	606-20-2				
N-DIOCTYL PHTHALATE	117-84-0				
1,4-DIOXANE	123-91-1	R	2B	V	
1,2-Diphenylhydrazine	122-66-7	R		V	Hydrazobenene
EPICHLOROHYDRIN	106-89-8	R	2A	V	Glycidyl chloride

(Continued)

TABLE 8-1. (CONTINUED)

Chemical Name as Listed	CAS#	NTP	IARC	VOCs	Some Synonyms (and Notes)
2-ETHOXYETHANOL	110–80–5				Cellosolve, ethylene glycol ethyl e
ETHYL ACRYLATE	140–88–5	R	2B		
ETHYLBENZENE	100–41–4			V	
Ethyl chloroformate	541–41–3				
Ethylene	74–85–1		3	V	Acetene
ETHYLENE GLYCOL	107–21–1				
Ethyleneimine	151–56–4				Aziridine
Ethylene oxide	75–21–8	R	2A	V	Epoxyethane
Ethylene thiourea	96–45–7	R	2B		ETU
Fluometuron	2164–17–2		3		
FORMALDEHYDE	50–00–0	R	2A	V	
FREON 113	76–13–1			V	Fluorocarbon 113, trichlorotrifluor
GLYCOL ETHERS (mono- and di-ethers of ethylene glycol, diethylene glycol, triethylene glycol)					
Heptachlor	76–44–8		3		
Hexachlorobenzene	118–74–1	R	2B		Perchlorobenzene
Hexachloro-1,3-butadiene	87–68–3			V	Perchlorobutadiene
Hexachlorocyclopentadiene	77–47–4			V	
Hexachloroethane	67–72–1		3	V	Perchloroethane
Hexachloronaphthalene	1335–87–1	R			
Hexamethylphosphoramide	680–31–9	R			
HYDRAZINE	302–01–2	R	2B		
Hydrazine sulfate	10034–93–2	R	2B		
HYDROCHLORIC ACID	7647–01–0				
Hydrogen cyanide	74–90–8				
Hydrogen fluoride	7664–39–3				

Name	CAS No.			Synonyms/Notes
Hydroquinone	123-31-9		3	
Isobutyraldehyde	78-84-2			
ISOPROPYL ALCOHOL*	67-63-0		1/3*	Isopropanol *(manufacturing = 1; s)
4,4'-Isopropylidenediphenol	80-05-7			
Isosafrole	120-58-1		3	
LEAD (and compounds)	7439-92-1	R*	2B	*(Lead acetate and lead phosphate)
Lindane	58-89-9	R	2B	
Maleic anhydride	108-31-6	V	3	Maleic acid anhydride
MANEB	12427-38-2			
MANGANESE (and compounds)	7439-96-5			
MERCURY (and compounds)	7439-97-6			
METHANOL	67-56-1		3	Methyl alcohol, wood alcohol
METHOXYCHLOR	72-43-5			DMDT, Metox
2-METHOXYETHANOL	109-86-4		3	Ethylene glycol methyl ether
Methyl acrylate	96-33-3			
Methyl tert-butyl ether	1634-04-4			
4,4-Methylenebis(2-chloro aniline)	101-14-4	R	2A	MBOCA
4,4'-Methylenebis (N, N-dimethyl) benzenamine	101-61-1	R	3	Michler's Base
Methylenebis (phenylisocyanate)	101-68-8		3	MBI, 4,4'-Methylenediphenyl disocy
Methylene bromide	74-95-3	V		Methylene dibromide
4,4-Methylenedianiline	101-77-9	R	2B	
METHYL ETHYL KETONE	78-93-3	V		MEK
Methyl hydrazine	60-34-4			
Methyl iodide	74-88-4	V	3	
METHYL ISOBUTYL KETONE	108-10-1			MIK
Methyl isocyanate	624-83-9			
METHYL METHACRYLATE	80-62-6		3	

(Continued)

TABLE 8-1. (CONTINUED)

Chemical Name as Listed	CAS#	NTP	IARC	VOCs	Some Synonyms (and Notes)
Michler's Ketone	90-94-8				
Molybdenum trioxide	1313-27-5				
Mustard gas	505-60-2	K	1		Moth flakes
NAPHTHALENE	91-20-3			V	
Alpha-Naphthylamine	134-32-7	K	3		CI 37265
beta-Naphthylamine	91-59-8	R	1		Fast Scarlet Base B (CI 37270)
Nickel (and compounds)	7440-02-0		1		
NITRIC ACID	7697-37-2				
Nitrilotriacetic acid	139-13-9	R			
5-Nitro-o-anisidine	99-59-2	R	3		Azoamine Scarlet K
NITROBENZENE	98-95-3			V	Oil of Mirbane
4-Nitrobiphenyl	92-93-3		3		
Nitrofen	1836-75-5	R	2B*		*(Tech grade)
Nitrogen mustard	51-75-2	R	2A		
Nitroglycerin	55-63-0				
2-Nitrophenol	88-75-5				o-Nitrophenol
4-NITROPHENOL	100-02-7			V	p-Nitrophenol
2-Nitropropane	79-46-9	R	2B		Isonitropropane
p-Nitrosodiphenylamine	156-10-5	R	3		
N,N-Dimethylaniline	121-69-7				
N-Nitrosodi-n-butylamine	924-16-3	R	2B		
N-Nitrosodiethylamine	55-18-5	R	2A	V	
N-Nitrosodiemthylamine	62-75-9	R	2A		
N-Nitrosodiphenylamine	86-30-6	R		V	
N-Nitrosodi-n-propylamine	621-64-7	R	2B		

Name	CAS number	R	Class	V	Synonyms
N-Nitrosomethylvinylamine	4549-40-0	R	2B		
N-Nitrosomorpholine	59-89-2	R	2B	V	
N-Nitro-N-ethylurea	759-73-9	R	2A		
N-Nitro-N-methylurea	684-93-5	R	2A		
N-Nitrosonornicotine	1654-55-8	R	2B		
N-Nitrosopiperidine	100-75-4	R	2B		
Octachloronaphthalene	2234-13-1				
Osmium tetroxide	20816-12-0				
PARATHION	56-38-2		3		
PENTACHLOROPHENOL	87-86-5				PCP, Dowicide 7, Woodtreat A
Peracetic acid	79-21-0				
PHENOL	108-95-2			V	Carbolic acid
p-Phenylenediamine	106-50-3		3		CI 76060
2-Phenylphenol	90-43-7		3		o-Xenol, Dowicide 1
Phosgene	75-44-5			V	Carbon dichloride oxide
PHOSPHORIC ACID	7664-38-2				
Phosphorus (yellow or white)	7723-14-0				
Phthalic anhydride	85-44-9				
Picric acid	88-89-1				Carbazotic acid, CI 10305
Polybrominated Biphenyls (PBBs)		R	2B		
Polychlorinated biphenyls	1336-71-4	R	2A		PCBs
Propane sultone	1120-71-4	R	2B		
beta-Propiolactone	57-57-8	R	2B		
Propionadehyde	123-38-6			V	
PROPOXUR	114-26-1				Baygon
Propylene	115-07-1		3		Propene
Propyleneimine	75-55-8	R	2B		
PROPYLENE OXIDE	75-56-9	R	2A	V	

(Continued)

277

TABLE 8-1. (CONTINUED)

Chemical Name as Listed	CAS#	NTP	IARC	VOCs	Some Synonyms (and Notes)
Pyridine	110–86–1			V	
Quinoline	91–22–5			V	
Quinone	106–51–4		3		
Quintozene	82–68–8		3		Pentachloronitrobenzene, PCNB
Saccharin*	81–07–2	R	2B		*(manufacturing)
Safrole	94–59–7	R	2B		
SELENIUM (and compounds)	7782–49–2	R*	3		*(selenium sulfide)
Silver (and compounds)	7440–22–4				
STYRENE (monomer)	100–42–5		2B	V	Cinnamene
Styrene oxide	96–09–3		2A		
SULFURIC ACID	7664–93–9				
Terephthalic acid	100–21–0				
1,2,2,2-TETRACHLORO-ETHANE	79–34–5		3	V	Tetrachloroethane
TETRACHLOROETHYLENE	127–18–4	R	2B	V	Perchloroethylene, Freon 1110
Tetrachlovinphos	961–11–5		3		
Thallium (and compounds)	7440–28–0				
Thioacetamide	62–55–5	R	2B		
4,4′-Thiodianiline	139–65–1		2B		
Thiourea	62–56–6	R	2B		
Titanium tetrachloride	7550–45–0				

Name	CAS	K	Class	V	Synonyms
Thorium dioxide	1314-20-1	K			
TOLUENE	108-88-3			V	Toluol
TOLUENE DIISOCYANATE (mixed isomers)	26471-62-5		2B		
Toluene-2,4-diisocyanate	584-84-9	R			m-Tolylene diisocyanate
Toluene-2,6-diisocyanate	91-08-7				
o-Toluidine	95-53-4	R	2B		
o-Toluidine hydrochloride	636-21-5	R			
Toxaphene	8001-35-2	R	2B		
Triaziquone	68-76-8		3		
Trichlorfon	52-68-6		3		Danex, chlorophos, chlorofos
1,2,4-Trichlorobenzene	120-82-1			V	
1,1,1-TRICHLOROETHANE	71-55-6		3	V	Methyl chloroform, trichloroethane
1,1,2-Trichloroethane	79-00-5		3	V	Vinyltrichloride
TRICHLOROETHYLENE	79-01-6		3	V	
2,4,5-Trichlorophenol	95-95-4				TCP, Dowicide 2
2,4,6-Trichlorophenol	88-06-2	R			Phenachlor, Dowicide 2S
Trifluralin	1582-09-8				
1,2,4-Trimethylbenzene	95-63-6			V	
Tris(2,3-dibromopropyl) phosphate	126-72-7	R	2A		
URETHANE	51-79-6	R	2B		Ethyl carbamate
Vanadium (fume/dust)	7440-62-2				
Vinyl acetate	108-05-4		3		

(Continued)

279

TABLE 8-1. (CONTINUED)

Chemical Name as Listed	CAS #	NTP	IARC	VOCs	Some Synonyms (and Notes)
Vinyl bromide	593-60-2		2A		
Vinyl chloride	75-01-4	R	1	V	Chloroethene, chloroethylene
Vinylidene chloride	75-35-4		3	V	1,1-dichloroethene
XYLENE (mixed isomers)	1330-20-7				
m-Xylene	108-38-3			V	m-Xylol, 1,3-xylene
o-Xylene	95-47-6			V	o-Xylol, 1,2-xylene, ortho-xylol
p-Xylene	106-42-3			V	p-Xylol, 1,4-xylene
2,6-Xylidine	87-62-7				o-Xylidine, 2,6-xylylamine
Zinc (fume/dust) (and compounds)	7440-66-6				
Zineb	12122-67-7		3		

Notes:

NTP—National Toxicology Program, *Fifth Annual Report on Carcinogens, Summary* (1989).

K = known carcinogen.

R = may reasonably be anticipated to be carcinogens.

IARC—International Agency for Research on Cancer.

   1 = carcinogenic.

2A = probably carcinogenic.

2B = possibly carcinogenic.

   3 = not classifiable as to carcinogenicity.

VOCs—EPA volatile organic compounds database.

The chemicals whose names are capitalized are believed to occur in consumer products.

have been shown to produce the latter two effects in humans. They include pesticides, metals, and organic solvents. Also ignored when focusing on carcinogens is the growth in and a new awareness of human sensitivity to certain chemicals that produce reactions ranging from headaches to severe depression.

DOES GOVERNMENT HELP WITH CONSUMER PRODUCTS?

As with toxic pollutants, government has been erratic and inconsistent in identifying and regulating toxic products. Part of the problem is inadequate resources. While the federal government probably spends less than $200 million per year investigating consumer products, industry is spending over $4 billion to advertise those same products and another $1 billion or more in research to develop new ones. But probably the most disturbing trait is that government regulators are widely seen to be more interested in protecting the industries they oversee than the public they serve.

There are other problems. First, every product or substance is evaluated exclusive of all others, even of other products that have the same toxic substance. This approach might be appropriate if there were only a few toxic products to contend with, but it is of questionable relevance when individuals are subject to multiple exposures to hundreds of chemicals in hundreds of consumer products. It is important to keep in mind that official decisions about chemical risks are based on toxicity *and* probable exposure. A chemical can be highly toxic but if it is assumed that the possibility of exposure is low, then the risk is judged low. For consumers this concept may be reasonable for chemicals that are not used in consumer products but less so for chemicals that are.

Second, except for actions covering pesticides, food additives, and drugs, product toxicity policies are not proactive. The government waits for some indication of a problem before it attempts to control toxic products. Even when the government is supposed to be proactive, as with pesticides, it cannot because complete and current data is not always available.

Third, the government often bases its regulatory decisions on tests that are paid for and controlled by the regulated industry. This policy saves the government a lot of money and time, but it opens the door to conflicts of interest. How widespread this problem is is not known, but it is clear from falsified generic drug data uncovered in 1989, and from the discovery several years ago that an independent testing laboratory in California had faked pesticide test results and done testing under improper conditions, that the government does not always get valid data on which to base regulations.

The government's ability to move against toxic products depends on the nature of the product. Food, cosmetics, and drugs are regulated by one system, pesticides by another, and all other products with hazardous substances by a third. Each system is run by a different agency, and each has its own rules and criteria for determining whether or not something is toxic (and who determines that) and how to respond. This diversity results—as in the case of wastes—in the same substances being allowed in some products but not in others.

Methylene chloride (TRI) is an interesting example. The scientific community is uncertain about its carcinogenicity. The National Toxicity Program rates it as a substance that can "reasonably be anticipated" to be a carcinogen. The IARC lists it as a "possible" human carcinogen. Acute reactions include loss of consciousness, confusion, and death. The Food and Drug Administration has banned methylene chloride from cosmetics but approved it as a contaminant of decaffeinated coffee. The EPA says that it is a toxic inert ingredient in pesticides. As such, it has to be named on the pesticide label, but the concentration does not have to be revealed.

The CPSC says that methylene chloride is a hazardous substance if present in concentrations greater than 1 percent and has found 13 classes of products that may contain it. Included are paint strippers, adhesive removers, spray shoe polish, adhesives and glues, paint thinners, glass frosting and artificial snow, water repellants, wood stains and varnishes, spray paints, and cleaning fluids and degreasers. The commission decided in 1987 that methylene chloride should be presumed to pose a carcinogenic risk to consumers but, because of uncertainty, has delayed ruling on a ban. That decision is not expected until 1992. Meanwhile, the commission and two industry groups have agreed to voluntary labeling of products with methylene chloride. As of September 1988, labels are to include a statement that methylene chloride has been shown to cause cancer in certain laboratory animals and to advise consumers to use the product only outdoors, if possible. One consequence of these labeling rules and the anticipation of a possible ban is the substitution of perchloroethylene (TRI). There is no action underway to take perchloroethylene off the market.

Pesticides

Pesticide products appear to come under a comprehensive regulatory system. If a formulator or manufacturer wants to sell a pesticide product in the United States, it has to be registered with the EPA. All such products (including disinfectants) have an EPA registration number on the label. Registration does not mean that the product works or that it is safe. Testing to determine the acute and, especially, the chronic effects of use has never caught up with the increasing number of pesticide products. *Consumer Reports* (Karpatkin 1988) succinctly summed up the problem of pesticides:

"There's too little information, too little testing, and too little action to remove hazardous pesticides from the marketplace."

When the comprehensive pesticide law was passed in 1972, the pesticides then in existence were "grandfathered" into the system. That means that they were accepted as is. No testing was required. The intent was that the government would require and review tests on new products while reviewing the old products over time. It hasn't worked because the regulatory system has been overwhelmed. There are too many products, and the process is too slow to keep up. As of 1986, there were 50,000 pesticide products with 600 active ingredients registered but without test reviews. The result today is that most pesticides in use have not been fully tested to determine their potential for causing chronic effects such as cancer, reproductive disorders, birth defects, and environmental damage.

In any event, pesticide testing programs only cover the active ingredients, those that control or kill a pest. Inert ingredients, which are intended to dissolve, dilute, propel, or stabilize the active ingredient, receive little official attention. Inerts can be the same problematic organic solvents found in many other household products. They can also be pesticide chemicals but, because they are not included in a product to kill pests, they are considered to be inert.

If that isn't enough uncertainty, the government does not attempt to assess the effect of multiple exposures—at and away from home—from a variety of products. One National Cancer Institute study in 1987 showed a link between leukemia in children and households that had used a variety of pesticides. A survey in Boston found that chlorpyrifos (the active ingredient in many home insecticides) was also used in public facilities, office buildings, retail stores and hotels, and the transit system.

Another critical point is the criteria under which the government can decline to register a pesticide and therefore keep it off the market. The EPA must make a finding that the product will cause *"unreasonable* risk to man or the environment, taking into account the economic, social, and environmental *costs and benefits* of use." We have stressed some key words in that phrase. *Unreasonable* is a judgment call. *Costs and benefits* means that the risks are balanced against the benefits of use; public health can lose out to private profits. Although some chemicals have been totally banned, more, such as chlordane, have simply been restricted from general (i.e., household) use.

Lastly, it is important to remember that the government does not test and evaluate pesticides. The pesticide industry does that for the government. Whether a new pesticide needs to be registered or an old pesticide reregistered, the manufacturer is responsible for the costs, the execution of tests, and the preparation of other required information. This obviously makes it possible for a company to withhold information it deems inappropriate or that might result in negative conclusions.

The key regulatory method for home pesticides is a strict set of labeling regulations. There is a good possibility that the label information will be an understatement because of the lack of government-verified tests and because the label data applies only to active ingredients. The evaluation of toxicity may only include acute effects.

Products with Hazardous Substances

The CPSC, under what is primarily a labeling statute, regulates products containing hazardous substances. Under another law, the commission sets standards to protect the public against "unreasonable risks of injury associated with consumer products." To ban or restrict a product, the commission must make a ruling either that labeling or that no feasible standard would adequately protect the public.

The CPSC's record in banning products with hazardous substances has been a checkered one. Since 1973 the commission has attempted to take products containing eight different carcinogens off the market but has only banned tris (a flame retardant in children's sleepware) and some uses of asbestos. Some investigations took so long that the CPSC discontinued action because the targeted use had virtually ceased due to other factors. Sometimes manufacturers have voluntarily removed products. Some substances have been under investigation for years: formaldehyde (TRI) emissions from pressed wood products, methylene chloride (TRI) in aerosols and paint strippers, paradichlorobenzene (TRI) in air fresheners, and perchloroethylene (TRI), a dry cleaning fluid. No decisions are expected soon, partly because of a lack of resources.

Products containing hazardous substances have to be labeled according to rules set by law and regulations. In general, it is left to the manufacturer to decide whether or not its product contains a hazardous substance. The regulatory definition is both technical and conditional. A substance is hazardous if it is toxic, corrosive, an irritant, a strong sensitizer, flammable or combustible, if generates pressure (through decomposition, heat, etc.), or if the substance "may cause *substantial* personal injury or illness" under conditions of use. The law excludes pesticides, foods, drugs, cosmetics, fuels, and tobacco and tobacco products. In addition, there are at least 38 regulatory exemptions for specific products. For instance, certain porous-tip marking pens do not have to comply with labeling regulations even though they may contain toluene (TRI), xylene (TRI), petroleum distillates, or ethylene glycol (TRI). The labeling rules for products with hazardous substances are similar to but not as inclusive or specific as those for pesticides.

In the last 10 years, the CPSC has worked closely with the industry it is supposed to regulate. One commissioner said in 1988 that the CPSC had translated the "spirit of reform" for less government interference, respect for individual choice, and simple common sense into a policy goal of more reliance on voluntary standards. That policy and the effect it has had on protecting the public have been strongly criticized by public interest groups such as the U.S. Public Interest Research Group (U.S. PIRG) and the Consumer Federation of America, as well as by members of the U.S. Congress. U.S. PIRG claimed in 1987 (Gilbert) that by deferring to industry's judgment and relying on voluntary standards, "CPSC has turned its responsibilities over to the industries it is charged with regulating."

Foods, Drugs, and Cosmetics

The third government agency to have some control over consumer products is the Food and Drug Administration (FDA). The FDA makes determinations as to whether or not food additives (direct and indirect ones such as packaging) and color additives, contaminants of additives, and environmental contaminants in foods and cosmetics are "generally recognized as safe." The FDA can ban a substance, set rules for safe use of an additive, require warning labels, and set tolerance and action levels. Under the "Delaney clause" of the law governing FDA actions, no additive is considered safe if it is a human or animal carcinogen. However, since 1982, additives that contain carcinogenic impurities have been allowed if the additive itself is not carcinogenic and there is a "reasonable certainty" that no harm will result from the contaminated additive.

Since 1950, the FDA has banned about half a dozen carcinogenic additives. The latest was in 1990 (red dye #3), but the ban applied only to cosmetics; food uses are exempted while the issue is studied for another 5 or more years. The use of methylene chloride to decaffeinate coffee has been allowed because of a minimal risk assessment, i.e., exposures too low to worry about. The FDA has allowed the use of at least four carcinogenic impurities in food packaging materials. During the 1980s, a number of cosmetic color additives were approved even though they were contaminated with known carcinogens. Four others were allowed because they were proved to be carcinogens only in animal tests, reversing a traditional interpretation of the law. Other cosmetic ingredients (such as vinyl chloride in aerosols, mercury compounds, zirconium, and chloroform) have been banned. These actions were all taken in the 1960s or 1970s; no others occurred until methylene chloride was banned in 1989. Foods, drugs, and cosmetics all have separate labeling rules. Compared with pesticides and hazardous substances, cosmetics labeling requirements are modest.

Environmentalist and consumer organizations have fought the FDA's attempts to weaken the Delaney clause regarding carcinogenicity. One of those groups, the Coalition for Consumer Health and Safety, also said in 1989 that it supported the development of third-party testing. The coalition believes that, to ensure accuracy in testing and results, tests should not be done by the FDA or by industry. Overall, the FDA's work is adversely affected by budgets that have not risen fast enough in the last few years to handle the agency's increased responsibilities. Meanwhile, the enforcement staff, whose job is to ensure that industry is adhering to regulations, has been cut 32 percent.

Almost everyone's attention, including the FDA's, centers on the agency's work on food and drugs at the expense of cosmetics. One reason is that consumers assume that cosmetics have to be pretested. Another reason is that the FDA's ability to act in the cosmetics area is very limited. The agency does not pretest cosmetics and does not have the authority to require manufacturers to do so. It relies on voluntary industry efforts to evaluate some 4,000 chemicals used to make over 40,000 formulations. In essence, the government's role regarding cosmetics is passive. The reality is that once in the marketplace, products are tested by consumers. The head of the FDA at the time, Dr. Frank E. Young, obtusely described the value of this consumer testing to the U.S. Congress in 1988: "we feel adverse reaction evaluation in cosmetics does provide a sentinel indication of problems." If the warning sounds are loud enough to get the attention of FDA officials, the agency must marshall its scarce resources to prove a product unsafe before it can be removed from the market. (It took the FDA over 4 years to gather enough evidence to ban methylene chloride.) At a congressional hearing in 1988, it was revealed that the National Institute of Occupational Safety and Health had checked 2,983 chemicals used in cosmetics against a list of reported toxics. Among the cosmetic chemicals were 884 reported to cause biological mutations, reproductive complications, acute toxicity, tumors, or skin and eye irritations. Only 56 of the 884 chemicals had been reviewed by the industry's Cosmetic Ingredient Review Board, and the board had discounted the toxicity of most of those due to evaluations of exposure. The board, by the way, operates on an annual budget of $750,000, provided by the cosmetics industry that generates $17 billion in sales each year. The relatively small budget has netted a review of only 300 chemicals in 10 years.

USE ONLY AS DIRECTED
AND DISPOSE OF PROPERLY

The companies that sell Americans billions of dollars worth of cosmetic, cleaning, pesticide, painting, and automotive products each year are accustomed to change. New products are introduced and existing products adjusted as sales figures and market analyses demonstrate that customer preferences are shifting. But the companies respond differently when consumer groups or movements and governments try to force product changes. When allegations are made that household products are toxic, the general response is to defend those products. Companies claim that consumers are being misinformed. Product advertising, on the other hand, which sells products on the basis of a host of values unrelated to product performance, is not misinforming consumers.

Surveys indicate that American consumers want nontoxic products and are willing to pay more for them. So far, this evolving attitude has been viewed favorably by industry as a niche market with some potential or suspiciously as a fad that may quickly disappear. The firms that have moved in to fill the opening are, so far, not the industry leaders but primarily small, innovative firms. Safer, Inc., Ringer Corporation, BioSafe, Zoecon, and many others now market alternative pesticides. A number of alternative products are imported into the United States: Ecover (Belgium) has a wide range of household cleaning products; Livos (West Germany) offers paints, cleaners, and polishes; and Auro (West Germany) sells paints. Because few American stores carry alternative brands, most are sold through mail order firms. Ironically, then, the introduction of green products into the U.S. market will cause an increase in mail waste (see Chapter 7).

Some established firms do quietly reformulate their products. Such is the case with Polaroid, which redesigned the battery pack for its cameras to eliminate mercury. Another example is Procter & Gamble, which years ago told all its suppliers to eliminate the use of toxic inks for packaging. Procter & Gamble has now gone public and become an industry leader in marketing environmentally friendly products in the United States.

The more common reaction to people's health and environmental concerns, especially by industry trade organizations, is a line of defense. Industry longs for the 1950s, when the public thought that chemicals would provide wholesale benefits without any risks. To defend their products, industry spokesmen make statements, such as the following (Etter 1988), that downplay the possibility of exposure: "every [household] product on the market performs the task required, *is safe when used as directed,* and poses minimal threat to human health and the environment when disposed of properly" (emphasis added).

When confronted with the fact that some products do contain toxics that can result in harm from use, industry responds that no substance is risk free. The benefits of these products, sold on the basis that they make life easier, more pleasant, and more sanitary, outweigh the potential harm. And, by and large, consumers agree. The pesticide industry may feel somewhat threatened by what one industry public relations person has called "environmental toxic terrorists," but they know that the markets (outisde of agriculture) for pesticides are continuing to grow. The public concern about toxic products is not yet showing up on the balance sheets of American corporations.

To prevent sales erosion, industrial groups defuse consumer concerns and regain consumer confidence in their products by funding lobbying groups and waste projects. The development of a new pesticide can cost $20 to $44 million, so reformulation is not the option of choice. It is more cost effective to pool together and fight enforced change. An industry coalition spent millions of dollars to combat the passage of California's Proposition 65 (Safe Drinking Water and Toxic Enforcement Act). It passed anyway and is now hailed (but not by industry) as the leading edge of a new environmental wave because its premise is that manufacturers should prove that their products are *not* toxic.

Plastics are today's miracle materials. In response to concerns about the possible releases of toxic chemicals from incinerators and landfills, the plastics industry spends millions of dollars each year. They have organized into such groups as the Council on Plastics and Packaging in the Environment (COPPE) and the Council for Solid Waste Solutions. The latter group planned to spend $8.5 million in its first 15 months of operation. The Society of the Plastics Industry has proposed a $150 million 3-year program to reverse growing negative public "perceptions" that result in polls showing that 72 percent of the public think that plastics are harmful to the environment. The countermessage these groups dispatch is that plastic products are important and contribute positively to the environment, and that plastics do not constitute a major volumetric problem in landfills and are not harmful when burned or buried in landfills. Despite the lack of acknowledgment of waste problems, the industry is supporting the recycling of plastics. It is not, however, publicly funding research to resolve controversial problems or find alternatives to certain plastics that might be harmful.

The Chemical Specialties Manufacturers Association (CSMA) is a trade association comprising the firms that formulate most household products, including detergents, pesticides, and disinfectants. They formed the Household Products Disposal Council in 1986 to (1) establish the credibility of the industry, (2) counter "consumer misconceptions" about the risk of household products and promote "consumer responsibility" for household waste, and (3) challenge legislative and regulatory actions on products or disposal restrictions. On this last point, they joined industry efforts against Proposition 65. Having lost that fight, the CSMA has filed a suit to invalidate Proposition 65 and is very active in lobbying against several state efforts to ban aerosol sprays and other products containing VOCs.

The centerpiece of the CSMA effort is the "Disposal: Do It Right" program, which provides the public with information on the disposal of household consumer products. The point stressed is that consumer products are designed with normal waste disposal systems in mind and, thus, can safely be sent down sewer drains or into incinerators and landfills. (Air disposal is not considered.) The CSMA does admit that the bacteria in sewers and septic systems, which are relied upon to break wastes down into safe compounds, can be overcome by excessive loads of substances. They also admit that some toxic materials may pass right through these systems, unaffected by the bacteria. They do not admit to any problems from incineration or landfills. They do admit that some materials can pose undefined disposal problems and that those should be saved for household hazardous waste collection days. The only materials so identified are old pesticides, gasoline, used motor oil, and ammunition. Interestingly, these are consumer products not made or no longer made by members of the CSMA.

A major flaw in the argument that each consumer product is designed for disposal in sewers and landfills is a lack of consideration for the cumulative effects of so many different products in each home and in so many products being disposed of at the same time. The "Do It Right" project also does not consider the health effects of using toxic products, such as inhalation of VOCs. Still, the CSMA announced to its membership in 1989 that the Household Products Disposal Council was successful because public perception was approaching the reality that "the main sources of toxicity from homes are used automobile batteries, old and used motor oil, and some outdated and banned pesticides."

To buttress the assertion that the benefits of use outweigh the risks, industry people often stress the uncertainty of the scientific data and of the process of determining toxicity. They use uncertainty to build a case *against* regulating or banning the use of specific chemicals. Thus they often ask: Can the results of animal studies be extrapolated to humans? Are the experiments biased toward conservative outcomes? Can conclusions based on ingestion be relevant when inhalation is the most likely route of exposure? And so on. They try to drive the determination of toxicity toward higher levels of proof. This is couched in terms of "science," but since perfect scientific data is extremely difficult to come by, the strategy only serves to delay decision making. And it has worked all too often.

Another strategy is the use of comparative analysis. In this approach, the risk of exposure to chemicals is compared with other health problems, such as cigarette smoking, poor diet, and alcohol and substance abuse. The conclusion drawn is that since these cause more deaths than can be attributed to chemicals in products and the environment, the public worries too much about the latter. This strategy glosses over the differences in certainty and the sense of control that the public uses to discriminate among these issues and ultimately frames the way they feel about them.

THE ART OF LABELING

The label is the point where the actions of industry and government combine to provide consumers with information about a product. Other than advertising, a negligible amount of government information, and work by public interest groups, it is the main way that positive or negative product information is communicated. The separate interests of government and manufacturers sometimes pile up and make for crowded labels with very small print (a result that is not exactly in tune with the selling of "convenience" products). The manufacturer uses the label to attract attention, convince people to buy, and provide directions for proper use. The government imposes constraints on product descriptions and requires safety or hazard information. Numerical analysis of the content, when available, is there because the government requires it to be.

Labels are part of product packaging, and packaging is primarily a marketing tool. The package is designed to project an image of reliability to gain consumer confidence. The simple additions of the word *toxic* can destroy confidence. A package is designed to make a product stand out from the competition and get potential customers to make a quick buy decision as they pass it in a store. Of dubious value are terms such as *organic, natural,* and *new and improved* that are defined by the manufacturer (under the sometimes watchful eye of the Federal Trade Commission). Confusion is already resulting from nomenclature such as *environmentally friendly* and *green products.*

For consumers concerned about product toxicity, the most objective and informative information may be on the back of the label in small print. The best, for those who can interpret it, is a label that has all the constituents listed in terms of percentage of total content. Unfortunately, this extensive information is rare. Companies always argue that such information is proprietary, but it is likely that consumers are the only ones not to have it. The government may require it but hold it secret. Formulators can break down a competitor's product and discover what it contains. The government's agreement to withhold product ingredient information seems to come from a position of being more protective of manufacturers than of consumers. Knowing a product's content is vital when accidents occur.

Product ingredient information on labels ranges from none to complete, with most labels falling in the middle. While some firms are more forthcoming than others, the extent of the list of ingredients is clearly a consequence of government regulations. If no regulation applies, no ingredients may appear. Pesticide producers list the percentage of active ingredients but lump all inerts, with one exception, into a single percentage because that is what the government allows. Products with hazardous substances, as interpreted by the manufacturer, may list only the names of the substances, not their amounts. Acute effects may be evaluated and chronic ones ignored. The most comprehensive listing of ingredients we have found was on some paint products, especially Duron paints, which listed both individual active and inert ingredients by percentage of content.

Cautionary or use instructions on labels are often determined by the government. But they are less than precise. Does "use with adequate ventilation" mean merely to open a window or to use the product with an exhaust fan? In reality, it depends on the product's contents and the sensitivity of the user. For a product that contains highly volatile methylene chloride, extreme measures are best and essential for users with heart problems. In a regulatory notice on methylene chloride, the CPSC commented that the phrase "use with adequate ventilation" has been applied to acute toxicity and is an insufficient notice for chronic toxicity, which may be induced by exposure levels below those for acute symptoms.

Labeling instructions can be unrealistic in practice. Users of highly toxic pesticides are told *not to breathe* vapors and, for less toxic pesticides, to *avoid breathing* vapors. First, the distinction between the two is somewhat obscure. Second, it is not always possible to know whether or not you are breathing or avoiding vapors. Either may become less feasible because of a trend in the pesticide industry to mask the unpleasant odors of chemicals.

A number of standard phrases are keys to relative toxicity even when ingredients are not listed. The phrase "Keep away from children" is not, however. It is pro forma, a requirement for all hazardous substances and pesticide products regardless of their relative toxicity. A standard phrase on both types of products is this disposal information: "Wrap tightly closed container and place in trash." The paper is there to absorb any leaks that might occur and injure a sanitation worker or end up in groundwater. So, although the phrase is not meant that way, it alerts one to toxicity. "Do not use with ammonia" means that the product is acidic. Mixing with ammonia will create a neutralization reaction that can be volatile. "Nontoxic when completely dry" and "Keep children away until completely dry" indicate that a product contains harmful VOCs.

For some products, manufacturers capitalize on public knowledge as a selling technique. Phrases such as "Contains no phosphorus" and "Contains no methylene chloride" are a recognition that the public is aware of some harmful substances. Cosmetic manufacturers use a variety of phrases that imply conditions that are not necessarily true. An "unscented" cosmetic may still contain small amounts of fragrances to mask unpleasant odors. *Hypoallergenic* only means less allergenic, and each manufacturer defines the basis of comparison. A hypoallergenic product can cause an allergic reaction.

We found a brand of aerosol paint with a bewildering array of information. Even though the label stated that the paint contained no lead, CFCs, methylene chloride, hexane, and methoxy- or ethoxethanol or their acetates, it also said: "Danger," "Extremely flammable," "Can Pressurized," "Harmful or fatal if swallowed," and "Vapor harmful." The reason was that despite all of those substances that were *not* in it, the product did contain acetone (TRI), propane, methyl ethyl ketone (TRI), xylene (TRI), toluene (TRI), methyl isobutyl ketone (TRI), and butyl alcohol (TRI).

WHICH TOXIC SUBSTANCES OCCUR
IN WHAT PRODUCTS?

Toxic substances permeate households. They are in the kitchen, bathroom, laundry room, hobby room, workshop, garage, and garden. They are in cleaners, adhesives, paints and polishes, pesticides, and automotive products. They are also part of packaging materials and the inks with which labels are printed. The uncertainty of exposure is high because there often is no practical, definitive way of knowing what chemicals are present due to inadequate or misleading labeling. Even when known chemicals are in a product, it can be difficult to understand the relative toxicity of each.

There are three kinds of toxic substances in household products: metals, stable organics and VOCs, and acids and alkalis. The possible exposure routes and toxicity of each differ somewhat. VOCs probably result in the most overall exposure to toxicity because they are present in so many products and because inhalation is involuntary and often unnoticed. Metals exposure from products may be the least probable (for adults) but represents a high source once product wastes are released into the environment. Pesticides may be the most toxic (both acute and chronic) household products. They are also the easiest category to avoid because of the many available alternatives.

Toxic Metals

Many metals are toxic, some more so than others. There is little dispute about their toxicity. The TRI list includes the following metals and their compounds: antimony, arsenic, barium, beryllium, cadmium, chromium, cobalt, copper, lead, manganese, mercury, nickel, selenium, silver, thallium, and zinc. Our discussion covers lead, cadmium, and mercury to illustrate the widespread use of toxic metals in household products. These metals are usually present in small quantities, but ingestion of even small doses can be highly toxic. Minute amounts tend to accumulate in body tissue over time and ultimately cause organ damage that can be irreversible and fatal. Systemic and chronic effects include damage to the liver, kidneys, and central nervous system, as well as cancer. There is epidemiological evidence that some metals can cause reproductive disorders or birth defects. Children who eat paint containing lead suffer from dysfunction of the central nervous system and many other problems that lead to learning disabilities.

There are millions of tons of toxic metals widely distributed among hundreds of thousands of products each year (see Tables 8-2a,b,c). Not all end uses are products typically found in households. Whether or not a consumer is liable to be affected by toxic metals during usage depends on how well encapsulated they are in the product. Unless a nickel-cadmium battery

TABLE 8-2A.   END USES OF CADMIUM, 1988

End Use	Amount (Metric Tons)	Percent of Total
Batteries	1,158	32
Coating and plating	1,050	29
Pigments	543	15
Plastics and synthetic products	543	15
Alloys and other	326	9
TOTAL	3,620	100

Source: Llewellyn (1988).

leaks, one is unlikely to become exposed. The mercury in a thermometer is secure unless the thermometer breaks. All of the metals used, however, eventually end up as wastes. Since metals cannot decompose, once present in the environment they can return to affect people. When burned in incinerators, metals become attached to particulates, vaporize, or end up in the incinerator ash. The vapors and some of the particulates are emitted into the air; the rest of the particulates and the ash end up in landfills, which means that they can leach into groundwater. Metals accumulate in water treatment plant sludges, which are sometimes used as fertilizer for food crops.

In 1987, 1.2 million metric tons of *lead* were used, most of it in storage batteries like automobile lead-acid batteries. Other consumer products that contain lead include paints and pigments, tin cans, and gasoline. In 1986 the permissible amount of lead additive in gasoline was reduced to 0.1 gram per gallon. Most food cans are welded today instead of lead soldered. Still,

TABLE 8-2B.   END USES OF LEAD, 1987

End Use	Amount (Metric Tons)	Percent of Total
Metal products	156,372	13
Storage batteries	953,598	78
Other oxides	68,094	6
Gasoline additives	na	na
Miscellaneous	52,323	4
TOTAL	1,230,387	100

Source: Woodbury (1987).

TABLE 8-2C. END USES OF MERCURY, 1988

Selected End Uses	Amount (Flasks)[a]	Percent of All End Uses
Paint	5,722	12
Electric lighting	891	2
Wiring devices and switches	5,102	11
Batteries	12,987	28
Instruments	2,233	5
TOTALS		
Selected end uses	26,935	na
All end uses	46,196	58

Source: Reese (1988).

[a]1 flask = 76 pounds

over 4 million cans produced in the United States in 1987 contained lead. It is unknown how many lead-soldered cans and cans containing goods are imported each year. Another imported source of lead is glazed ceramic ware that has not been properly fired.

Very low concentrations (a few parts per million) of lead can severely and persistently affect the learning capability of children by causing deficits in central nervous system functioning. In 1978 the government banned the sale of household paints with more than 0.06 percent lead. The ban also applied to paint used on toys and other items intended for use by children. Lead is still used extensively in pigments. Over 79,000 metric tons of lead pigments were produced in the United States in 1987. Chrome yellow (lead chromate) and moly orange (lead molybdate) pigments are used in paints and printing inks. Pigments, which also contain chromium, may be used in packaging labels, as well as to color plastics. Some plastics [such as polyvinyl chloride (PVC) wire and cable insulation] also contain lead as a stabilizer.

There is no current substitute for the lead-acid battery. There is no way to know which printed ink or which plastic article contains lead, although yellow and red plastic articles likely contain lead pigments. Another source of lead in households is drinking water. It is picked up from lead pipes and lead solder in copper pipes. An estimated 40 million people drink water that doesn't meet government standards for lead content.

A high concentration of *cadmium* is acutely toxic; chronic low exposures can damage kidney tissue. Cadmium may be a human carcinogen and may cause birth defects. Exposure often occurs from cadmium in the environment; one major source is food that has been grown in soils containing cadmium. Over 3,600 metric tons of cadmium were used in 1988. All end uses of cadmium include consumer products. Cadmium-coated or -plated metals are used in electronic equipment, automobile parts, washing machines (to resist the corrosive effect of detergents), and as components in hardware. Rechargeable nickel-cadmium batteries are sold individually, but most (75 to 80 percent) are encapsulated in appliances, tools, toys, and so on. Almost all of the cadmium pigments produced end up in plastics (PVC, polystyrene, polypropylene), the balance in inks, paper, and paints. Cadmium is present as a stabilizer in plastics (mainly PVCs). If a household item contains plastic or a battery, there is a good chance that it also contains cadmium.

Despite its unquestioned toxicity, there are no U.S. restrictions on its use. Sweden banned most uses of cadmium in 1985 with many exemptions that gave its industry until 1992 to find replacements. In 1990 Sweden attempted to get Western European countries and Japan to ban cadmium (and other similar materials) but was unsuccessful. Substituted was a proposal by the U.S. government to study ways to limit exposures.

Mercury use totaled 3.5 million pounds (almost 1,800 tons) in 1988. The toxicity of mercury varies depending on its form: elemental metal or inorganic or organic compound. All forms can be absorbed through the skin. Effects range from dermatitis to central nervous system and organic dysfunction; systemic poisoning can lead to death.

The largest single end use is batteries, primarily alkaline, although carbon-zinc batteries also contain mercury. Since 1982, the total amount of mercury used for batteries has been cut in half due to improved technology. Other consumer products containing mercury are pigments, paints, electrical and electronic equipment, and instruments (such as thermometers). Mercury is used in latex paints as a preservative to prolong shelf life and to prevent mildew once applied. Exterior latex paint may contain an organic mildewcide instead of mercury.

As for most metals, identifying specific products that contain mercury can be difficult. Mercury can be avoided to some extent by using carbon-zinc batteries, but at the cost of generating more waste, since alkaline batteries last considerably longer. (In Chapter 7 we present a comprehensive analysis of battery usage.) Button cell batteries have the highest mercuric oxide content; eventually they may be replaced by other systems such as zinc-air batteries. Batteries without mercury are available in Europe and Australia.

Toxic Organics

Toxic organic substances are found in products as diverse as cleaners, nail polish remover, paint thinner, glues, pesticides, and antifreeze. Most—but not all—organic substances are made from petroleum and natural gas. Pine oil, used in many general-purpose cleaners, comes from the resin of pine trees. Another "natural" organic is turpentine. Whether synthetic or natural, organics are composed of carbon and hydrogen and a variety of other chemicals, such as chloride, fluoride, bromide, and phosphate.

The true chemical makeup of organics in consumer products is often hidden behind such terms as *solvents, petroleum distillates, volatile organic compounds (VOCs),* and *mineral* or *petroleum spirits.* A solvent is technically any substance capable of dissolving another substance to form a solution. While water can serve as a solvent, solvents are generally organics. Some common solvents are listed in Table 8-3 by groups, another way that people talk about organics. VOCs are a class of organics that vaporize at room temperature and pressure. Petroleum or mineral spirits are refined petroleum distillates with certain properties (such as flash point and volatility) that make them suitable as solvents in paints, varnishes, and others. *Petroleum distillates* is a general term for hydrocarbon fractions varying from petroleum ether to lubricating oils. Included are kerosene, mineral seal oil, naphtha, gasoline, mineral spirits, and stoddard solvents.

TABLE 8-3.   SOME COMMON SOLVENTS

Group	Substance[a]
Hydrocarbons	
Aliphatic	pentane, hexane, naptha, Stoddard solvent, kerosene
Aromatic	BENZENE, TOULENE, XYLENE, NITRO-BENZENE
Oxygenated solvents	
Alcohols	METHANOL, ETHANOL, ISOPROPYL ALCOHOL
Ketones	ACETONE, METHYL ETHYL KETONE
Ethers	Diethyl ether, isopropyl ether
Esters	ethyl acetate
Glycols	ETHYLENE GLYCOL, CELLOSOLVE
Halogenated solvents	
Chlorinated	METHYLENE CHLORIDE, CHLORO-FORM, TRICHLOROETHYLENE, PER-CHLOROETHYLENE, CARBON TETRA-CHLORIDE, 1,1,1,-TRICHLORO-ETHANE (methyl chloroform)
Fluorinated	fluorocarbons, freons

Source: Adapted from Purin et al. (1987).

[a]TRI chemicals are capitalized.

The use of so many different generic terms and the lack of specificity make product toxicity difficult to judge. Like other ingredients in household products, organics can be fatal—certainly harmful—if ingested (depending on the dose) and are almost always irritants if they come in contact with eye tissue. Some organics can be absorbed through the skin, and those that vaporize are easily inhaled. Some are very ignitable, which makes them dangerous to use near any flame. Solvents, PCBs, and organochlorine pesticides have been shown in epidemiological studies to cause abortions or birth defects. Many organics are known or suspected human carcinogens.

The acute effects of low concentrations of a few organics have been singled out by the CPSC for special labeling rules. They include:

- Diethylene glycol and mixtures containing 10 percent or more are harmful if swallowed.
- Ethylene glycol (TRI) and mixtures containing 10 percent or more are harmful or fatal if swallowed.
- Benzene (TRI): mixtures containing 5 percent or more may cause blood dyscrasias through inhalation of vapors or ingestion.
- Toluene (TRI), xylene (TRI), and petroleum distillates: mixtures containing 10 percent or more may be aspirated into the lungs and result in chemical pneumonitis, pneumonia, and pulmonary edema; inhalation can result in systemic injury.
- Methyl alcohol (methanol/TRI) and mixtures containing 4 percent or more may cause death and blindness from ingestion; inhalation is also harmful.
- Turpentine (including gum turpentine, gum spirits of turpentine, sulfate wood turpentine) and mixtures containing 10 percent or more: ingestion may cause systemic poisoning; aspiration into lungs may cause pneumonitis, pneumonia, and pulmonary edema.

These represent only a few of the many organics in use. Because their uses are ubiquitous, it is not possible to break down the end uses, as we have done for metals. The EPA has a database that identifies 320 VOCs found in outdoor or indoor air. Table 8–1 indicates which TRI chemicals are VOCs, according to that database. One report has identified over 60 different VOCs used in consumer products and says that the more common solvents include ethanol, isopropyl alcohol (TRI), kerosene, propylene glycol and propellents in aerosols such as isobutane, butane, and propane. Another study estimated the percentages of VOC emissions by consumer products in California as follows: personal care (30 percent), household (18), pesticides (6), aerosol paints (12), auto and industrial (32), and miscellaneous (2).

Dioxin in Products

Dioxin, a chlorinated organic chemical present in consumer products, is getting more attention today. The chemical family has over 100 compounds, including 2,3,7,8-tetrachloro-p-dibenzodioxin (TCDD) and 2,3,7,8-tetrachlorodibenzofuran (TCDF). Dioxins are found in paper products—such as writing and printing papers, facial and toilet tissues, towels, disposable diapers, sanitary napkins and tampons, food packaging, and many others—because it is formed during the paper bleaching process. Dioxin can be avoided by simply not buying and not using bleached paper products. At least two brands of brown coffee filters, Natural Brew and Melitta, are readily available today.

In Canada, when low levels (0.04 and 0.75 parts per trillion) of dioxin were detected in milk packaged in plasticized, bleached paperboard cartons, the industry agreed to reduce the amounts in that product. A very high percentage of paper products in Sweden are unbleached as a result of consumer pressure. Meanwhile, the U.S. government is still studying the issue in cooperation with U.S. pulp and paper mills. An assessment by the industry concludes that the risk to consumers is minor, but the study covers only two of the many possible compounds (TCDD and TCDF) and does not consider the cumulative effects on one person who is using a variety of products containing dioxins.

Plastic Materials

Depending on one's perspective, plastics can be either the chief cause of all MSW problems today or a major reason for advances in living standards. The truth is less clear-cut, although both positions have some merit. As the use of plastic materials grows, so does the waste they generate. There are major unresolved issues about the toxicity of plastics, however, and the present debate is not clarifying those issues. There are several reasons for this situation. Among them are that people tend to generalize too much and that the industry deflects rather than confronts public concerns.

Plastics (or polymers) are not a material. They are a class of materials. The basic raw materials—resins—are synthesized from petroleum and natural gas. The polymers are made by linking groups of resin molecules together. Thus, styrene, propylene, vinyl chloride, and ethylene become polystyrene, polypropylene, polyvinyl chloride, and polyethylene. During both the resin and polymerization processes, other chemicals and metals are added. The result is hundreds of types of plastic materials with varying properties: flexible or rigid; transparent, opaque, or colored; easy to tear or as strong as steel. Plastics are so variable that they can be engineered to have special properties to meet the requirements of specific end uses.

Because of this variability, it is inaccurate to say flatly that plastic products and packaging are toxic. Some may be; some may not. The information available to verify which ones are toxic and which aren't is lacking. Uncertainty is high because the plastics industry has chosen to deflect people's concerns about the potential toxicity of plastics and their impact on the environment by funding recycling operations and producing so-called biodegradable plastics. Environmentalists have already started to debunk the myth of degradability as a waste solution. Recycling may prove to be a limited solution, especially since plastic recycling is not continuous but results instead in only one loop (from bottle to fence post, for instance). Another choice the industry could make is to publicly support scientific research on the consequences of plastics manufacturing, use, and disposal.

It is quite accurate to say that the production of plastic materials exposes workers to many toxics and results in a host of toxic wastes. As well as those listed above, other TRI chemicals such as benzene, carbon tetrachloride, diazomethane, and cadmium and chromium compounds are used to make plastic packaging materials. What is not well understood is whether or not the toxics used to make plastics can leach out of them during use and while in landfills. There is uncertain evidence that chlorides (in PVCs) contribute to dioxin formation in and releases from municipal incinerators. It is more certain that metal additives end up in air emissions and ash residues. There is evidence that addititves in plastic microwavable food trays and films may leach into food during heating (see Chapter 7). In some European countries, public pressure has resulted in the elimination of PVC packaging for food.

History is replete with examples of wondrous, highly acclaimed materials that have turned out to be harmful. Whether or not plastics fall into this category is not well understood. This suggests that until more is known, people can choose to avoid plastic products and packaging. Unfortunately, plastics are ubiquitous, so that avoidance may mean not simply choosing among different versions of products but giving up certain products altogether.

Dyes and Pigments

Much is said about products containing toxic metal pigments and the harm they may cause during consumer use and, more probably, once thrown away. Little, however, is said about the organic synthetic dyes used to color yarns and textiles for clothing, upholstery, toweling, carpets, and other household products. Organic dyes are used in cosmetics and food as well. For the latter, the FDA is supposed to ensure safety. For other uses, there is no counterpart agency. For instance, Solvent Yellow 3 (o-aminoazotoluene) is not approved for use in foods, drugs, or cosmetics but is used to color shoes and other wax polishes.

Organic chemical dyes number in the hundreds; 90 million pounds of them were produced in 1987. The TRI list contains over 30 organic dyes. (Some can be identified by the "C.I." for Colour Index prefix or by a color name in the "Synonym" column of Table 8–1.) As with other classes of organic chemicals, their toxicity varies. Some are known human carcinogens; some are suspected carcinogens. Acute toxicity ranges from low to high. Some are allergens. As with chemicals in general, complete information about their toxicity is not known. The EPA has recently made an agreement with eight companies to begin testing one dye—C.I. Disperse Blue—because of its unknown effects on workers and the environment.

Consumer use of dyed fabrics and items may or may not be harmful. The information is sketchy. The EPA assumes, for instance, that consumer exposure to C.I. Disperse Blue, used almost exclusively in polyester fibers, is low because it is held tightly within the fiber structure. The chemical O-anisidine hydrochloride (TRI), an animal carcinogen, is used in the manufacture of dyes. This is what the National Toxicology Program (NTP) said in 1989:

> The primary routes of potential human exposure to O-anisidine hydrochloride are inhalation and dermal contact. According to [the CPSC], residual traces of O-anisidine may be present in some dyes manufactured from O-anisidine and in the final consumer product. Exposure even to trace amounts may be a cause for concern. No data are available on the actual levels of O-anisidine in final consumer products.

The NTP makes similar statements about a number of chemicals, such as P-cresidine (TRI), which is used to manufacture 11 dyes for textiles; 3,3-dichlorobenzedine (TRI), used to manufacture pigments for printing ink, textiles, and plastics; and 2,4-diaminotoluene (TRI), an intermediary in the production of dyes used to color silk, wool, paper, and leather, as well as some fibers, wood varnishes, and stains. In 1971 2,4-diaminotoluene was banned from hair dye formulations.

Can dyes or dye contaminants enter the body and result in acute or chronic exposure? In some cases, skin absorption of dyes or their contaminants has been tested. Dimethylbenzidine-based dyes and pigments, for instance, were not absorbed "to any substantial degree" when tested in rabbits. These kinds of conclusions leave people in the same old situation. How much information and scientific certainty are enough? Is there a real danger? Skin absorption studies do not deal with situations such as children chewing on fabrics.

The toll on the environment from consumer use of dyed products is unknown. Dyes and residual chemicals that wash out of fabrics end up in the sewer system. These substances and their components may be released when wastes are incinerated and may leach under certain conditions from landfills. Buying products dyed with toxic chemicals makes the consumer part of the toxic waste-generating chain that starts with the chemical factory and textile mill. Unfortunately, except in very limited situations when products are claimed to be colored with natural dyes or not to contain dyes (Cheer detergent), consumers have few alternatives.

Acids and Alkalies

Acids and alkalies (strong bases), which produce acute toxic reactions, have long histories of use in households. Lye (sodium or potassium hydroxide), a strong base, has been the traditional ingredient in drain and oven cleaners. Both products may also be made from acids, such as sulfuric or hydrochloric acid. Chlorine bleach (sodium hypochlorite) is an alkali.

Acids and alkalies corrode. This ability to eat away at materials is what makes them attractive ingredients in many cleaning and polishing products. The weaker they are and the more diluted strong ones are, the less likely acids and alkalies are to cause serious damage (and the less cleaning power they may have). The government has banned liquid drain cleaners that contain more than 10 percent lye unless they comply with special packaging rules. Acids and bases can be neutralized by mixing them together, but the reactions can be violent. Acidic products are usually labeled with the advice not to mix them with a product containing ammonia. Ammonia reacts with water to form ammonium hydroxide, a strong alkaline solution that reacts violently with acids.

A chemical's pH indicates how strongly acidic or alkaline it is. A pH of 7 on a 14-point scale is neutral, with 14 being the most alkaline and 1 the most acidic. The pH scale is logarithmic, which means that each number is a multiple of 10. For instance, a pH of 8 is 10 times as large as a pH of 7. (This is similar to earthquake measurements, in which a quake of 6 on the Richter scale is 10 times greater than one of 5.) Hand soap has a pH of about 9.5. Oven cleaners containing lye have a pH of about 13. A dilute solution of strong hydrogen chloride (pH 1) is 100 times as acidic as weak dilute acetic acid (vinegar) (pH 3). Lemon juice has a pH of 2.

Household Cleaners

Ever since manufacturers discovered that Americans will buy specialized products, the numbers and varieties of cleaning products for household has proliferated. Once, households relied on soap, ammonia, baking soda, and lye. Now there are bathroom cleaners, kitchen cleaners, tile cleaners, toilet bowl cleaners, rug cleaners, furniture polishes and waxes, floor waxes, bleach, detergents, fabric softeners, oven cleaners, drain cleaners, abrasive cleaners, metal cleaners, spot removers, hot tub cleaners, pool chemicals, jewelry cleaners, pet cleaners, VCR tape cleaners, upholstery cleaners, window cleaners, disinfectants, and air fresheners. More recent additions include microwave oven cleaners and cleaning wipes for the bathroom, windows, and furniture.

The household products cleaning industry comprises about 200 companies marketing over 5,000 brands of about 32 major types of products. The industry shipped an estimated $9.2 billion worth of products in 1988. Somewhat in awe, Carol Dawson (1988), a member of the CPSC, stated:

> "One could not imagine a home without soaps, detergents, insecticides, and a myriad of cleaning and disinfectant products. Simply put, these products are essential to the everyday life of the average American."

Evident ongoing changes in the industry do not reflect much concern about the toxicity of products. One major shift has been a partial return to multifunctional and general-purpose cleaners. This is a recognition that as product specialization increases, convenience decreases. It is easier to buy one box or bottle of cleaner that will do a variety of jobs. Laundry detergents now bleach, reduce static cling, and soften fabrics, as well as clean. Household cleaning may be coming full circle. Manufacturers have used the development of specialized products to drive consumers to new heights of concern about cleanliness. Now consumer demands for convenience may be driving a willingness for less cleanliness. Manufacturers are trying to hold on to specialized product markets by introducing new, more convenient forms such as toweling impregnated with cleaning solutions that cater to the throwaway ethic. Boxes of Spiffits (Dow Chemical Company) come in five different selections; they clean windows and bathroom, clean up grease, scour dirt, and polish furniture. Disinfectant versions are made by Lysol and Pine-Sol.

There is one thing most household cleaning products have in common: they are made from synthetic chemicals derived from petroleum and natural gas or other sources (such as pine oil). Their toxicity in use (or misuse) depends, then, on the form of the petroleum derivative or the strength of the acid or alkali ingredient. One general—but imperfect—guide to toxicity of cleaning products is the claims made by the manufacturer. The more powerful the manufacturer makes its product seem, the higher may be the potential toxicity. In advertising and labeling, manufacturers use terms like *extra strong, powerful, high speed,* and *super strong* to encourage consumers to buy. Although these terms are ill-defined, consumers can view such language as reasons *not* to buy.

Alternative products are appearing in the market (Table 8–4). Many have originated in Europe, where consumer demand for green products is higher than in the United States. Ecover, a Belgian company, distributes its "eco-

TABLE 8–4. SOME GREEN PRODUCTS

Company/Product Name	Types of Products
Ecover	Laundry powders and liquids Fabric conditioner Dishwashing liquid Floor cleaner Toilet cleaner Fabric softener
Kleer II	Concentrated gel for dishwashers
Life Tree	Concentrated laundry liquid
Citra-Solv	General-purpose cleaner
Neo Life Green	General-purpose cleaner
Simple Green	General-purpose cleaner
Livos	Furniture and floor waxes Cleaners Leather and shoe polishes Paints and coatings Art materials
Ivory Soap flakes	Laundry powder
Bon Ami	Soap Soap cleaning powder (abrasive)
Easy-Off Non-Caustic Formula	Oven cleaner
Pango Modelo Brevettato	Mechanical drain unclogger
Auro	Paints and coatings
3M Safer Stripper	Paint stripper
Aubrey Organics	Skin and hair care products
Aveda	Hair and Skin care products Cosmetics
Tom's of Maine	Shampoo Toothpaste Mouthwash Deodorant
Citra Hand Cleaner	Cleans off grease, paint, adhesives, ink, and tar

(Continued)

TABLE 8-4 (CONTINUED)

Company/Product Name	Types of Products
Natural Cedar	Air freshener spray
Carpet Stuff	Rug deodorizer
Air Therapy	Air freshener
Ringer	Hand cleaner
	Pet shampoo
OUT!	Pet odor eliminator
	Pet stain eliminator
Safer	Lemon pet shampoo

logically safe" cleaning products in the United States. Ecover produces a whole line of cleaners: laundry powders and liquids, fabric conditioner, dishwashing liquid, floor cleaner, toilet cleaner, and so on. Green products are not cheap. Ecover's laundry powder comes in a 105-ounce size for $17.95, which is 17 cents an ounce. By comparison, a regular laundry detergent costs about 10 cents per ounce (for a 42-ounce box) and as little as 4 cents per ounce for a 125-ounce box. Green products can be over-priced.

Home-mix alternatives to consumer products are made from water and vinegar, lemon juice, and baking soda (sodium bicarbonate). These are weak acids and alkalies or diluted versions of strong ones. An advantage of home mixes, compared to commercial products, is that users know exactly to what they are exposed. There are no hidden inert ingredients, no petroleum-derived solvents. Table 8-5 is a list of some of the many home-mix alternatives.

Detergents

An estimated 15 billion pounds (7.5 million tons) of detergents end up in the nation's waterways each year, almost half of them laundry detergents from households. Other products include automatic dishwasher and liquid dishwashing detergents and hand soaps. Throughout the 1980s, the consumption of laundry detergents alone increased by more than 4 percent per year—much faster than population growth. Americans now consume one-half of the world's output of the major detergent ingredient. It appears that cleanliness is next to wastefulness. Although the ingredients in detergents can be toxic to humans, their impact on water resources may be of even greater potential harm.

Some of the complex hydrocarbon compounds in detergents ultimately

TABLE 8–5.   SOME HOME-MIX ALTERNATIVES

For or To	Try
General cleaner	Mix of baking soda in water Hot water, soap and borax
Surface cleaner	Mix of vinegar, salt, and water Washing soda and water (1/2 cup per bucket); doesn't work on aluminum
Stubborn spots	Rub with 1/2 lemon dipped in borax-rinse; dry
Bathroom/tile	Baking soda and water
Mold and mildew	Concentrated solution of borax or vinegar and water
Clean windows/glass	Mix 1/4 cup vinegar in 1 quart warm water Mix 1 or 2 tablespoons lemon juice in 1 quart of water Remove residual wax from cleaners with alcohol; clean with 50/50 solution of white vinegar and water
Clean oven	Baking soda and water Baking soda, salt, and water
Clean glass oven door	Wet sponge dipped in baking soda
Clean toilet bowls	Toilet brush and baking soda or mild detergent
Clean coffee pot	Vinegar solution
^aUnclog drain	Plunger Flush with boiling water, 1/4 cup baking soda and 2 ounces vinegar
Garbage disposal odors	Baking soda Grind lemon rinds
Air freshener	Ventilation Set vinegar out in open dish Simmer cloves and cinnamon in boiling water Open box of baking soda in refrigerators or closets
Disinfectant	Mix 1/2 cup borax in 1 gallon of hot water
Polish brass	Use Worchestershire sauce Use 50/50 mix of salt/flour with a little vinegar Use water in which onions have been boiled

(Continued)

TABLE 8-5 (CONTINUED)

For or To	Try
Polish silver	Soak in boiling water with baking soda, salt, and piece of aluminum
Polish copper	Lemon juice or hot vinegar and salt
Shine chrome	Rub with newspaper; or rubbing alcohol; or white flour in dry rag Rub with baby oil and soft cloth
Clean linoleum floors	Solution of 1 cup white vinegar mixed with 2 gallons of water
Polish linoleum floors	Club soda
Wood floors	Soapy water to clean and soft cloth to shine
Clean rug/upholstery	Dry cornstarch sprinkled on surface; vacuum up
Remove spots	Club soda, lemon juice, or salt Immediate cold water Cornmeal and water paste; brush when dry
Floor/Furn polish	1 part lemon juice, 2 parts olive or vegetable oil
Laundry clothes	Soap products; boost with washing soda
Laundry bleach	1/2 cup sodium hexametaphosphate per 5 gallons of water
General bleach	Borax or washing soda
Automatic dishwasher	In soft water: 50/50 borax/washing soda In hard water; lowest-phosphate detergent cut with baking soda
Pool chemicals	Ozone or ultra violet light system

[a]To prevent clogging pour down boiling water once a week.

break down into carbon dioxide and water as they move through sewers and into septic tanks and water treatment systems, although intermediate products such as ethanol are also produced. Compounds with nitrogen form nitrates. Up to 50 percent of detergents can be phosphates. Phosphates have been identified as a pollutant in water bodies because they reduce the dissolved oxygen content, producing an environment that encourages the growth of algae.

A detergent cleans by moving the dirt on a fabric or a dish into the wash water and keeping it there. Detergent raw materials are synthesized from chemicals such as benzene, ethylene oxide, and propylene and from natural oils such as coconut and palm oils. Primary ingredients are surfactants (surface active agents), builders (sequestrants), and additives. Powdered laundry detergents are 15 to 20 percent surfactants and 23 to 55 percent builders. The formulation of automatic dishwasher detergents and liquid dishwashing detergents is similar. Of the three, dishwashing detergents are considered the gentlest and dishwasher detergents the harshest.

There are more than 400 different synthetic surfactants. The most frequently used are linear alkylbenzene sulfonates (LAS), alcohol ethoxylates, alkylphenol ethoxylates, alcohol ether sulfates, and alcohol sulfates. These and other surfactants are also used in personal care products such as soaps, shampoos, cosmetics, and toiletries. (Soap, the original surfactant, is now used principally in toilet bars.) With so many chemicals, general statements about toxicity are inappropriate. Small doses of some surfactants can cause low to high toxic effects when ingested; some are skin irritants. The Department of Commerce calls LAS one of the safest surfactants on the market because it is rapidly biodegradable, is nontoxic in the concentration used, and does not accumulate in the environment.

Builders are added to help the surfactants work by lowering the calcium and magnesium content of water (i.e., to make hard water soft). The principal builder is sodium tripolyphosphate, where allowed. About 12 states have banned or restricted the use of phosphates in household *laundry* detergents. For those markets, phosphates have been replaced with inorganic compounds such as soda ash, sodium silicates, sodium aluminosilicates, and citric acids (in liquid detergents). These substances can be more caustic than phosphates; in general, builder chemicals are skin and eye irritants and can be moderately toxic when small doses are ingested. Comprehensive data on the environmental fate of these alternatives is not available. Although phosphate use is declining in household laundry detergents, dishwasher detergents and institutional detergents still contain phosphates. Liquid laundry detergents (over 40 percent of the market) do not contain phosphates because the compounds are not stable in liquid form.

There is a trade-off between performance and environmental friendliness. In a trade advertisement on additives, the AKZO Chemical Company says: "Making detergents gentle to the environment while retaining cleaning and conditioning properties is one of the major challenges faced by detergent manufacturers." Depending on the claims of a detergent, current additives may include bleaches (sodium hypochlorite, sodium perborate, and hydrogen peroxide), bleach activators (tetraacetylethylenediamine), antiredeposition agents (cellulose ethers), antistatic agents, and fabric softeners (quaternary ammonium compounds). Millions of pounds of optical brighteners or fluorescent whitening agents (principally stilbene derivatives) were added to detergents in 1988 to make clothes *appear* cleaner. Other additives are toxic metals and VOCs (in handwashing and liquid detergents). Chlorine bleaches in detergents can react in wastewater with organics to form trihalomethanes, creating a problem for wastewater treatment facilities.

Detergent manufacturers have faced two major reformulations because of the water pollution that detergents create. The first change occurred because the most widely used surfactant was not biodegradable. LAS was substituted because it does decompose in water. However, that is not to say that all of the components in a detergent product biodegrade. Liquid Tide and Cheer bottles have a long list of ingredients (including solvents such as ethyl alcohol and propylene glycol). The labels say: "The surfactants and enzymes . . . are biodegradable." While the surfactants may constitute the largest percentage of the ingredients, the statement implies that the balance are *not* biodegradable.

How some surfactants biodegrade is not as well known. For instance, high levels of nonylphenols have been found in sewage sludge and in effluents from municipal treatment plants. Nonylphenols are used in the production of nonionic ethoxylated surfactants, some of which end up in detergents. (Nonylphenol is also an inert ingredient in pesticides.) The EPA signed a consent order in February 1990 under which a group of chemical manufacturers agreed to conduct a testing program to better understand the chemical and environmental fate of nonylphenols in aquatic environments.

The detergent industry and its suppliers are still searching for a "perfect" replacement for phosphates. According to the industry, the performance of detergents that don't have phosphates is not as good as that of detergents with phosphates. (*Consumer Reports* says that the difference is only slight.) In lobbying against bans, the industry also argues that consumers will end up paying more for detergents without phosphates. The trade-off is, of course, improved water quality. In the Chesapeake Bay area of the East Coast, phosphates were banned or restricted by Maryland (in 1985), the District of Columbia (in 1986), and Virginia (in 1988). The bans have had a dramatic effect, reducing phosphorus levels by 45 to 50 percent from sewage plants in Maryland and Virginia. The plants also save money (over $550,000 a year in Maryland) because less has to be spent to control phosphate levels.

Because of the wide range of ingredients in and formulations of detergents, their overall human toxicity varies. The CPSC labeling guidelines for detergents list six different permutations of acute toxicity, corrosivity, and eye and skin irritations. Where not banned, people can avoid phosphates by using liquid laundry and some liquid dishwasher detergents. They can avoid some of the additives in detergents because laundry detergents are now being formulated without perfume and dyes. Detergents are avoidable by using soap, the precurser of detergents. Soap (such as the Ivory brand) is discounted by some because it does not clean as well as detergents (as measured by today's high standards). If consumers can lower their standards a bit, using soap products will lower personal exposure to and environmental loading of petroleum-based surfactants. Soap also eliminates the need for fabric softeners. Detergents clean so well they leave fabrics feeling scratchy. Fabric softeners solve this problem by thinly coating the fibers. Fabric softeners also reduce the buildup of static electricity produced by the tumbling action on synthetic fibers in dryers. Using soap does not eliminate the static problem. At the very least, consumers can reduce consumption of detergents by lowering the amount used per washing machine load.

Oven and Drain Cleaners

Detergents, with their myriad ingredients, make it difficult to balance the pros and cons of use. With oven and drain cleaners, it is quite clear. The best advice is to avoid them. They are two of the most hazardous household products. Over 2,000 people were injured seriously enough by lye-based drain cleaners in 1988 to require a trip to a hospital.

Because they both usually contain strong acids or alkalies, the toxic effect of those cleaners is highly acute and local. Their corrosiveness—which can cause severe damage to skin, eyes, and lungs—is what makes them work. These products, when in liquid form, may also contain petroleum distillates, which can acutely affect lung tissue. Aerosol oven cleaners are the most dangerous. Not only do they contain a VOC as a propellant, but the spray action makes it difficult to avoid being exposed to the active ingredients.

Substitutes for oven cleaners include noncaustic versions (one brand is Easy-Off Non-Caustic Formula) and home-mixed solutions (see Table 8–5). Preventive, regular oven cleaning will obviate the need for anything other than a mild cleaner. Regularly used ovens can get to a state of equilibrium in which spilled items are baked off the surfaces. Self-cleaning ovens are designed to do the same, although at higher than normal baking temperatures.

Clogging of drains can be prevented by a weekly flushing with salt and water. For unplugging a drain, mechanical devices can be effective. A simple plumber's helper costs only about $2 and can be used over and over again. A plumber's snake or drain auger (from $7) can chew its way through a clog. A device called the Pango Modelo Brevettato, which delivers a blast of air, is sold by Brookstone stores and catalogs.

Disinfectants

The annual sales of disinfectant products, along with deodorizers (air fresheners), totaled $600 million in 1988. Disinfectants, which only temporarily reduce household germs (bacteria and fungi), are pesticides and, therefore, must adhere to the pesticide testing and labeling regulations (see below). However, prior to a 1988 revision of the law, disinfectant products did not have to be tested for chronic toxicology or environmental fate. Reviews of new test data may not be completed by the government until the mid-1990s.

Like pesticides, disinfectants contain a small amount of active ingredient mixed with solvents. Reported ingredients include diethylene or methylene glycol (TRI), formaldehyde (TRI), various phenol compounds (TRI), bezalkonium chloride, sodium hyprochlorite, and quaternary ammonium compounds. Disinfectants can be toxic, especially those containing glycols, phenols, and other VOCs; some are corrosive. In addition, phenols may cause problems for wastewater treatment plants. The ecological toxicity of quaternary ammonium compounds is uncertain. (They were canceled in 1973 for use as a sanitizer in poultry drinking water.) In a trade journal, industry sources (Binenstock 1989) were credited with saying that "even though they are, of necessity, somewhat toxic, disinfectants greatly improve the quality and safety of life. In any cost/benefit analysis . . . that would be a persuasive argument."

Disinfectant ingredients, as well as parabens, are used as preservatives in many household and almost every personal care product. Their purpose is to extend the product's shelf life by retarding the development of bacteria. Extending the shelf life reduces the distribution costs of products. Consumers may pay less for preserved products, but it is questionable which is more harmful—the bacteria or the chemicals. Antibacterial preservatives have been found to be a major cause of allergic reactions to cosmetic products. These products are not labeled as pesticides because they are not sold as such.

Because disinfectants provide only temporary relief from germs, if they are not used repeatedly, little is accomplished. But the repeated use of a disinfectant means exposure to the pesticide's active ingredient and VOCs. The prudent choice seems to be to avoid them altogether. People who are highly sensitive or allergic to certain bacteria and fungi may have to consider sterilization to achieve effective relief.

Lysol brand products have 90 percent of the disinfectant market. The active ingredient in many Lysol products is O-phenylphenol (TRI). The compound was monitored in an EPA study of residual household pesticides in indoor and outdoor air. It was found in relatively significant concentrations in indoor air. The newest Lysol entry is Bathroom Touch-Ups, sheets of paper toweling impregnated with disinfectant. This product, which Lysol says "cleans, disinfects, deodorizes," and "leaves your bathroom smelling fresh," is marketed to satisfy consumer desires for cleanliness and convenience. Advertisements say: "just wet, wipe and flush away." The ease of use may encourage people who don't already disinfect to start doing so, as well as hold on to customers who seek disinfecting convenience. The product caters to the throwaway ethic, which we cover in Chapter 7. These single-use products (like paper towels) increase the quantity of household waste. For a consumer who already uses a liquid disinfectant and applies it with a paper towel, Touch-Ups may reduce packaging waste because the plastic bottle is eliminated. Reduction depends, however, on how much disinfecting is possible with a box of Touch-Ups versus a liquid.

Another new product on the consumer market is liquid "germ-fighting" hand soaps. Plain soap, these products imply, no longer gets hands clean enough. Liquid Dial is billed as "the ultimate weapon in germ warfare." Softsoap is for relief in "the daily fight against dirt and germs." The active ingredient in the Dial product is 2,4,4'-trichloro-2'-hydroxydiphenyl ether (triclosan). Ethers generally affect the nervous system, and this chemical is rated as being moderately acutely toxic. Antibacterial soaps, although they may contain pesticide chemicals, are not registered with the EPA. Washing hands with regular soap does not kill bacteria but is quite effective at removing them and, of course, using soap eliminates the uncertainty of these new products.

Air Fresheners/Deodorizers

Despite their name, air fresheners do not freshen. They mask (or cover up) existing odors in the air. They coat nasal passages with an oil film or decrease the sense of smell with a nerve-deadening agent. Solid forms of air fresheners have a VOC content of 5 to 10 percent; aerosol forms, of 90 to 99 percent. More important, the active ingredient may be paradichlorobenzene (TRI), a probable human carcinogen.

A superior freshening method is to let in outside air, although it too may contain toxics. Still, this will help to reduce indoor air pollution in houses that are so tightly sealed that they do not have adequate air exchange. Certain houseplants—*Draceana massangeana, Spathiphyllum,* and Golden Pothos—and flowering plants (Gerbera daisy and chrysanthemum) will help to purify indoor air.

Some Specialty Cleaners

Specialty cleaners are a class of at least 13 different kinds of cleaners that generated $1.5 billion in sales in 1988. According to some market analysts, they are the most active segment of the household cleaning products market.

Most *glass cleaners* include ammonia or vinegar as an active ingredient. They may also include surfactants. The inert ingredients are most likely solvents, such as isopropyl and butoxy-ethanol. Their toxicity depends on specific surfactants and solvents used, which differ among brands. Ammonia vapors can damage lungs. Glass cleaners are easy to avoid, though. Windows can be cleaned with water alone if they aren't too heavily soiled. Home mixes of vinegar and water clean as well as most commercial products and are at least 10 times cheaper to use. *Toilet cleaners* with ingredients such as hydrochloric or oxalic acid and calcium hypochlorite are corrosive and will irritate eyes and skin. Some may contain the pesticide paradichlorobenzene (TRI).

Now that single-purpose products conflict with consumers' desire for convenience, the industry has developed "special" *general-purpose cleaners.* They are liquids that dissolve, emulsify, or suspend dirt, and their principal active ingredients are solvents (such as pine oil and ethanol) and ammonia. They can be highly alkaline. General-purpose cleaners are designed to be used full strength as spot cleaners or diluted for floors and walls. At full strength, some of them can mar surfaces. A cleaner that is less strong may not be as efficient in cleaning. The trade-off for efficiency can be, of course, higher-potential toxicity.

Waxes and Polishes

There are polishes for nails, shoes, floors, and furniture, and there are polishes for metals like silver and brass. With sales of $300 million in 1988, polishes and waxes make up the smallest segment of the household cleaning market. They generally contain petroleum distillates. Specific ingredients can include ammonia, isopropanol (TRI), diethylene glycol, or nitrobenzene (TRI). Products containing more than 10 percent petroleum distillates or diethylene glycol are defined as hazardous substances by the CPSC and must follow special labeling requirements. Metal polishes can be acidic. Wood polishes may contain phenol (TRI). Nail polish and polish remover contain acetone (TRI).

Bleaches

The basis for bleaches are chlorine (as sodium hypochlorite) or borate (as sodium perborate). Sodium perborate is used in powdered bleach products and is less toxic than sodium hypochlorite. An alternative to either is washing soda, a by-product of the baking soda production process. It is available in most stores in the laundry products section.

Products with sodium hypochlorite concentrations of over 10 percent are supposed to be labeled "poison" because they can burn human tissue. Clorox bleach ingredients are sodium hypochlorite (5.25 percent) and inerts (94.75 percent). The label has a warning that the product can cause "substantial but temporary eye injury," is harmful if swallowed, and may irritate skin.

A fragrance has been developed to mask the chlorine and other odors in household bleaches. This seems to be a step in the wrong direction, since malodorousness can help prompt customers to be cautious when using corrosive bleaches.

Automotive Products

These products are made from various organic compounds and are chemically treated with additives. Toxicity, then, depends on what organic compound or compounds are used. Gasoline itself contains benzene (TRI), a known human carcinogen, and other TRI toxics such as 1,3-butadiene, toluene, and xylene. Higher concentrations of benzene, toluene, and xylene are found in high-octane gasoline. Most people focus on the contribution that gasoline makes to VOCs and, ultimately, to the greenhouse effect. Few consider that exposure to the toxic fumes occurs whenever one pumps gasoline at a gas station that does not have vapor recovery nozzles.

There are few nontoxic substitutes for automotive products. Two toxic products of note are antifreeze, made from ethylene glycol (TRI), and windshield cleaning solutions with methyl alcohol (TRI), also known as methanol. Products containing more than 4 percent methanol are required to have the following label statements: "Vapor harmful" and "May be fatal or cause blindness if swallowed." Ethylene glycol products are also subject to special labeling rules. A new antifreeze, made from propylene glycol, is being marketed in Canada and may move into U.S. markets.

The very popular WD-40 aerosol product is used in thousands of households, both in garages and throughout the house. It is labeled with the word "Danger" but lists only one ingredient, petroleum distillates. The signal word "Danger" and the ingredient listing imply that WD-40 is more than 10 percent petroleum distillates. Additional warnings on the product refer to extreme flammability, swallowing, and inhalation problems.

Since there are few alternatives, many automotive products can be avoided only by not having an automobile or by having all servicing done by professionals. Neither is an optimal solution. The latter option simply transfers the risks of automotive products. Not having a car is an appropriate prevention reaction to global warming but may be practical only for people living in cities or areas with extensive public transportation.

But consumers may be able to reduce their exposure somewhat by careful selection of products. We found three engine degreasers with different kinds of labeling; two contained petroleum distillates, and the third had no product content information. One with petroleum distillates had the signal word *danger,* which implies that the concentration is greater than 10 percent. The manufacturer called it a "heavy duty" degreaser. The other petroleum distillate degreaser had a *warning* signal, thus, its concentration was below 10 percent. For acute toxicity, then, the second of the two reduces exposure to petroleum distillates. The third product—with the mysterious contents—had a *caution* signal (legally equivalent to the warning on the second product) but a note that the CPSC had certified it as nonflammable. The caution label implies that the product is a strong sensitizer or irritant. None of these differences among the three products deal with potential chronic exposure.

Paints and Coatings

This broad category of products includes not only paints but also preservatives, removers, brush cleaners, thinners, lacquers, and varnishes. About $11 billion worth of such products are sold each year, half of them for commercial and household use. The amount sold to individuals for household use was 200 million gallons in 1987.

Over 1,000 raw materials (solvents, pigments, resins, and additives) are used to formulate paints; most come from petrochemicals. A study conducted in 1986 found over 300 toxic chemicals in household paints; half of them were assessed as probable carcinogens. VOC emissions from household and commercial paints and coatings in California have been estimated to account for 18 percent of all solvent product emissions. Highest exposures occur during application and as paint dries, but fumes can continue to be released for months or years afterward.

Paints are either solvent- or water-based. The solvent content of conventional paints is 40 to 50 percent by volume. Lacquers, however, have the highest solvent content, at 80 percent or more. By regulation, lead (a pigment) content has to be less than 0.06 percent. Many different solvents are used in paints; the ones on the TRI list include toluene, xylene, butyl alcohols, methyl ethyl ketone, acetone, methyl isobutyl ketone, ethylene glycol, and various chlorinated solvents. Additives include mercury compounds as preservatives and mildewcides. Fungicide uses of mercury were banned in 1976; its use in paint was one of the few exceptions allowed. In 1990 paint manufacturers agreed to voluntarily eliminate mercury from interior latex paints after a child had an acute toxic reaction to fumes inhaled after his home was painted. Mercury will remain in exterior paints.

Using water-based (latex) instead of solvent-based paints avoids releasing and inhaling VOCs, but exposure to other toxic substances in paints remains. Equipment used to paint with latex can be cleaned with water (instead of toxic solvents). Easy cleanup may decrease the tendency to use brushes and rollers only once, thereby reducing waste as well. One drawback of latex paints is that they may not perform as well as solvent-based paints on surfaces that tend to get wet (bathrooms, kitchens).

Wood preservatives are technically pesticides and are often reported to contain pentachlorophenol (TRI). But in 1984 the chemical was canceled under the pesticide regulatory system, and consumer wood preservative products are not supposed to contain it (or creosote) any longer. It is possible for consumers to buy wood products that have been treated with pentachlorophenol, however. Under a voluntary industry program, consumers are supposed to be notified about possible exposures to pentachlorophenol.

Of the many specialty coatings available, paint thinners are solvents and rust paints can contain methylene chloride (TRI), petroleum distillates, and toluene (TRI). Paint removers/strippers contain methylene chloride (TRI), a probable carcinogen, which easily vaporizes. The fumes can ignite explosively and produce an acutely toxic gas that can destroy metal. The label of one such product, which contains acetone (TRI), toluene (TRI), methanol (TRI), petroleum distillates, *and* methylene chloride (TRI), says "Risk to your health depends on level and duration of exposure. Reports have associated repeated or prolonged occupational overexposure to solvents with permanent brain and nervous system damage." Some manufacturers are now substituting perchloroethylene (TRI) in strippers to avoid such explicit labeling.

A category of paints that doesn't get much attention is art and hobby supplies. These materials have generally been exempted from hazardous substance labeling regulations despite the fact that they can be highly toxic. Part of this toxicity is due to pigments, part to a variety of solvents. Art materials can contain the following TRI substances: arsenic, asbestos, lead, formaldehyde, antimony, cadmium, chromium, manganese, or mercury. Art paints are exempt from the regulation that keeps the lead content of household paint below 0.06 percent. A new labeling law, which goes into effect in November 1990, requires cautionary statements on art materials that have the potential to produce chronic effects.

One way to reduce exposure to paint products is to reduce the number of paint applications by buying the most durable latex paint. Another is to buy alternative paints made by Auro and Livos, two West German firms. Their ingredients include pine resin, linseed oil, chalk, india rubber, and mineral-derived pigments. The Livos line includes paint, wood preservatives, shellac, spackles, adhesives, thinners, and children's art materials. 3M is marketing a paint stripper called Safest Stripper that does not contain methylene chloride.

Pesticides

Pesticides are meant to kill, hopefully selectively. The risks posed by the use of pesticides are widespread and highly uncertain. One source of pesticide exposure is food. But, as discussed in Chapter 6, only half of the pesticides produced annually are for agriculture. The other half are used in industry and to rid houses, gardens, lawns, and buildings of unwanted pests, weeds, bacteria, and fungi.

Depending on what products are included, the consumer pesticide market has been estimated at $1.0 to $2.3 billion for 121 to 300 million pounds of insecticides, herbicides, fungicides, and rodenticides. Pesticide chemicals are also used in disinfectants, lice shampoos, pet flea collars and sprays, shower curtains, paints, bedding, and food packaging. People don't normally think of mothballs as pesticides, but they are and can consist of the TRI chemicals napthalene or paradichlorobenzene.

Consumers tend to be overusers of pesticides. Surveys have found that households use more pounds of pesticides per acre than do farmers; in one survey, the rate was four times more. Because they want quick reactions from pesticides, consumers buy $550 million worth of chemical cockroach sprays a year that provide a visible kill but cause the roaches to build natural defenses, so that each year more and more insecticide is needed.

The pesticide market is growing despite increasing efforts to publicize the environmental costs. Why? Because, according to Alan Caruba (1989), a pesticide industry public relations counselor:

> the public does not want to live with pests and vermin—flies, ants, cockroaches, mice, rats. Nor do they want termites destroying their biggest private investments—their homes. This is why pesticide manufacturers continue to exist—in fact, more than exist: the market for pesticides is not just good, it is growing.

Another reason for growth in the consumer market is a maturing of the agricultural market. Agricultural firms are coping with that potential decline by developing pesticides for home use.

Active Ingredients

The industry claims that today's pesticides are not as toxic as yesterday's because they don't persist as long in the environment. That only tells part of the story. The government has removed many pesticides (in some cases, only certain forms of the pesticides were removed) from the consumer market by classifying them as "for restricted use," which means that they can be applied only by commercial firms. But the toxicity of the remaining products is highly uncertain. Of 50 active ingredients in consumer pesticide products, Consumer's Union reported in1987 that 66 percent have been inadequately tested for carcinogenicity, 72 percent for mutagenicity, 62 percent for teratogenicity, 64 percent for adverse reproductive effects, and 98 percent for neurobehavioral toxicity.

Many of the restricted pesticides are organochlorines (chlorinated hydrocarbons) that have been replaced in the consumer market by other classes, primarily organophosphates and carbamates. DDT was banned in 1972; now only two organochlorines, methoxychlor (TRI) and endosulfin, are in general use. Organophosphates include malathion, diazinon, chlorpyrifos, parathion, and acephate. Some carbamates are carbaryl (TRI), propoxur (TRI), and aldicarb. Many carbamates are moderately to highly toxic; some may be carcinogenic, teratogenic, and/or mutagenic. Organophosphates do break down on plants within weeks or months but can persist for up to a year in soils. Aldicarb use was banned on Long Island, New York, because it was found to have contaminated groundwater, an indicator of long-term persistence. Before they break down, organophosphates can be more acutely toxic than organochlorine pesticides; they are the cause of most hospitalizations due to pesticides each year. Carbamates are less acutely toxic than organophosphates but can be converted into carcinogenic nitroso compounds in the stomach. When pesticides are tested, it is the active ingredients—not the breakdown products or metabolites—that receive most attention.

As with all other toxic chemicals, the exposure routes of pesticides include inhalation of vapors or dust and accidental oral ingestion. Pesticides, however, are modified versions of nerve gases, which means that they can be absorbed through the skin. It also means that they generally affect the central nervous system. The toxic effects can be either acute or chronic, but more is known about the acute effects. The acute toxicity of pesticides varies widely even within chemical classes. Based on the dose required to cause acute fatality, some pesticides are rated as having moderate or low toxicity. That means that a relatively large amount must be taken orally or absorbed through the skin to cause death. Skin absorption is deceptive, however. It is not possible to know when it is occurring, since there is no warning sensation. DEET (n,n-diethl-m-toluamide), a common ingredient in insecticide repellents (and flea sprays), is applied directly to skin and is rapidly absorbed by the body. It has been shown to cause serious side effects, but none of the tests provided the government by formulators have dealt with chronic exposure.

There are many pesticides on the consumer market. That fact, plus the uncertainty and variability of their toxicity (even on a chemical class basis), makes a comprehensive discussion impossible. Instead, we present the following pertinent, but sketchy, information on some of the most widely used pesticide chemicals:

- *2,4-D [dichlorophenoxyacetic aicd]* (TRI).   There are more than 1,500 formulations of this popular weed killer, 3 million pounds of which are spread around homes each year. It irritates lungs, the stomach, and mucous membranes. It can injure the liver, kidneys, and nervous system. The cancer risks are still uncertain, although various U.S. and European studies have shown a link to lymph node cancer. The IARC lists all chlorophenoxy herbicides as possible carcinogens.
- *Diazinon.*   This is a widely used insecticide, especially for lawn care. It affects the central nervous system and is highly toxic when ingested or exposed to skin. It is highly toxic to birds. The mutagenic, neurological, and reproductive toxicity in humans are unknown.
- *Carbaryl* (TRI).   This is an insecticide for gardens and pets. It may cause birth defects in dogs; chronic exposure causes loss of appetite and weight and weakness in humans. It also causes mutations in experimental animals. Carcinogenic, reproductive, and neurotoxicity data are incomplete.
- *Methoxychlor* (TRI).   This is an organochlorine insecticide that affects the central nervous system. There is high uncertainty about its toxicity, since all data is from the 1950s and 1960s. It has low acute toxicity.
- *Chlorpyrifos.*   This insecticide has been registered for sale for 24 years, but tests are lacking regarding its potential to cause cancer, mutations, or nerve damage. It is known to be extremely toxic to fish and birds. It is used in many insect sprays, including pet sprays, and is injected into the ground around houses to control termites. Its use is expected to increase due to the halt in manufacturing of chlordane.
- *Malathion.*   This product controls garden insects and household pests. It has low to moderate acute toxicity via skin and moderate to high toxicity through ingestion. Its chronic effects—whether or not it is carcinogenic, mutagenic, causes birth defects, or is a neurotoxic or reproductive toxic—are unknown.
- *Simazine.*   This an an herbicide used in ponds and swimming pools. A California study says that 8 of 10 earlier studies have incomplete data. All of its chronic toxicity effects (carcinogenicity, mutagenicity, neurotoxicity, birth defects, and reproductive toxicity) are unknown.
- *Captan.*(TRI).   This is a widely used commercial and household fungicide that has been classified as a probable human carcinogen since 1984. In 1989 the EPA decided, based on available information, that captan is an experimental mutagen but not a teratogen; that, while it does cause reproductive effects in animals, the risk to humans is acceptable; and that in terms of hypersensitivity, it is a mild irritant and a moderate skin sensitizer.
- *Dichlorvos* (TRI).   This is found in more than 800 products, including

insect sprays and pet flea collars and sprays. It is used in hanging pest strips that continuously emit vapors over 90 days or more. It is a possible carcinogen, mutagen, and may be neurotoxic.

An EPA study (1990) of nonoccupational exposure to 32 pesticides included all of the above except simazine. When air measurements were taken (in Jacksonville, Florida, and Springfield, Massachusetts), diazinon, chlorpyrifos, and dichlorvos were among those for which the highest mean concentrations were found indoors. The pesticide with the highest concentration in Jacksonville was propoxur (TRI), an experimental carcinogen that is widely used for indoor pest control, particularly cockroaches and flies, as aerosols and foggers, and in baits.

Both the EPA's pesticide study and the agency's decision regarding captan are interesting examples of how multiple exposures to either a variety of chemicals or to a single one are not considered when making risk determinations. In the study, the risks for the pesticides were individually calculated. Of 13 pesticides for which cancer risks and of 19 for which noncancer risks were determined, none individually exceeded the EPA's benchmark of 1 excess cancer in 1,000,000 or a noncancer hazard index of 1. The study essentially concluded that overall risks from exposure to 32 pesticides, then, were low or negligible. If the cancer risks are added up, however, a total risk of 5 in 1,000,000 is obtained, and that puts a different spin on the study. The number is well within regulatory limits. Even that number is probably low. Three pesticides for which data on carcinogenicity is lacking (chlorpyrifos, methoxychlor, and malathion) and one (dichlorvos) for which data confirms animal carcinogenicity were not included in the EPA's estimate of cancer risk.

After completing an 8-year review of captan, the EPA decided in 1989 to cancel some agricultural uses. The cancellation does not apply, however, to 24 food crops, any seed treatments, or any nonfood uses. (Captan is used as a preservative or protectant for awnings, draperies, bedding, and leather goods, and in soaps, cosmetics, paints, paper, wallpaper flour paste, and plastic.) Household pesticide use was not banned. The EPA estimated, individually, that the cancer risk from captan in home garden use and in a dozen or so nonfood exposures ranged from 1 in 10 billion to 1 in 10,000 (pet shampoos). When challenged on some of its conclusions by the Natural Resource Defense Council and the California Rural Legal Assistance Foundation, the EPA admitted that the risk from using cosmetics containing captan is unknown and recommended that the FDA study the matter further. Even given this and other uncertainties and not counting possible captan exposures from food, when the individual risks are added up, the range narrows down to between 1 in 100,000 to 1 in 10,000—again, well within normal regulatory limits.

Inert Ingredients

Inert does not mean benign. In fact, the risk from inert ingredients in pesticides may be higher than that from active ingredients because of dilution. Inerts can be up to 99 percent of the total content of a pesticide product. They can be chemicals, such as carbon tetrachloride (TRI), that have been banned as active pesticide ingredients, current pesticide chemicals, or toxic solvents such as petroleum distillates, methylene chloride (TRI), or xylene (TRI), used in over 2,000 pesticides.

Through a disputed process in 1987, the EPA divided all 1,200 inerts then in use in pesticides into four categories. The lists were revised in late 1989. List 1 contains the most assuredly toxic inerts. There are about 40 chemicals on the list; over half of them also appear on the TRI list, and half of those are known or suspected carcinogens. List 2 has over 60 chemicals that are candidates for testing because their structure or properties suggest toxicity. Twenty-four of them are on the TRI list; among them are familiar product ingredients like xylene and toluene. List 3 contains 800 to 900 chemicals of known toxicity but unknown risk, and List 4 consists of 273 inerts of minimal concern.

The outcome of listing inert ingredients does not mean that the worst ones will be removed from pesticide products. Nor does it mean that consumers will be given much more information about inert ingredients. For List 1 chemicals, manufacturers have to revise their labels to read: "This product contains the toxic inert ingredient———." The percentage of that inert and of the rest of the inerts, as well as their names, can stay hidden. List 2 chemicals may eventually be tested. Probably the most egregious inaction involves List 3 chemicals, those of highest uncertainty. The EPA does not expect to test or require formulators to test these chemicals, 30 percent of which have never been made public because they are proprietary.

Alternatives

The alternatives to current pesticides are almost as varied as the chemicals currently in vogue. They range from less toxic synthetic and natural chemicals to nontoxic biological and mechanical controls, barriers, and traps. Solutions vary depending on whether use is inside or outside of homes. Some are discussed here; others are presented in Table 8–6.

Changed aesthetics and improved information can reduce pesticide use. Deep green manicured lawns are highly sought but are an unnatural state. One of the costs is a high use of insecticides (diazinon) and herbicides (2,4-D). Simply accepting a lawn with several species of plants can eliminate the use of pesticides. It will also cut down measurably both the time it takes to maintain such a lawn and the cost of doing so. Commercial firms that treat lawns have access to more toxic pesticides than do individuals, and

TABLE 8-6. SOME ALTERNATIVES TO SYNTHETIC CHEMICAL PESTICIDES

For	Try[a]
Multiple pests	Pyrethrum[b] sprays (Hargate,[c] Ringer Insect Attack Traps with floral and sex scent lures (SureFire) Introduce colony of predators such as ladybugs, praying mantises, lacewings Insecticide soap (Safer, Ringer Attack) Protective netting (Reemay, Agronet, VisPore) Diatomaceous earth (DE) Wrap for tree trunks (Tree Skin)
Orchard pests	Visual attractant with sticky surface (Red Sphere, White Rectangle)
Roaches	Boric acid powder spread in appropriate spots; cheap, lasts for years; special boric preparations also sold Hydroprene growth regulators (Gencor)
Aphids	Traps with floral and sex scent lures (SureFire) Introduce colony of ladybugs Insecticide soap with citrus (Ringer Aphid-Mite Attack)
Yellow jackets	Traps with floral and sex scent lures (Surefire, Rescue)
Grasshoppers	Infect with spores of *Nosema locustae* (microbe) (Ringer Grasshopper Attack)
Ants	Insecticidal soap Essential oils concentrate; repels by scent (Ant-Free) Pour line of cream of tartar, red chili powder, paprika, or dried peppermint at point of entry
Fleas	Insecticidal flea soap (Safer, Indoor Flea Guard) Methoprene growth regulators; May include synthetic pyrethroids,[b] however (Precor) (Zeocon) Powders made from pyrethrum[b] (Natural Animal Flea Powder, POW) Herbal shampoo and flea dip Herbal flea powder (Green Ban) Herbal flea collar (Shoo) Prevent by feeding pets brewers yeast or vitamin B (Bits Anti-Flea vitamins)
Flies	Flypaper; purchase or make by painting a board or surface bright yellow and cover with Tanglefoot.

(Continued)

TABLE 8-6   (CONTINUED)

For	Try[a]
	Fly traps (Big Stinky Fly Terminator) Screens as barrier to prevent entry
Japanese beetle	Trap with sex attractant (Bag-A-Bug, Surefire, Catch Can)
Lawn grubs	*Bacillus popilliae* infects grubs with Milky Spore disease (Ringer Grub Attack)
Lice	Products containing pyrethrins[b] (Triple-X, A-200 Pyrinate, Rid)
Moths	Line closets with cedar Cedar balls, blocks, and chips Cembra Essence added to shellac; applied to chests/closets/drawers (Livos) Hebal sachet (Moth Away)
Soil dwellers	Beneficial nematodes (BioSafe, Scanmask)
Worms/catepillars	*Bacillus thuringiensis* (BT), a bacterial insecticide
Slugs	Diatomaceous earth (but is not selective) Lava pellets with herbal formula; repel slugs (Slug Off)
Slugs/snails	Put beer (as bait) in a container (Slug Bar) Special bait (malt formula) with trap (Snailer, Slug Saloon)
Insect repellant	Herbal liquid (Green Ban)
Nematodes	Chitin-protein products promote growth of beneficial bugs (Safer ClandoSan) Plant marigolds nearby
Gophers/moles	Vibrating probes set into soil (Gopher Deflector, Mole Mover, Go'pher It II)
Rodents	Lava stones soaked in herbal formula; repel (Rodent Rocks)
Mice	Reusable baited traps (Havahart)
Weeds	Soap (Safer, SharpShooter) Pull weeds Cover soil with weed control fabric

TABLE 8-6   (CONTINUED)

For	Try[a]
Fungi	Minerals (quartz, bentonite clay, coral lime) concentrate (Sil Ka Ben) Soap (Safer); also moss/algae killer available Microbes that inhibit and suppress growth (Ringer Dis-Patch)

The names in parentheses are sample product brands that can be purchased through one of he following companies: The Necessary Trading Company (703) 864-5103; Gardener's Supply (802) 863-1700; Ringer (800) 654-1047; Gardens Alive! (812) 623-3800.
Pyrethrins is the family name for insecticides derived from natural botanical sources. Pyrehroids are closely related synthetic compounds.
This is an aerosol spray; the propellant is carbon dioxide.

the GAO reported in 1990 that the government does not have an effective enforcement program against false claims of safety claims.

Misinformation about cockroaches leads Americans to support half of the consumer pesticide industry. Use of insecticide sprays is a vicious circle. Spraying does not stop roaches from breeding. The more spray used, the more that has to be used because roaches build natural defenses against them. Unlike fleas, bedbugs, lice, mites, scorpions, and spiders, roaches are not a public health problem. So, while roaches are unpleasant creatures, they are not normally a hazard. Alternative controls are boric acid powder placed in strategic locations, plugging of entry cracks, and removal of trash and clutter. Boric acid is so long-lasting that it can be placed behind walls as a house is built to prevent roach problems for years.

Other household insects can be controlled in various ways. Flies can be caught, either in traps or on flypaper. Soap and water are effective against ants. Fleas can be treated with insecticidal flea soap, a product safe enough to use on pets, on rugs, and in bedding. Soap applications have to be repeated often, though.

New developments in insecticides are insect growth regulators (IGRs) such as methoprene for fleas and hydroprene for cockroaches. IGRs have been developed because of the growing ability of insects to achieve resistance to traditional pesticides. Many IGRs are based on insect hormones and act by interfering with key aspects of an insect's physical development such as molting and metamorphosis. IGRs can be inherently selective (apply only to specific species) and are not quick killers; they can take days to work. The lack of speed, however, means that insects can track IGRs back to nests, widening the effect of a single application. The substances are instable and thus are not persistent in the environment. However, although methoprene is not considered to be carcinogenic, its neurotoxicity has not been determined. Another new form of insecticide is a chitin inhibitor that affects the outer covering (exoskeleton) of an insect.

Gardeners who want to avoid pesticides start by learning integrated pest management (IPM), just as commercial organic farmers do. That means taking a systems approach to gardening. It includes using cultural controls (such as maintaining healthy plants through proper soil treatments), mechanical controls, natural chemicals, and botanical controls. The Necessary Trading Company (a gardening mail order firm) suggests the following IPM steps: get to know your enemy; use good bugs; trap bugs; use gentle, selective controls; and, lastly, resort to botanical insecticides. But, first, use *prevention* through good management techniques like providing healthy soil and plants, appropriate varieties and timing of planting, and crop rotation.

Insecticidal soaps are a form of mechanical control. They contain potassium salts of fatty acids that penetrate soft-bodied insects (such as aphids, mealybugs, spider mites, and whiteflies) and cause death by dehydration. (A soap has also been developed as an herbicide.) Other mechanical controls are horticultural (dormant) oils that place an airtight seal over insects and suffocate them. Diatomaceous earth (DE) ruptures an insect's protective coating, causing dehydration. DE is not selective, so it will kill beneficial insects along with pests. Traps, such as a board or card painted bright yellow and coated with stickly tanglefoot, can attract whiteflies, aphids, and flies.

Natural (or botanical) chemicals are not necessarily nontoxic. In fact, some are more acutely toxic than some synthetic chemicals. Botanical chemicals (like nicotine, rotenone, ryania, pyrethrins, and sabadilla) are extracted from plants and tend to break down very rapidly once applied. This means that, to be effective, they must be applied often but also that they are unlikely to contaminate soil, water, or food crops. Nicotine and rotenone, however, are more acutely toxic to humans than many synthetic pesticides; a very small dose taken orally can be fatal. On the same basis, ryania and pyrethrins are moderately toxic; sabadilla is relatively nontoxic. The high acute toxicity of nicotine and rotenone means that extreme care has to be taken when applying them. The toxicity to animals of botanical pesticides varies widely. A caution, however: the chronic toxicity of some botanical chemicals is not well understood.

Biological controls—"bugs against bugs"—one of the oldest forms of pest control, are experiencing a revival. They tend to be selective, which makes them very safe but complicated for consumers to use. A variety of predatory insects (ladybugs and praying mantids) and soil microorganisms are available. BioSafe is a nematode delivery system that destroys soil-dwelling insects; a similar system for household cockroaches is being developed.

Aerosol Packaging

The aerosol spray can is not toxic, but the propellant that makes it work may be and the can is a form of excessive packaging. Not buying products in aerosol cans can reduce exposure to toxics, eliminate one source of VOCs that contribute to global warming, and reduce a household's quantity of waste.

This popular package contains a mixture of the product and a propellant (gas) under pressure. When the button is pressed, the propellant forces out a fine (aerosol) spray of the mixture. The propellant separates from the product by vaporizing and ends up in the air; the product, hopefully, lands on the target, but much is wasted. Aerosol cans are labeled with instructions not to inhale the fumes but, unless users close their mouths and hold their breath, avoidance is difficult to achieve.

CFCs, once the major aerosol propellant, were banned for most products in the United States in 1978. They are still widely used in the rest of the world and for about 2 percent of U.S. aerosols (certain medical and pharmaceutical products). Current propellants are liquefied gases like hydrocarbons (propane, isobutane, and N-butane), dimethyl ether, and compressed gases (mainly carbon dioxide, nitrous oxide, and nitrogen). Emerging as propellants are hydrochloro- and hydrofluorocarbons (HCFCs). HCFCs, however, as discussed in Chapter 5, are already receiving attention as inappropriate alternatives to CFCs.

An estimated 2.9 billion aerosol units were filled in 1988, a growth of 7 percent over 1987. These numbers tell the industry that there is a continuing consumer preference for aerosols despite environmental concerns expressed in a 1989 poll in which 75 percent of the respondents favored a ban on household aerosols. Industry refutes the environmental concerns by reminding the public that U.S. aerosol products are "ozone friendly" (i.e., do not contain CFCs) and by suggesting that since consumer products are not a *major* contributor of VOCs to the atmosphere, they ought not to be banned.

The health issue—direct inhalation of the propellant—is rarely addressed. The industry's Aerosol Education Bureau brochure on safe use of aerosols does say, "There are different chemicals of different strengths in various products, so a reasonable precaution is to keep [the] spray away from [the] face or eyes. Whenever heavy spraying is necessary, do this in a well-ventilated room." This brochure calls aerosols "a must in modern living" and lists no fewer than 61 types of household aerosol products.

Consumers don't have to wait for aerosols to be taken off the market. They can simply not buy them. There are plenty of alternatives without giving up the products or even switching brands. Most products packaged as aerosols are also sold as solids or liquids in a pump spray package in which the force causing the spray is finger action on the pump (rather than the gas under pressure in an aerosol). Pump sprays, while easier to control than aerosols, tend to be less fine and forceful. The loss of these properties is a small price to pay to eliminate what may be a major source of VOC releases in homes.

THREE WAYS TO REDUCE UNCERTAINTY AND TAKE CONTROL

It isn't easy to take control over toxic products, but consumers don't need chemistry degrees. There is a need, though, for consumers to take on the responsibility of becoming better informed. We suggest three ways that can help. They are presented in order of increasing difficulty; that is, the amount of knowledge needed to practice pollution prevention increases with each level. A person can pick the level at which he or she is most comfortable and stay there. Alternatively, a person can work through increasingly sophisticated levels. Using combinations of levels results in the most control. The levels are:

1. Use a household hazardous waste list.
2. Read labels for positive and negative signals.
3. Get to know chemicals by name.

Because of the lack of public information on the true content of all consumer products, all three levels have major imperfections. But, given the paucity of information, they are reasonable approaches.

Begin by taking an inventory of the toxic products already found at home. Most people will be amazed with the number and variety they have in stock. Consumers can test the three levels using their own inventory and asking three questions: How many household hazardous waste products do I own? How comfortable am I with the information provided by the labels? How many TRI chemicals do I have stored throughout my house, garden, and garage?

Level 1:   Using a Household Hazardous Waste List

The first step toward control is to narrow down the list of products in which one can expect to find toxic substances. The most concise form of this information is a *household hazardous waste* list. Many organizations and some government agencies now prepare and make these lists available, usually free. We have consolidated a number of lists into Table 8–7. There are two drawbacks to this level. First, because product formulations can differ among manufacturers of the same kind of product, not all versions contain toxic substances. Avoiding all products means ignoring some good ones. That is a consumer choice. If enough consumers use that power, manufacturers may start to realize the benefits of standing out from the pack by letting consumers know how their products differ in content from those of their competitors.

The second drawback is that household hazardous waste lists are aimed at getting people to take their toxic product residuals to a household collection place. That action is important once these products are in the home. But our message is different. Prevention means not buying these products in the first place to reduce both the health effects of use *and* the environmental and health damage of disposal. Another problem with household hazardous waste lists is that some of the information is irrelevant for making purchasing decisions. The lists often contain toxic products that may be stored in homes but are not on the market today.

Level 2:   Reading Labels

Once people know the kinds of products that may contain toxics, discretionary skills increase by taking the time to *thoroughly* read labels as part of making a purchasing choice. This weeds out the bad and locates the good products. There are two types of quick information on labels: signal words and precautionary statements. Despite this information, reading labels is not foolproof.

Locating the Good Ones

If consumers' desire for green products grows and is sustained, manufacturers will become more active in marketing safe products and in telling consumers why existing products are safe. Already some stores are adding shelf notices near products that they have decided are environmentally friendly.

The word *nontoxic* already appears on some products; its frequency should grow. Beware, however: there is no standard definition of *environmentally friendly* or *nontoxic*. Their use may become as confusing as that of the words *natural* and *organic* became when attached to food products. Without an accompanying ingredients list and the ability to interpret it, there is no way to know for certain what *nontoxic* means.

Some countries have designed label symbols to highlight good products. To qualify, products must pass tests and meet certain criteria. Until that

TABLE 8-7. PRODUCTS ON HOUSEHOLD HAZARDOUS WASTE LISTS

Category	Types of Products	Reasons for Toxicity		
		Organic	Metal	Corrosive
Cleaners	Surface cleaner	X		X
	Toilet bowl cleaner	X		X
	Disinfectant	X		
	Air freshener	X		
	Oven cleaner			X
	Drain cleaner			X
	Abrasive cleaner			X
	Window cleaner	X		X
	Wax	X		
	Floor & furniture polish	X		
	Metal polish	X		X
	Detergents	X	X	
	Chlorine bleach			X
	Rug & upholstery cleaner	X		X
	Mothballs	X		
	Ammonia-based cleaner			X
Automotive	Antifreeze	X		
	Transmission fluid	X		
	Brake fluid	X		
	Used oil	X	X	
	Batteries		X	X
	Degreasers	X		
Paints & coatings	Oil-based paint	X		
	Water-based paint	X	X	
	Rust paint	X		
	Thinners & turpentines	X		
	Strippers	X		
	Wood preservatives	X		
	Stains/finishers	X		
Miscellaneous	Photographic chemicals		X	X
	Adhesives	X		
	Art supplies	X	X	
	Pool chemicals			X
	Batteries		X	
	Shoe polish	X		
	Nail polish remover	X		
Pesticides	Insecticides	X		
	Herbicides	X		
	Pet care products	X		
	Fungicides	X		
	Disinfectants	X		

kind of system is available in the United States, consumers should be somewhat leary of marketing claims. At the most, without analyzing the contents, a product labeled nontoxic should be considered less toxic than its neighbor on the shelf. When a label says that a product is "nontoxic as defined by the Federal Hazardous Substances Act," this means that it does *not* have "the capacity to produce personal injury or illness to man through ingestion, inhalation, or absorption through any body surface." That sounds definitive, but the statute deals primarily with acute rather than chronic effects.

Routing Out the Bad Ones

Reading for negative information can be more definitive than looking for positive signals. However, negative information appears on products only when the government requires it, and there are unknown numbers of toxic products on the market that have fallen through the regulatory cracks. So, the lack of a negative signal on a product does not mean that it is safe or nontoxic.

Pesticide products and products containing hazardous substances are required to have the word *danger, warning,* or *caution* on their label. For pesticides, the meaning of these signal words is precise; for hazardous substances it is not, and the words *warning* and *caution* are equivalent. A precautionary statement that helps to define the risks of use is included on both kinds of products, and a warning to "Keep out of reach of children" must be on both.

First, some details on pesticides, which are split into one of four acute toxicity categories. Table 8–8 lists the signal word, hazard indicators, and precautionary statements by toxicity category. It also shows the toxic effects to be expected from each category. Even though the signal word for Categories III and IV is the same, only products in Category III are required to have a precautionary statement.

It is worthwhile to note how the three precautionary statements vary, especially in the first line: Category I (*fatal*), II (*may be fatal*) and III (*harmful*). Pesticide labels also include a statement about environmental toxicity, if any, as well as any physical or chemical hazards from use, such as flammability (the statement varies, depending on whether the container is pressurized or not).

It may be the manufacturer, rather than the government, that decides whether or not a product contains a hazardous substance. For such products, the signal words are *danger* for products containing toxic, extremely flammable, or corrosive substances; *danger* and *poison* for those containing highly toxic hazardous substances; and *warning* or *caution* for all other hazardous substances (an irritant, a strong sensitizer, or inflammable). Unlike the situation for pesticides, the latter two signal words are equivalent.

TABLE 8-8.   TOXICITY INDICATORS ON PESTICIDE LABELS

	Acute Toxicity Category		
	I	**II**	**III**
SIGNAL WORDS:	*Danger* (and *poison*)[a]	Warning	Caution
HAZARD INDICATORS[b]			
Oral LD_{50}	Up to and including 50	50 to 500	500 to 5,000
Inhalation LC_{50}	Up to and including 0.2	0.2 to 2	2 to 20
Dermal LD_{50}	Up to and including 200	200 to 2,000	2,000 to 20,000
Eye effects	Corrosive; corneal opacity not reversible with 7 days	Corneal opacity reversible with 7 days; irritation persisting for 7 days	No corneal opacity; irritation reversible within 7 days
Skin effects	Corrosive	Severe irritation at 72 hours	Moderate irritation at 72 hours
PRECAUTIONARY STATEMENTS			
	Fatal (poisonous) if swallowed (inhaled or absorbed through skin). Do not breathe vapor (dust or spray mist). Do not get in eyes, on skin, or on clothing.	May be fatal if swallowed (inhaled or absorbed through skin). Do not breathe vapors (dust or spray mist). Do not get in eyes, on skin, or on clothing.	Harmful if swallowed (inhaled or absorbed through skin). Avoid breathing vapors (dust or spray mist). Avoid contact with skin (eyes or clothing).

[a] Poison (with skull and crossbones) has to be added to labels that qualify for category I because of oral, inhalation, or dermal effects.

[b] Oral and dermal LD_{50}s are measured in milligrams of the substance per kilogram of body weight.

Approximate doses: A few drops to a teaspoon for Category I

A teaspoon to 1 ounce for Category II

1 ounce to more than 1 pint for Category III

Inhalation LC_{50}s are measured in milligrams of substance per liter of lung capacity.

Only *principal* hazards must be identified and phrases used include *Harmful or fatal if swallowed, vapor harmful, flammable,* or *skin and eye irritant.*

Other than food products and the new art materials labeling, the only other government labeling requirement is for cosmetics. But a warning statement does not have to appear unless a cosmetic is packaged as an aerosol or is a feminine deodorant spray, a foaming detergent bath product, or a coal tar hair dye (which may pose a risk of cancer).

To sum up the negative signal words, a product labeled:

- *danger* and *poison* contains a hazardous substance that is highly toxic or is a pesticide that is highly toxic by the nature of the internal reactions it causes.
- *danger* is a pesticide that is highly toxic (skin or eye effects) or contains a hazardous substance that is toxic, extremely inflammable, or corrosive.
- *warning* is a pesticide that is highly to moderately toxic or contains a hazardous substance that is flammable, an irritant, or a strong sensitizer.
- *caution* is a pesticide that is moderately or slightly toxic or contains a hazardous substance that is flammable, an irritant, or a strong sensitizer.

Keep in mind, however, that these ratings of relative toxicity may only consider a product's ability to cause an acute reaction. Some products may have a precautionary statement that indicates that they may be a carcinogen or cause other chronic effects.

Level 3:   Getting to Know Chemicals by Name

The bottom line is that it is specific chemicals that make products toxic. The most definitive way to practice pollution prevention, then, is to learn the names of toxic chemicals. This is also the most difficult method because there are hundreds of thousands of chemicals. Even if a person could recognize the important ones, few products contain complete lists of ingredients.

To cope with the thousands of chemicals, we have selected the TRI list of chemicals as a starting point. The list appears in Table 8-1, and throughout this chapter we have identified TRI chemicals when they appear. Still, a list of over 300 chemicals can be daunting. A first step might be to learn to recognize those that are considered carcinogens, or the VOCs, or those thought to be in consumer products (capitalized on our TRI list).

To make this level more complicated, few chemicals have only one name. Most have many synonyms, some of which may be trade names. Each chemical on the TRI list has at least three synonyms; some have up to 30.

An example is dichlorvos, an insecticide. It is also called 2,2-dichlorovinyl dimethyl phosphate, DDVP, and Vapona. Multiple synonyms means that two similar products may have the same ingredient but appear to be different. The only definitive way of distinguishing one chemical from another is to use the Chemical Abstracts Services (CAS) number. The thorough Duron paint label, mentioned earlier, included the CAS number for some ingredients. That made it quite easy to use a TRI list arranged by CAS number and discover that one of those ingredients (ethylene glycol) as listed.

The usefulness of this level of control is, of course, limited to products with ingredient lists. In some cases, manufacturers voluntarily list them. Most ingredient lists, however, are required and defined by the government. As noted above, those include pesticides, hazardous substances, and cosmetics. For pesticides, only each active ingredient must be named and its percentage content listed. All inert ingredients are shown as a total and are not individually identified unless an inert is on the EPA's Inerts List 1. Then it has to be identified by name. The active ingredients are listed by chemical name and common name (if there is one).

If a product has a hazardous substance, that substance has to be identified on the label (by common or chemical name), but there is no requirement for the percentage content of either the hazardous substance or any other ingredients. This means that you can't tell whether the named hazardous substance makes up 1 or 99 percent of the product; dilute concentrations of some substances can be highly toxic.

On cosmetics, each ingredient has to be named in descending order of predominance. There are exceptions, of course. Trade secrets are not listed. Substances that happen to be included in a product because they were part of the manufacturing process do not have to be named if they are present in *insignificant* levels and do not have any technical or functional effect (i.e., are inert). Cosmetics do not have percentage breakdowns of ingredients.

Some product manufacturers give a customer service telephone number (often a toll-free 800 number) on their labels. This service can be used to discover what chemicals are in a product and to get information about those chemicals. For a number of years, manufacturers have been required under occupational health and safety regulations to make product safety sheets (called *Material Safety Data Sheets,* or *MSDSs*) available for any chemical used in the workplace. Now some manufacturers provide them for consumer products as well.

SUMMARY

Consumers have a choice in regard to household toxic products. They can choose to believe that government policies and industry practices protect their health and that of the planet and continue to buy these products. Or they can choose to question those policies and practices, to recognize the inevitable high uncertainties that accompany the use of toxic products. Toxics use reduction—pollution prevention—is the response to that choice.

The best consumer is an informed consumer. Environmental and public interest groups help immeasurably to fill some of the gaps left by government programs. However, major reform is needed on labeling; the goals ought to be simplicity and consistency. Consistency means that *all* household products are subject to the same system rather than *some* being subject to differing systems. Simplicity demands easily recognized and understood negative or positive symbols or codes. And, despite manufacturers' reluctance to admit product toxicity, consumers have a right to have that information, regardless of whether or not the government decides that the possibility of exposure is low for the average person.

The consumer product right-to-know concept includes (1) acute toxicity, (2) chronic toxicity, (3) physical hazards of use, (4) environmental fate, and (5) for completeness, environmental wastes produced during manufacture. In a positive so-called *eco-logo* system like West Germany's *blue angel* or Canada's *environmental choice* program, a product has to meet a set of minimum criteria to win approval to use the logo. The certification is done by an independent board or institute. An official American logo would eliminate the confusion (and deception) that inevitably result from manufacturers using the words *green, nontoxic,* or *environmentally friendly* on their own. Such terms lack consistent meaning.

Conversely, in a negative marking system, a numerical code (or bar scale) could indicate the outcome in each of the five categories. If a simple rating—say, from 1 to 5—was used, a potential consumer could quickly compare similar products. A question mark could indicate that data cannot substantiate a conclusion. Consumers could read this system positively, by choosing a product with low numbers, or negatively, by avoiding products with high numbers. The set of numbers would also allow trade-offs based on individual value systems. A consumer may decide, for instance, to use a product that has high acute human toxicity but a positive environmental fate.

No system of labeling is truly effective without a continuous consumer education campaign. The combination would improve the marketplace. If consumers shun products that explicitly tell the bad news, then manufacturers will be driven to develop and market products that can attain low values in the numerical rating system. Conversely, if consumers flock to products with a positive eco-logo, then manufacturers will scramble to add products that will gain that stamp of approval.

Whatever its merits, an American eco-logo or systematic product coding may be a long way off. Consumers must contend with a system that almost always makes them responsible and, at the same time, not fully informed.

As an official of the CPSC, charged with protecting the public from harmful products, said (Dawson 1988): "The effectiveness of some products depends on formulations using ingredients which, *improperly used,* could be harmful." Unfortunately for the consumer, government and industry tend to apply this attitude too broadly. It may not be possible to formulate a product that, when accidentally swallowed, would not cause some acute toxic reaction. At the other extreme, however, are products that contain substances that, by their nature and with even proper use, are difficult for consumers to avoid. Examples include inhalation of VOCs and pesticide absorption through skin.

The high uncertainties of chronic toxicity and the known statistics on consumer injuries from acute exposure to consumer products suggest that more industry and government responsibility is needed. Until it is provided, consumer responsibility includes not just proper usage but taking control through prevention. Pollution prevention means less consumption of toxic products and products suspected of being toxic.

Taking control doesn't have to mean total elimination and fundamental restructuring of one's lifestyle, and it certainly does not lead to a lower standard of living. But there are a number of products that are worthy of avoidance. At the top of the list are pesticides (and related products like disinfectants) and oven and drain cleaners. For both categories, there are a host of alternatives or alternative products. VOCs are so ubiquitous that avoidance may be more difficult. A major reduction in exposure is possible, however, by carefully weighing, on a product-by-product or functional basis, the need for a product versus the possible risks and costs involved. For some products there are no alternatives at this time. In those cases, the need for the function that a toxic product serves can be questioned.

What seems clear is that American society has substituted many toxic chemicals in countless everyday products for time and physical work. Industries have succeeded in marketing unnecessary—even frivolous—amenities. Consumer habits have become fixed. Now, however, for a safer home and planet, every consumer should reevaluate his or her purchasing habits and needs from a new perspective.

If pollution prevention is necessary for planetary health, then toxic use reduction in the household is also necessary. If people are at all concerned about their personal health and want to minimize food and water contamination with toxic chemicals, then they need to reexamine countless purchases that are just as threatening as contaminated food and water, not to mention toxic waste sites and industrial plants. If consumers send clear signals to manufacturers, toxic products will eventually give way to safe products. Green product labeling laws are needed to serve industry and consumers.

9

NO TIME TO WASTE

Something is missing. That something is an environmental strategy, a sense of knowing where we want to go and how we want to get there. Largely unnoticed in the past 20 years of the environmental movement is that the U.S. government and other major institutions have zigzagged clumsily, without a clear, detailed, commonly understood and supported *strategy*. It is difficult to make real progress without a good plan, without spelling out the exact short- and long-term objectives to be met and the means to be used in the process. Protection from pollution has suffered because nearly all actions have been reactive, ad hoc, uncoordinated, and often contradictory. Developing an environmental strategy provides many opportunities, especially for bringing pollution prevention to the attention of the public.

Pollution prevention should become the keystone of a formal environmental strategy for the United States and for the world. No other concept says so much with so few words. If something is potentially harmful to health and the environment, do not make it in the first place. The closest the U.S. government has come to this strategy is a statement of national policy included in the 1984 amendments to the Resource Conservation and Recovery Act, the basis for the hazardous waste regulatory program: *"The Congress hereby declares it to be the national policy of the United States that, wherever feasible, the generation of hazardous waste is to be reduced or eliminated as expeditiously as possible."* That is a very good statement and a good start. A key problem, however, is that the statement applies only to federally defined hazardous wastes, a rather narrow slice of the whole waste and pollution pie. And, of course, a policy statement is significant only if it leads to tangible actions, which this policy statement has done to only a limited extent.

CONSEQUENCES OF A LACK OF STRATEGY

Without a finely honed strategy, success is not only unlikely, it is difficult even to assess what success is. But from narrow perspectives, an absence of strategy can be desirable. A lack of strategy, for example, may benefit politicians and bureaucrats, who can claim success for nearly all results. Government has minimum accountability because lack of strategy means an absence of criteria by which the public can judge the worth of actions. Ambiguity and flexibility fill the vacuum created by a lack of strategy, allowing economic self-interests to maintain their status quo and resist change. With unclear objectives and priorities, it is possible to delay uncomfortable, inconvenient, and threatening changes, even though they benefit society as a whole. The absence of an environmental strategy has contributed to unnecessary tension between economic health and environmental health.

The lack of an environmental strategy means that among the plethora of actions, none take precedence. Among the various kinds of actions being used, including pollution prevention sometimes, many people and organizations resist giving more support to some than to others. It seems reasonable to advocate a balanced set of actions. Alternatively, people can point to preventive actions as a sign that all is not wrong and that the system is moving incrementally in the right direction, which it may be. But incrementalism will not rescue the planet in time. Incrementalism does not ensure protection for the many. Incrementalism helps protect the economic wealth of the few. By definition, environmental incrementalism means that considerable resources are spent in zigzag directions. Without a strategy, in other words, human, financial, and technical resources are not allocated to a clear path, and the environmental protection results are uneven. Limited past preventive actions have not had much effect on the complex system of waste and pollutant production. Just as there is much talk in the American chemicals industry about waste reduction, enormous amounts of money are spent on hazardous waste incinerators. A few good preventive actions can be, and often are, easily offset by less sound measures. It is like an overweight person who diets several days a week, binges on the others, and then is surprised that there is no net weight loss.

The absence of an environmental strategy inevitably causes economic inefficiency and, as time has revealed, ineffectiveness in protecting *both* public health and the natural world. The last point illustrates another important consequence of not having an environmental strategy. Sadly, environmental programs have been unable to establish the relative priorities of protecting public health versus protecting natural resources, both animate and inanimate. Lacking a strategy, the environmental protection system has naturally drifted into emphasizing human health. Facing the need to limit spending for government programs, for example, the EPA has routinely ignored protection of the planet's natural resources. After all, would voters and taxpayers support spending money on protecting animals, plants, and wetlands if it meant lessening or delaying protection of human lives? Should not the question be: How much money do we need to spend to protect both? The focus on protecting human health ignores the scientific truth that ultimately human health is eroded when the natural world is degraded. This helps explain why many conservation groups have devoted themselves to protecting resources, in contrast to groups focused nearly exclusively on protecting people from toxic wastes, for example.

There is, in fact, no real choice. But the absence of strategy pits one worthy goal against another. Federal agencies line up against one another and against the public. The EPA emphasizes protection of human health, and the Interior Department has power to influence natural resource conservation. Conservation often loses out to development objectives and the environment to other economic interests. Private environmental groups become constituents of either one of these agencies. Without a strategy, no one can efficiently coordinate the responsibilities and efforts of diverse government agencies. The Energy and Defense departments end up creating terrible environmental problems, for example. Both departments must make an enormous cleanup effort because of poor past practices, and both are major generators of environmental waste and therefore have established pollution prevention programs. That of the Defense Department is sophisticated and substantial. But that of the Energy Department is less so. Indeed, a consultant (Varnado 1990) for the Energy Department said in February 1990 that "DOE is in disarray with regard to its waste minimization program." This consultant also noted that "the scope of EPA's pollution prevention goes far beyond DOE's waste minimization efforts by including all forms of pollution."

Developing a strategy would do more than resolve the relative priorities of protecting human health versus protecting natural resources. It would also help establish priorities among different environmental problems. To be sure, government bureaucrats have often complained that it is impossible for them to be judged successful by the public because they have not been told which problems are more important than others. Without a strategy, resources provided for individual environmental problems may be insufficient to achieve real success. Resources are too split up, especially as each year brings more problems to the fore-front. An especially important need is to distinguish among different *scales* of environmental problems. This is not to say that local health and environmental damages do not merit attention, but rather to acknowledge that global and other larger-scale problems pose a qualitatively different problem. When the natural equilibrium or ecological carrying capacity is destroyed on a global scale, the impacts are more severe, more irreversible, more inescapable, and less amenable to traditional pollution control solutions than when it happens on a smaller scale.

However, in thinking of taking actions to solve global problems, it is imperative to understand what Richard F. Tucker (1990), the President of Mobil Oil Corporation, described as the need to devise "microscale solutions that have macroscale consequences: catalysts and processes that fine-tune manufacturing to minimize or eliminate unwanted byproducts, computer modeling to optimize every manufacturing or production system, new molecular structures that pass harmlessly through the environment." In other words, technological implementation of all pollution prevention tactics remains local in scale even when attacking global environmental problems.

An environmental strategy must pay attention to scale and to setting priorities, in part, based on scale. However, giving high priority to global problems such as ozone layer depletion and global warming should not diminish government or public interest in addressing smaller-scale problems such as toxic wastes and pesticides. Developing an environmental strategy, however, does not merely offer a way to divide up the money pie more sensibly, as if some environmental problems are more important than others; it has the potential to make the pie larger so that all environmental problems can be solved. U.S. environmental spending, although high, is still small compared to that of other needs. In 1990 only about 1 cent of the average American tax dollar will have been spent on the environment, compared to 24 cents for defense and 14 cents for interest on the federal debt.

By developing a strategy, the whole range of environmental problems can be brought into clearer perspective and the need for more resources to solve them can be seen more effectively. It is always more difficult to defend an increased commitment to an area that lacks a well-thought-out strategy. Strategy has always been an issue in foreign policy and national defense, for example, but strangely, not for environmental protection. Moreover, an explicit environmental strategy offers the opportunity to recognize other related areas. For example, environmental strategy needs to address directly global population growth, economic growth, Third World debt, international trade agreements, and educational policy. And environmental strategy development can help place the environment on the agenda of top decision makers as an equal partner with other policy areas, which is definitely not the case now. The absence of a strategy inevitably has contributed to making the environment a second-class concern relative to others, especially economics. An environmental president would make development of an environmental strategy his top priority. The development of an environmental strategy for the United States could be a job given to the Council on Environmental Quality or a newly established presidential commission.

To sum up, the lack of a U.S. environmental strategy means that government, industry, and individuals are not routinely pulling in a common direction with common goals and using common methods. By recognizing the absence of a strategy, it is possible to have a public debate that transcends traditional narrow, often conflicting and competing interests. While establishing a strategy removes the flexibility that many people find desirable, it also provides an opportunity for society to join together, find common ground, and devote itself to achieving a collective good with efficiency, effectiveness, and equity. Finally, developing an environmental strategy would help individuals and industrial companies better understand their responsibility and role in achieving local and global protection of health and the environment, and it would greatly assist international environmental efforts.

POLLUTION PREVENTION AS THE BEST STRATEGY

In this book, we have tried to show why and how pollution prevention should be the centerpiece of an environmental strategy that addresses all environmental problems, in all environmental media, caused by all sources worldwide. Of special importance is that pollution prevention is much more critically needed for large-scale environmental problems, such as global warming, acid rain, and destruction of the protective ozone layer, which are caused by countless sources of environmental wastes. Pollution prevention has the intrinsic benefit of being simple in concept and holistic and comprehensive in application. Pollution prevention is especially attractive because it is parsimonious and robust; that is, it is a simple yet powerful concept that also makes common sense. People can understand the pollution prevention imperative: eliminate or minimize *all* waste outputs and pollutants. Moreover, it is difficult to see how any other approach can successfully cope with the increasing global population.

The global population has doubled since 1950, and today's 5.3 billion people could become 10 billion energy- and material-consuming, waste-producing individuals by 2025. Stemming population growth cannot be ensured. It also seems impractical and foolish to believe that peaceful methods will limit the aspirations of most people of the world to achieve a higher standard of living through industrialization and consumption. Exporting the current American lifestyle based on high personal consumption in general and especially on single-use, disposable products spells doom. Increasing per capita production of pollution and waste militates against pollution prevention countermeasures to do just the opposite.

Applying pollution prevention only to industrialized nations, which now produce the largest amounts of waste, will not ultimately be enough to ensure planetary health, although without setting an example, all hope is lost. If pollution prevention is not applied as soon as possible to the developing nations, they will adopt more of the old, inefficient, polluting technologies and products that the United States and other industrialized nations have used for so long. Alternatively, developing nations will buy traditional end-of-pipe pollution control technologies, which often fall into disrepair, instead of addressing the source of the problem. In the exploding, heavily polluted metropolitan Mexico City area, only 30 percent of the industries have antipollution equipment, and much of that equipment is insufficient or inoperative. Should Mexican industries buy expensive, ineffective pollution control equipment for polluting processes or new clean manufacturing technologies? It's all in the numbers. Billions of additional people overload planet Earth with unmanageable, disequilibrating amounts of wastes and pollutants unless pollution prevention is a global strategy.

However, accepting the primacy of pollution prevention does not mean abandoning traditional pollution control or end-of-pipe actions or the government regulatory and legal systems designed to ensure their implementation. Not all waste and pollution can be eliminated, either immediately or in the long run. What is absolutely crucial, however, is to recognize the *primacy* of pollution prevention in the hierarchy of environmental options. Pollution prevention should be the first choice and the option against which all other options are judged. The burden of proof imposed on individuals, companies, and institutions should be to show that pollution prevention options have been thoroughly examined, evaluated, and used before lesser options are chosen. The current preoccupation with recycling, for example, needs to be reoriented in order to promote true pollution prevention measures. The basis for evaluating recycling must shift from waste management options like landfills and incinerators, which are worse, to pollution prevention, which is better. Recycling perpetuates many patterns of production and consumption which are based on toxic chemicals and which inevitably produce manufacturing wastes. Recycling itself also poses environmental threats, as the Canadian government recognized when Environment Canada (1990) said that "it is important to remember that there are economic and environmental costs associated with waste collection and recycling processes."

To have an environmental strategy based principally on a commitment to pollution prevention is to take the first and most necessary step to ensure that pollution prevention will be fully used in both the short and the long term. The implications of doing so for constructing an environmental strategy are several. First, it means that the general but universal goal is to strive for an absolutely minimum amount of waste and pollutant generation from every source. This, in turn, means that governments and persons shift their attention from deciding what "safe" or "acceptable" levels of wastes and pollutants are to targeting *any* level of waste and pollutant production for elimination or reduction. No amount of waste or pollutant generation should be seen as inevitable, acceptable, necessary, economically tolerable, or safe.

The object should not be to prove with scientific certainty that a waste or pollutant will result in damage to human health or the environment before it can be eliminated. The presumption should be that any waste or pollutant is potentially harmful and preventable. All too often, research is called for instead of action. On global warming, Senator Timothy E. Wirth (1990) has dramatized the need to act preventively:

> The question is not whether the Earth will warm but how much and how soon . . . our top scientists are telling us that they fear a dramatic, unprecedented and perhaps catastrophic warming. We are gambling with the climatic underpinnings upon which all societies depend for survival. We must develop a comprehensive national energy policy that makes reduction of greenhouse gases and air pollutants a national objective.

With a pollution prevention strategy, human intelligence and creativity as well as science and technology can focus on preventing, eliminating, or reducing the production of all wastes and pollutants. That, and not the management or treatment of wastes and pollutants, is the prime objective of a prevention-based strategy. Zero waste outputs on the planet is, of course, an abstraction, but it defines an ideal state. The planetary system must strive to move toward zero waste. It is either that or moving in the opposite direction, toward producing more waste, which is exactly where the world is today—heading in the wrong direction, toward planetary disequilibrium and destruction. To their credit, some major corporations, such as 3M and Monsanto, have openly spoken of the need to strive for zero waste generation. Polaroid also has major reduction goals.

The second necessary action is to establish more specific objectives and priorities on all levels. For example, it is necessary to set numerical targets for reduction and a timetable to achieve them. Global, national, state, industry, company, and individual reduction targets for specific wastes and pollutants are feasible and ultimately necessary. The scope of the environmental problem dictates the scale of the necessary prevention and reduction targets. Attacking the loss of the protective ozone layer and global warming, for example, clearly require *global* prevention objectives. Other problems are different. MSW can be reduced at the community, household, and commercial establishment levels. Pesticide reduction can be structured at the federal and state levels.

Reduction targets must be ambitious if they are to drive the existing system to widespread pollution prevention. We have noted how frequently U.S. studies have concluded that a 50 percent reduction is feasible for a certain class of waste. Comprehensive studies have also shown the economic and technical feasibility of very ambitious pollution prevention strategies. In 1989 the Canadian Council of Ministers of the Environment announced a federal government waste initiative aimed at cutting industrial waste generation by 50 percent by the year 2000. Sweden has committed to banning environmentally dangerous products such as alkaline batteries in 1990. Notable among national studies is the work of the Dutch government.

Their highly sophisticated and detailed analyses of different environmental strategies make U.S. efforts, for example, seem slovenly, amateurish, and backward. In their 1989 report *National Environmental Policy Plan—To Choose or to Lose,* the Dutch concluded that over the next 20 years "it is not impossible to combine a doubling of consumption and production with reductions in emissions of 70 to 90 percent [and 20 to 30 percent for carbon dioxide]." Indeed, the preferred strategy which emphasizes pollution prevention over traditional end-of-pipe controls, entails a lot of societal change but also offers many advantages. Adverse effects on the Dutch economy are said to be unlikely, especially if other countries pursue the same strategy. Moreover, in terms of international industrial competitiveness, the Dutch see the longer-term potential: "There may also be good reason for giving some sort of lead in matters relating to the necessity for cleaner means of production and calling for changes in production and investment: what initially looks like a handicap might later turn out to be an economic headstart."

For the United States, a president with vision should see adoption of a federal pollution prevention strategy as a key means to make American industry a leading contributor to solving global environmental problems. Industries that will ultimately capture the largest world market shares will be the ones selling the most innovative clean technologies and products. In Chapter 4 we presented considerable data to show that American industry produces proportionately more hazardous waste than industries in comparable industrialized nations. In February 1990, the OTA (1990) concluded that "American manufacturing has never been in more trouble than it is now." Part of that trouble—a part generally ignored—is that American industry produces too much environmental waste and pollution. Japan generates about 20 pounds of industrial hazardous waste per person a year compared to about 4,000 pounds in the United States. A more competitive industry that produces higher-quality goods should and can be one that also produces less environmental waste.

All told, the measure of being an environmental president or a nation that shows real leadership is adoption of a vigorous pollution prevention strategy. Ambitious national pollution prevention goals for the United States would help focus consumers' attention on the problem of household trash. A worthy goal is to reduce the U.S. municipal wastestream by at least 2 percent per year. Without this kind of effort, no amount of recycling, new incinerators, and new landfills will provide a satisfactory answer to America's garbage problem. In summing up the MSW situation for Congress, the Environmental and Energy Study Institute in January 1990 said: "Most experts agree that in order to tackle our current and anticipated garbage disposal capacity problems, the United States must reduce the amount of garbage we generate." But the political system is not listening to the experts. Manufacturing, packaging, and waste management companies, as well as environmental groups, are focusing on recycling.

The EPA and its Administrator, William K. Reilly, often speak of a 25 percent reduction goal for MSW by 1992. However, a little research quickly shows that this goal includes recycling. Hardly any effort goes to cutting the generation of waste by changing products and packaging. As part of the regulations to limit air emissions, the EPA came up with the idea of forcing municipal waste incinerator facilities or communities to separate a minimum of 25 percent of trash for recovery, an amount that incinerator companies said was too high and that environmental groups said was too low. Clearly, the EPA's and nearly everyone else's priority is to promote recycling. A number of states have established numerical goals for cutting the disposal of MSW. However, here too, nearly all of them combine reduction and recycling and rely on recycling. One exception is New York; its 1988 law stipulates 40 percent recycling and 10 percent waste reduction by 1997, which is only about 1 percent per year reduction.

Another important example of the EPA's preference for recycling over pollution prevention is its February 1990 report, required by Congress, on plastic wastes. U.S. production of plastics leaped from 3 billion pounds in 1958 to 57 billion pounds in 1988—an average growth rate of 10.3 percent annually. When it came time to examining ways to deal with the problem of plastic waste, the EPA (1990) cautioned that "source reduction actions need to be carefully examined." The concern was that substitutes for for plastics might not result in net benefits. But when examining recycling, the EPA made no mention of environmental problems with recycling facilities.

Another necessary and feasible goal is to cut in half the U.S. production of industrial hazardous waste by the year 2000. Massachusetts adopted a comprehensive Toxics Use Reduction Act in July 1989 that includes a state-wide goal of reducing toxic waste by 50 percent by 1997, using true pollution prevention measures. Cutting the use of pesticides overall by 5 percent a year could be another specific goal. Such goals would put public welfare above private profits and would show the commitment of American society to pollution prevention. After seeing how well voluntary goals worked, the government could consider stronger measures.

Turning a prevention-based environmental strategy into action must entail constructing a host of policies and programs to ensure that obstacles are overcome and incentives provided. Several of the most critically needed government actions are briefly discussed below. Many individual actions have already been proposed in this and other chapters. The current government structure does not support pollution prevention and must be changed. Peoples' and companies' actions are reflective of and constrained by government policies and programs. Changes in government that people want to support and advocate must be identified. Truly meaningful changes in governmental policies and programs will not become reality unless many individuals become convinced of the need to stress pollution prevention and work to have government adopt it fully. The ideas presented here are meant to be illustrative of what is critically needed and what appear to be the most productive based on past experiences, but they are not an exhaustive inventory of all possible actions.

LEADERSHIP, ADVOCACY, AND ORGANIZATIONAL COMMITMENT

The commitment of government to a goal or strategy is revealed by the position given to it in the bureaucracy and the amount of funding for it. Pollution prevention has not yet achieved much status. There is a pollution prevention office at the EPA, but it is at a low level and receives little funding. In 1986, the OTA pointed out that less than 1 percent of government environmental funding went to pollution prevention. That situation has improved only a little. The OTA and other groups have suggested that a high-level pollution prevention office at the EPA, with its own Assistant Administrator (or Assistant Secretary if the agency becomes a cabinet department), is called for. But neither the EPA, industry, environmental organizations, nor Congress has given this idea much support.

The result is that the EPA remains focused largely on traditional end-of-pipe solutions and, although Congress has shown a definite increase in interest in and support for pollution prevention, funding for pollution prevention remains token. Even though the current EPA Administrator has shown more interest in pollution prevention, it has not yet surfaced as a genuine high priority as a theme used by the EPA to link all environmental problems together. The EPA seems unable to switch gears, from sluggishly regulating small parts of the waste and pollutant universe, bit by bit and on the basis of arguable scientific evidence, to quickly promoting the elimination or reduction of *all* wastes and pollutants by asserting the primacy of the pollution prevention principle.

But the EPA's lack of commitment to the primacy of pollution prevention over other options is a mirror image of American society. It is nearly impossible to find the word *prevention* in the indexes of countless books on environmental subjects. One exception is the 1990 book *Environmental Policy in the 1990s—Toward a New Agenda.* But even though the authors Norman Vig and Michael Kraft, say that *"pollution prevention* or reduction and elimination at the source must guide future environmental policy,'' they reach an incorrect conclusion. They allege that ''it is evident that prevention is the most difficult course to follow since it requires alterations in human behavior and technology over a relatively short time that are likely to be economically costly, scientifically uncertain, and politically unpopular.''

Although pollution prevention faces opposition and obstacles, it is *not* economically costly, *not* scientifically uncertain, and certainly *not* politically unpopular to the vast majority of people who have much more to gain than to lose. The information and analysis available today make it *evident* that pollution prevention is profitable, practical, and politic.

The "Valdez Principles" created in 1989 by the Coalition for Environmentally Responsible Economies, received praise as an important statement of an environmental ethic or code of conduct for corporations. But those principles did not articulate the importance and primacy of pollution prevention. One of the principles is: "Reduction and disposal of waste. We will minimize waste, especially hazardous waste, and wherever possible recycle materials. We will dispose of all wastes through safe and responsible methods." This is a good statement and one which, only a few years earlier, would probably not have been made. Yet, in this statement and in the other nine principles, there is no clear message that the first responsibility of corporations is to reduce or eliminate all environmental wastes at their sources in manufacturing, toxic chemicals in products, and postconsumer garbage.

Another example of giving short shrift to pollution prevention is the work of the Worldwatch Institute. In its *State of the World 1990,* the universal applicability of pollution prevention is missed and nearly all the emphasis in regard to solid wastes is on recycling, although it does say: "Although early moves away from the throwaway society are concentrating on recycling, sustainability over the long term depends more on eliminating waste flows."

It can be argued that a major federal pollution prevention initiative must be made outside of the EPA in order to integrate the activities of all federal agencies and to overcome the EPA's historic focus on inefficient end-of-pipe control measures. Perhaps the President's Council on Environmental Quality is the appropriate place to at least supervise and coordinate efforts by the EPA and other departments, such as Energy, Defense, and Commerce. Consider, for example, the difficulty of changing the policies of the U.S. Postal Service to impose higher costs for sending junk mail in keeping with the primacy of pollution prevention. Or consider the clean coal program of the Energy Department, which is largely end-of-pipe oriented and should have been called the "cleaning" coal program because it does not emphasize use of clean (low-sulfur) coal or clean process technologies. Some federal agencies are themselves major waste producers. For example, in November 1989 the General Accounting Office noted that the Department of Defense generates over 400,000 tons of hazardous waste annually from processes used primarily to repair and maintain weapon systems and equipment.

The Council on Environmental Quality could also serve as a place where U.S. progress in pollution prevention is measured. A special annual report could tell the American people whether or not specific categories of wastes and pollutants were produced in lesser or greater quantities. If federal pollution prevention goals were established, then the Council could assess progress in meeting them. In energy, the measurement of progress has been based on a sound concept. On any level, company or national, progress is measured in terms of *decreasing* energy use per unit of output or GNP. A good measure has helped everyone learn that economic growth has been possible with greater energy efficiency.

A good measure of progress is crucial for the environmental area. By itself, growth in GNP reveals nothing about the quality of life. Indeed, the production of environmental wastes and harm to health and the environment easily increase economic activity. That, however, is no reason to continue allowing the production of environmental wastes. Companies and nations must show that less environmental waste is being produced on some logical basis, such as per capita, per worker, per unit of company or industrial output, per unit of GNP, or some other measure. Improving environmental quality requires that economic growth be contingent on producing less environmental waste per unit of economic activity. Producing less environmental waste is equivalent to using less planetary assets and reducing the traditionally hidden environmental costs of making and consuming goods and services. *Environmental efficiency must become as compelling as energy efficiency.*

What is needed is true advocacy for pollution prevention so that difficult programs, such as pollution taxes and government requirements for industrial pollution prevention plans and achievements, can be launched. There is a remarkable opportunity for leadership that has not yet been seized by any senior official in the U.S. government or at the international level—or, for that matter, by an environmental organization.

POLLUTION AND WASTE TAXES

There is no more fundamental way to offset the externalization of environmental costs and the lack of incentives to practice prevention and reduction than to impose *high* taxes on pollutants and wastes. There are two options. If pollutants or wastes are managed so that their amounts are reliably known, then it is feasible to tax them directly. But when products lead to direct discharges into the environment, especially when they result from many sources, then it is more feasible to tax a raw material manufacturing input or a product because the production of waste is inevitable and predictable. In the first category, for example, are industrial toxic wastes and government-regulated water and air releases. In the second category are ozone-depleting chemicals (now taxed), pesticides, disposable products, excessive product packaging, petroleum, and coal. A substantial pollution tax on fuels and energy sources is essential to combat inefficient and excessive use of fossil fuels in the United States and their undeniable contribution to global warming.

In theory, high enough taxes can compensate for artificially low market prices which do not account for environmental costs shifted to other parties, places, or times. Increasing the costs of producing waste should cause a decrease in waste production. It is a challenge to overcome opposition to imposing high pollution taxes sufficiently to change behavior. To overcome opposition to high taxes, broad public support for pollution prevention is necessary. Opposition comes from many expected and unexpected quarters. Those who are taxed logically may oppose pollution taxes, but a more subtle source of opposition is those in the business of waste management or pollution control. The following short news story from *Resource Recovery Report* (1990) illustrates the effectiveness of increasing waste management costs on causing waste producers to reduce waste and the ironic response it inspired:

> The Rhode Island Solid Waste Management Corp. is considering lowering its tip fee by $10 to $49/ton to generate more waste. Commercial waste has declined after a recent fee increase and RISWMC seeks to avoid a $5 million loss of revenues.

However, there is another important function for pollution taxes that can help create broad support. Pollution taxes can raise funds dedicated to implementing activities to support pollution prevention. For example, by placing such revenues in a special fund, it would be possible to support R & D to develop safe alternatives, new products, and new manufacturing processes. Providing technical assistance to industrial facilities on a grand scale would also be possible. With the government helping industry in these two ways, the imposition of taxes would have a positive effect on the supply side of industry.

Pollution tax monies could also support education and training activities and public education campaigns to promote pollution prevention, especially through smart consumer purchases. In this way, the demand side of the equation could be moved in the right direction. Dislocations and hardships on companies and communities caused by chemical or product bans could also be offset by government assistance. For example, if a pesticide manufacturing plant went out of business, the government's special fund could help to locate another type of "clean" business there or help people in the community directly. Economic assistance to companies, communities, and workers must be a key ingredient of any serious U.S. conversion to pollution prevention. Without it, workers, industries, and communities which fear economic disaster will use their political muscle to fight necessary preventive actions.

Nowhere is this conflict seen more clearly than in areas around polluting coal-burning power plants and coal mines producing high-sulfur coal. A southern Illinois coal miner, Don Baldwin (1990), threatened by the switch to low-sulfur coal, summed it up simple: "Environmentalists want a perfect world, and they don't care if my family goes hungry." On the other side was an environmentalist who saw the choice as being between "a world with cleaner air" or a world that continued to use "the air as a sewer." Politically and humanely, it is necessary to expedite the conversion to pollution prevention by assisting those people, companies, and communities that are asked to shoulder the burden of the historic transition. It is not enough to say to such affected parties: you lived well for decades by fouling the world, now pay the price. It is necessary to remove the acts of violence to the planet by removing the economic necessity and opportunity to use such acts as the only means of survival.

Considering the enormous quantities of wastes and pollutants produced, it is feasible to generate enormous sums of money for a National Pollution Prevention Trust Fund. For the United States, placing a fee or tax of one-tenth of 1 cent on each pound of household, commercial, and institutional MSW produced would generate about $400 million annually. Such a tax could be collected efficiently through municipal waste collection agencies and waste management facility operators. Alternatively, municipal agencies and facilities could choose to raise more money by charging households and other waste producers more through the taxes or fees currently collected. To overcome the use of average taxes which do not take into account wide variations in household waste generation, government agencies should consider innovative ideas. For example, local governments can allow collection of a standard container (which should be small) and charge more money for collections above a base level. This is the approach taken by Seattle, and it has been successful.

A one-tenth of 1 cent tax per pound on federally defined industrial hazardous waste would generate about $1 billion annually. The government currently collects extensive information from industrial waste generators. This makes tax collection quite feasible. Another feasible alternative would be to use the federal TRI system to tax industrial waste generation. Since these TRI figures are for pure toxic substances, unlike the often dilute aqueous liquids in the hazardous waste universe, a tax of 10 cents per pound is more reasonable. This would raise about $2 billion annually. Aside from hazardous wastes and toxic substance production, billions of dollars more could be generated by taxing many other categories of waste outputs and common commodities such as petroleum and coal. It would be simple to collect a pollution tax from the producers of petroleum and coal. It would also be easy to add a pollution tax on gasoline sales to consumers, which is needed anyway to spur fuel savings.

Sensible as taxing waste seems, a waste-end tax to help support the Federal Superfund program was narrowly defeated by Congress in 1986. Although many companies and industry trade associations supported the proposal, many other industrial interests successfully opposed it. This event demonstrated how difficult it is politically to tax openly acknowledged, unwanted waste generation. About 30 states have imposed such waste taxes, and their programs have functioned well. However, in nearly all cases, the amount of taxation is too small to affect decisions made by industry. But these taxes have raised substantial amounts of money to run many different kinds of state environmental programs—but not pollution prevention ones in particular.

It is self-evident, reasonable, and logical that establishing a federal pollution prevention trust fund that would raise over $1 billion annually could help transform America. If used solely to promote pollution prevention, the money would provide an enormous resource to implement a federal prevention-based environmental strategy. The U.S. commitment to pollution prevention would be tangible. As the nation moved forward with prevention, pollution tax rates could be raised to increase the economic incentive for remaining waste producers. There is no injustice to tax heavily waste-intensive industries and products. Since there is no benefit from environmental wastes, there should be no opportunity for low taxes to be viewed as a way to justify pollution by paying for the right to produce it. Pollution taxes should *not* be seen as an acceptable cost of doing business. The only real injustice is to avoid pollution taxes and to keep them at a token level that does little to help the national and international conversion to clean technologies and processes. For these reasons, the level of taxes mentioned above should be automatically increased by no less than 50 percent every 5 years to keep building the economic incentive to reduce waste generation and to maintain revenue generation to help the United States work on the remaining sources of environmental wastes.

With a large base of revenues, the United States and other industrialized nations could also provide assistance to developing nations to help steer them toward prevention. Indeed, it is only fair that wealth be transferred on a massive scale from the nations that have polluted the planet to those which have done less damage and which still have time to limit their contribution to planetary degradation.

Another way of implementing the pollution tax concept is to integrate it into international trade. Products which are exported and which produce pollution where they are manufactured or used should be burdened with a special pollution tariff. The least preferred products could become less competitive in world trade if the tariffs were sufficiently large. This is especially important for developing nations, which often emphasize export industries to generate income in desired foreign currencies but which also cut corners environmentally. Nations and industries that spend more money to practice pollution prevention should derive a benefit from doing so.

Lastly, the lack of a green political party in the United States is a handicap. Unlike most other nations, environmentalism in the United States depends on environmental organizations and the news media to force bold political action. While this system is often effective, it does not offer direct political opportunities to individuals, unless they are given a chance to vote on specific measures. The latter action has been extremely effective in California, where the voters—not the politicians—have enacted very progressive environmental legislation through state ballot initiatives. The interesting question is: would a majority of Americans vote for significant pollution taxes which their elected officials are unlikely to pass?

BANS OF CHEMICALS AND PRODUCTS

Barry Commoner has been instrumental in identifying the most effective generic action to protect human health and the environment: the banning of specific chemicals and products. Bans are a sure way of eliminating the production of a particular source of waste or pollution. A prevention-based environmental strategy would explicitly endorse the need for government bans on dangerous chemicals and products. This would make it easier for the federal government to ban chemicals and products and to join international agreements on chemical bans.

This governmental action triggers opposition for several reasons. On principle, many people find it intrusive and distasteful because it removes individual freedom of choice. However, individual freedom must often be limited in order to protect society as a whole. Second, companies naturally oppose losing the opportunity to make products in which they have invested money to develop and market and from which they make profits. Putting companies and whole industries out of business, with the corresponding impacts on workers and communities, cannot be dismissed or ignored. But neither should such concerns stop bans that are necessary for effective protection. For negative economic impacts, government should consider compensation and assistance rather than letting industry delay or block a ban.

Third, consumers tend to fear the loss of products for which no immediate substitutes or alternatives may be available. There is little historical evidence that this is a significant problem. For indispensable products, government should help finance crash programs to develop safe alternatives. Destruction of the planet's protective ozone layer illustrates the need to institute bans aggressively and comprehensively. Necessary bans on ozone-depleting chemicals should not be framed in terms of concerns about losing refrigeration and air conditioning. The real issue is helping safe alternatives quickly enter the marketplace. Innovative incentives need to be considered. For example, the government could provide a one-time income tax credit for the purchase of certified ozone-safe refrigerators and air conditioners.

Bans will remain a critical solution to many problems, especially when there is relatively little time to solve the problem, when profits from the existing products are high, and when environmental damages are known but their costs are externalized. Bans, however, are not the perfect solution either, as experience with pesticides and ozone depleters has shown. A banned chemical or product may not be replaced by a truly safe substitute. This problem can be solved by providing a higher priority to examine safe substitutes.

Moreover, bans may not be comprehensively applied. A single nation's product ban may be symbolic but ineffective. It will not stop the global transport of pollutants. Earth's atmosphere is a pollution pipeline, delivering toxic perils planetwide. Pollution prevention cannot stop delivery of what is already in the pipeline, but it can stop new entries. But since the pipeline picks up and delivers worldwide, more international agreements on chemical bans are essential.

INFORMATION FOR CONSUMERS ON LABELS

Consumers cannot be expected to use their power in the marketplace until they are provided with reliable and user-friendly information on the consequences of their decisions. There is enormous historical evidence that consumers change their behavior when given information that is meaningful and useful to them. All sorts of lifestyle changes verify that wide-scale change is not only feasible but that it can happen very quickly. The 1986 requirement for much of U.S. industry to report the release of some toxic substances (the TRI) and to make that information directly available to people is generally recognized to have had an immediate impact on industry. Generators of industrial wastes and pollutants often came to understand for the first time how inefficient their operations were. The concerned public, as well as environmental activists, were handed a potent tool with which to draw attention to the polluting ways of companies. Armed with that back-end information, people and environmental organizations, as well as state environmental agencies, have been able to put much more pressure on companies to demonstrate their commitment to front-end waste reduction.

In reality, most Americans have had no awareness whatsoever of the data on industrial facilities. But information on *facilities* is minor compared to the potential impact of providing the public information on *specific products*. Every person is a consumer, and even relatively small shifts in market share send potent signals to industries to change their products quickly.

Theoretically, every consumer product could be labeled or identified in a way that would tell a consumer the relative environmental benefit of the product from a pollution prevention standpoint. Three kinds of information need to be obtained and incorporated in a simple, easily understood symbol or code: (1) the amount and hazard of the waste produced during manufacture, (2) the threat to consumer use because of toxic components in the product, and (3) the amount and hazard of the waste produced as a result of using and disposing of the product. This complex life-cycle information must be reduced to numbers, colors, or images, or a combination of them. Like international signs for drivers on roads, product standards could become immediately understood. Such a system would also facilitate placing pollution taxes on products. It is also necessary to have a trusted entity, perhaps a quasi-government agency, ensure that such information about a product is accurate. Misleading and inaccurate information on product labels is a well-known problem.

Perhaps the greatest irony about greatly improved product labeling requirements is that proponents of market incentives rarely embrace it. In truth, of course, the most potent market force is an informed consumer. But people in government and industry rarely focus on product labeling. They prefer to advocate market-oriented policy instruments like tradable emissions permits which allow some inefficient, waste-producing companies to resist change by buying the right to pollute from other companies that produce less than their allowable limit. This redistributes pollution but does not necessarily produce a net reduction of it. Esoteric economic theories and regulatory actions need to give way to commonsense requirements for useful information provided directly to consumers on product labels.

PREVENTION IMPACT STATEMENTS

The environmental impact statement (EIS) has been generally recognized as an important instrument to force attention to environmental *effects*. The idea that it is prudent to analyze the potential environmental impacts of major projects and actions has intrinsic merit. But the EIS is mostly an end-of-pipe or back-end concept, although it has often been preventive by causing the cancellation of some projects. More often, however, mitigation of environmental effects by pollution control or waste management measures is the outcome. A pollution prevention or waste reduction impact statement would shift the focus from analysis and anticipation of *effects* to *sources* of environmental problems.

This idea means that all government actions would have to be assessed for their impact on pollution prevention and waste reduction. Doing so would eliminate many actions which would inevitably result in more pollution and waste. As an example, consider that the EPA routinely takes actions which make it easier to dispose of industrial wastes, such as deregulating hazardous wastes and granting permits to operate landfills, injection wells, and incinerators. Such actions do not promote waste reduction, and they should be allowed only after the EPA can certify that they cannot be replaced by pollution prevention actions. Local governments could also implement this idea. For example, granting local building permits or licenses to operate a business could be made contingent on detailed assurances that the maximum level of waste reduction will be employed. Through such measures by government agencies, private companies would be motivated to conduct waste reduction audits and make tangible commitments to future actions through formal and publicly accessible waste reduction plans.

FOREIGN AID

Foreign aid or loans from international lending institutions (or, for that matter, domestic loans and government aid) should not be provided unless the recipients can demonstrate that the money will be used only for activities consistent with the pollution prevention imperative. This measure is critical, considering the desire for economic growth and industrialization in developing countries. Capital investment must be directed to projects which cut existing rates of waste generation or can ensure that future rates are as low as is technologically feasible. However, there is a long history of ignoring the environmental impacts of foreign aid and international loans. Even today, the environmental services and equipment industry has been able to take advantage of the growing environmental interest in developing countries by selling traditional end-of-pipe technology. While this is an improvement over the discharge of raw wastes and pollutants, it is inferior to ensuring that clean technologies are used to the maximum extent practical.

GOVERNMENT PURCHASES

Government agencies are major consumers and have enormous influence in the marketplace. With a prevention-based environmental strategy, all government purchasers would be required to favor goods and services that produced the least amount of wastes and pollutants. This would motivate manufacturers and service providers to describe accurately their generation of wastes and pollutants, to conduct waste reduction studies, and to minimize their production of wastes. Government purchasing authorities could be allowed to pay a modest premium for environmentally clean products and services. Contractors operating government facilities could be required to demonstrate active waste reduction programs. The success of such programs should help determine award fees from government agencies. This would provide a profit motive for inventors and entrepreneurs to commercialize new products and services. This premium could be justified economically because it would ultimately save money to address problems due to environmental wastes.

REGULATORY FLEXIBILITY

An often overlooked need is to prevent the current end-of-pipe environmental system from stealing resources from more beneficial pollution prevention activities. Current legal requirements often force companies to spend time and money on conventional pollution control and waste management solutions rather than preventive measures. This happens because it can be a lot easier for companies to purchase control equipment or build waste management facilities than to change their manufacturing technologies and operations or products. Regulatory systems need to be flexible enough to allow waste generators to commit to waste reduction projects that offer more public protection in the long term. However, government regulatory agencies must have enough expertise to evaluate proposals for some delays in regulatory compliance, for example, to ensure that companies have a firm commitment to a definite, well-planned pollution prevention project.

PROGRESS REQUIRES CITIZEN ACTIONS

Protection of human health and natural planetary assets is a responsibility borne by individuals whose actions affect other persons in their own communities and other nations, as members of societies on planet Earth affecting the well-being of other societies, and as members of a generation whose collective actions will affect the well-being of future generations. A private citizen is a manager of the environment. Adopting a pollution prevention strategy institutionalizes the responsibility of every person to do everything reasonably possible to prevent pollution as a direct or indirect consequence of their own activities. Even though attention focuses increasingly on large-scale or global environmental problems, the responsibility of individuals in contributing to these problems is not diminished.

But individuals typically feel uncertain about the negative impacts of their "small" contributions to big environmental problems and equally uncertain about the positive impacts of changing their behavior. Personal justification for taking pollution prevention actions does not have to be based on asking "Does my small contribution really matter? If I stop using paper towels hourly or stop buying a few of the many toxic products I normally use, will it make a difference?" A better question is: "If a lot of other people do as I do, will there be harm or benefit to people's health and the ecosystem?'' The issue is not one of finding *the* solution, but of each of us feeling a part of the solution rather than the problem. Environmentally friendly behavior and consumption are easy concepts for persons to understand and act on. At issue is not perfect personal behavior, or whether every ''right'' action is taken, but rather whether a person's total pattern of living is a significant step in the direction of prevention and whether that shift is continuous.

It is useful to think of three strategic levels of human actions: (1) individual self-education, conviction, commitment, and practice; (2) additional education and practice through sharing information and experiences with family, friends, co-workers, and local community groups; and (3) explicit political activity to influence public policy and the agenda of private organizations.

Step 1 has the objective of changing *patterns* of behavior. Practicing a little pollution prevention is like being a little ethical, honest, racially tolerant, or hard-working when it's convenient and easy. Such behavior is better than nothing, but it's not good enough. It is fundamental to the notion of pollution prevention that many seemingly small actions must be taken, starting at the individual level. The pattern of new behavior must encompass all of one's activities at home, at work, and at play. The goal is not simply to feel good but to accomplish something. This requires finding and creating opportunities, overcoming obstacles, and assessing progress. Until new behavior patterns are solidly in place, it takes great discipline to fight old, waste-intensive habits. And, of course, most people will still not be able to stop generating all pollutants and wastes. But every one of us can know whether the total impact of our lifestyle is better than it once was and whether it is getting better from a pollution prevention perspective.

Step 2 is also crucial because, by sharing information, understanding, and experiences with other people, one's own behavior pattern will be reinforced and other people will become converted to pollution prevention. Seeking fellowship from like-minded persons will nurture, sustain, and expand one's own pollution prevention practices. Hearing why other people are not interested or enthusiastic is also important to develop effective countermeasures, including getting better facts about benefits.

Not everyone will take step 3 because not everyone feels comfortable trying to move institutions and organizations, nor does everyone have the time or opportunity—as in voting—to do so. But many people want and like to act in a more political way to bring about wide-scale change. Literally every type of organization can be used to achieve pollution prevention. People need to think of advocating the practice of pollution prevention in the groups to which they now belong, including religious groups, labor organizations, civic and business associations, professional societies, parent-teacher associations, garden clubs, and sports groups. As an example, just think of the national impact of replacing the use of disposable utensils at thousands of daily meetings and conferences with reusable serving ware.

The message must be translated into universal action: pollution prevention is a *necessity* for individual and collective well-being and survival. There is no other comparable, intelligent choice. Enough evidence is in; 100 percent certainty about the exact cause and magnitude of environmental problems is unnecessary and impossible. The natural equilibrium and life-sustaining capabilities of planet Earth have been greatly disturbed by the darker side of modern science and technology, as well as by greed and stupidity. There is a limit to the pollution that the planet can assimilate without being permanently changed. Restoring equilibrium and habitability requires a new prevention-based strategy. Pollution prevention does not threaten our economic prosperity or standard of living, although there are bound to be winners and losers as new processes and products replace old ones. And of course, the style of living must change for many people in affluent nations.

The larger truth is that without a full commitment to pollution prevention, global economic well-being is threatened. Everyone's standard of living *will* decrease if more and more money has to be spent struggling to clean up and react to a hostile and inhospitable environment. Without a massive shift to environmentally clean processes and products, the chief uncertainty is when health and environmental damages will take their full toll on people's health and planetary assets. In a thought-provoking article entitled "Reflections After the Cold War," Richard J. Barnet (1990) concluded: "The threat to human life posed by man's destruction of the environment far surpasses any threat posed by increases in the weapons stockpiles." He noted that the future of the United States is imperiled by "it's permissiveness with respect to environmental destruction" more than by the threat of war. Ordinary people also seem to have internalized this knowledge and, therefore, there is fertile ground for pollution prevention to take hold and blossom. Indeed, a 1989 survey by the Roosevelt Center for American Policy found that environmental problems ranked first as "top priority threats" for 47 percent of people, compared to 46 percent for the spread of nuclear and chemical weapons, and 15 percent or less for the spread of communism, the Persian Gulf conflict, Defense Department waste and corruption, and terrorist attacks.

If common sense prevails, pollution prevention will be used as the basis for an international environmental strategy. Prevention on merely a local or national scale cannot protect people from some of the most severe global environmental risks. There is no safe spot on planet Earth anymore. Only by executing a global prevention strategy tenaciously will planet Earth's future, and ours, be assured. For the most part, it is not technology which must be created, but rather the will to act preventively. All nations must replace the search for a middle ground between ecology and economy with a pollution prevention strategy that puts ecology first and that rigidly requires economic growth to be based on the use of environmentally clean technology and clean products. Pollution prevention is an explicit, environmentally driven strategy compared to the unclear goal of sustainable economic development. A prevention strategy requires that we get out priorities straight: planet Earth first, economics second.

There has been too much denial of environmental problems and too much lethargy in solving them because of purported threats to the economy or the standard of living. Pollution and waste generation is a shortsighted, immoral, and illusory path to economic prosperity. The sooner the transition to pollution prevention occurs, the sooner we will achieve economic prosperity that is in harmony with the natural world and the best human values. The governments and industries of affluent and polluting nations have an opportunity and a responsibility to lead all nations to the pollution prevention strategy, by example and by providing assistance. Individuals who have come to understand the meaning of pollution prevention and the urgent need to adopt it as a governing principle of society have an opportunity and a responsibility to lead governments and industries to the proper strategy. There is no time to waste because this is no time to waste.

REFERENCES

CHAPTER 1

Brower, Michael, 1990. *Cool Energy—The Renewable Solution to Global Warming.* Cambridge, MA:Union of Concerned Scientists.

Darman, Richard G., 1989. As quoted in The Decade by the Numbers. *Washington Post,* December 11.

Erikson, Kai, 1990. Toxic Reckoning: Business Faces a New Kind of Fear. *Harvard Business Review* 90(1):118–126.

Frosch, Robert A., and Nicholas E. Gallopoulos, 1989. Strategies for Manufacturing. *Scientific American* 261(3):144–154.

Grizzle, Charles L., 1989. As quoted in For EPA, War on Pollution Strikes Home. *Washington Post,* December 12.

McGinnis, J. Michael, 1988. National Priorities in Disease Prevention. *Issues in Science and Technology* 5(52):46–52.

OECD, 1989. *OECD Environmental Data Compendium, 1989.* Paris: Organisation for Economic Co-operation and Development.

Picker, Michael, 1989. Detoxifying Industry—The Easy Way Out. *L.A. Weekly,* December 1.

Pope John Paul II, 1990. Message of His Holiness Pope John Paul II for the Celebration of the World Day of Peace, January 1

Postel, Sandra, 1987. *Defusing the Toxics Threat: Controlling Pesticides and Industrial Waste.* Washington, D.C.: Worldwatch Institute.

Silverstein, Michael, 1989. Facing a Huge Environmental Debt. *Chicago Tribune,* November 25.

Singer, S. Fred, 1990. The Answers on Acid Rain Fall on Deaf Ears. *The Wall Street Journal,* March 6.

U.S. Office of Technology Assessment, 1986. *Serious Reduction of Hazardous Waste: For Pollution Prevention and Industrial Efficiency.* Washington, D.C.: U.S. Government Printing Office.

U.S. Office of Technology Assessment, 1989. *Coming Clean: Superfund Problems Can Be Solved . . . 1989.* Washington, D.C.: U.S. Government Printing Office.

Washington Post, 1989a. Frogs, Toads Vanishing Across Much of World. December 13.

Washington Post, 1989b. '40s Plutonium Emissions Concealed. December 19.

Washington Post, 1989c. Low-Level Radiation Causes More Deaths Than Assumed, Study Finds. December 20.

CHAPTER 2

Ancker-Johnson, Betsy, 1989. Quoted in Sustainable Development—Will American Industry Pay Heed Soon Enough to Take the Lead? *The Conservation Exchange* 7(4):1, 6-7.

Anderson, Robert A., 1989. Quoted in Sustainable Development—Will American Industry Pay Heed Soon Enough to Take the Lead? *The Conservation Exchange* 7(4)1, 6-7.

Brower, Michael, 1990. *Cool Energy—The Renewable Solution to Global Warming.* Cambridge, MA: Union of Concerned Scientists.

Buried alive. 1989. *Newsweek,* November 17.

Davis, G.R., 1989. *Energy Technologies for Reducing Emissions of Greenhouse Gases.* Paris: Organisation for Economic Co-operation and Development.

Flux, Mike, 1987. Quoted by John Elkington, *The Green Capitalists.* London: Victor Gollancz, Ltd.

Gibbons, John H., 1988a. John Gibbons: Pursuing the Conservator Society. *EPRI Journal* 13(8):22-28.

Gibbons, John H., 1988b. The Conservator Society: A New Imperative. *Issues in Science and Technology* 5(1):30-31.

Irvine, Sandy, 1989. Consuming Fashions? The Limits of Green Consumerism. *The Ecologist* 19(3):88-94.

Lander, C.W.M., and R.J.M. Maas, 1990. Economics of Sustainability in a Policy Process. Paper presented at The Economics of Sustainable Development conference, January 23-26, Washington, DC.

Lawlor, Anthony, 1990. Letter in *Time,* January 8.

Leathers, James C., 1989. Quoted in Sustainable Development—Will American Industry Pay Heed Soon Enough to Take the Lead? *The Conservation Exchange* 7(4):1 6-7.

Masciantonio, Philip X., 1989. EPA and the Environment: An Industry Perspective. *Forum for Applied Research and Public Policy* 4(1):94-96.

Muratore, Stephen, 1985. Stewardship is Enough: Ecology as Inner Priesthood. *Epiphany* 6(1):26-34.

New Zealand, Ministry for the Environment 1989. Resource Management Bill. December.

Pezzey, John, 1989. Quoted in Sustainable Development—Will American

Industry Pay Heed Soon Enough to Take the Lead? *The Conservation Exchange* 7(4):1 6-7.

Pollution Engineering. 1989. Burrrp. November, p. 49.

Ruckelshaus, William D., 1989. Toward a Sustainable World. *Scientific American* 261(3):166-174.

The World Commission on Environment and Development, 1987. *Our Common Future.* New York: Oxford University Press.

CHAPTER 3

Delcambre, Ryan, 1988. Quoted in Waste—An Ounce of Prevention. *Chemical Engineering.* August 15, pp. 34-35, 37.

Earl, Anthony, 1989. Key Notes. *Water Environment & Technology* 1(11):432.

Goldberg, Ray A., 1989. Quoted in Luring 'Green' Consumers. *New York Times,* August 6.

Harwood, Joseph E., 1988. Waste Reduction Pollution Prevention Initiatives: Industry's Need for Pollution Prevention. In *Waste Reduction—Pollution Prevention Progress and Prospects within North Carolina,* pp. 55.1-55.2. Raleigh, NC: North Carolina Pollution Prevention Program.

Higgins, Thomas E., 1989. *Hazardous Waste Minimization Handbook.* Chelsea, MI: Lewis Publishers.

Hollod, Gregory, 1990. Perspective on a National Waste Reduction Initiative. In *Waste Reduction: Research Needs in Applied Social Sciences.* Washington, DC: National Research Council.

Huisingh, Donald, 1989. Waste Reduction at the Source: The Economic and Ecological Imperative for Now and the 21st Century. In *Management of Hazardous Materials and Wastes: Treatment, Minimization and Environmental Impacts,* ed. S.K. Majumdar et al., pp. 96-111. Pennsylvania Academy of Science.

Karras, Greg, 1989. Pollution Prevention: The Chevron Story. *Environment* 31(8):4-5.

Lafferre, Thomas H., 1989. Stewardship—The New Environmental Paradigm: A New Challenge for Chemical Engineers. Paper presented at In Our National Interest . . . Pollution Prevention for the 1990s: A Chemical Engineering Challenge. American Institute of Chemical Engineers. December 4-5, Washington, DC.

Little, Linda W., 1988. Rethinking Waste Management: From "Hole in the Ground" to Holistic. In *Waste Reduction—Pollution Prevention Progress and Prospects within North Carolina,* pp. 56.1-56.6. Raleigh, NC: North Carolina Pollution Prevention Program.

Mulligan, Bill, 1989. Quoted in Growing 'Greener'—Corporate America Comes Around to the Environmental Cause. *The Conservation Exchange* 7(4):4-5.

National Research Council, 1990. *Waste Reduction: Research Needs in Applied Social Sciences.* Washington, DC: National Research Council.

Nichol, David A., 1989. Quoted in Luring 'Green' Consumers. *New York Times,* August 6.

Quarles, John R., Jr., 1975. The Need for Action Grows. *Proceedings, 1975 Conference on Waste Reduction,* U.S. Environmental Protection Agency, April 2–3. Washington, DC: U.S. Government Printing Office.

Ruffins, Paul, 1989. Black America: Awakening to Ecology. *Washington Post.* December 24.

Steinmeyer, Dan, 1989. Waste Prevention—What Can We Learn from the Success of Energy Conservation? Paper presented at In Our National Interest . . . Pollution Prevention for the 1990s: A Chemical Engineering Challenge. American Institute of Chemical Engineers. December 4–5, 1989, Washington, DC.

U.S. Office of Technology Assessment, 1989. *Facing America's Trash.* U.S. Congress. Washington, DC: U.S. Government Printing Office.

Vig, Norman J., and Michael E. Kraft, 1990. *Environmental Policy in the 1990s—Toward a New Agenda.* Washington, DC: Congressional Quarterly Press.

Wall Street Journal, 1990. Advertising—New Agency Devoted to the Environment. February 22.

CHAPTER 4

Craig, J.W., and J.L. Warren, 1988. EPA's Latest Hazardous Waste Data. *Waste Age* 19(10):75, 84.

Diaz, L.F., and C.G. Golueke, 1985. Solid Waste Management in Developing Countries. *BioCycle* 26(1):46.

Franklin Associates, Ltd., 1988. Characterization of Municipal Solid Waste in the United States, 1960 to 200 (Update 1988). Report prepared for the U.S. Environmental Protection Agency.

Freedonia Group, 1989. As reported in Hazardous Waste Management: Planning to Avoid Future Problems. *Industrial Launderer* 40(11):28.

Hazmat News, 1990. 1992: The European Community Prepares for Environmental Unification. 3(1):26, 28–35.

Haznews, 1990. Europe's Environmental Industry in 1992. January, pp. 14–16.

Keen, R.C., and N. Jhaveri, 1989. The Development of Hazardous Waste Inventories in Asian Countries. Paper presented at the Pacific Basin Conference on Hazardous Waste, April 1989, Singapore.

Moscow Plays Ketch-Up, 1990. *Washington Post,* February 1, p. A18.

Murarka, I.P., 1987. *Solid Waste Disposal and Reuse in the United States.* Boca Raton, FL: CRC Press.

Netherlands, 1989. *National Environmental Policy Plan—To Choose or to*

Lose. Ministry of Housing, Physical Planning, and Environment. The Hague: SDU Publishers.

New South Wales, 1988. *Annual Report 1988–1989.* Metropolitan Waste Disposal Authority.

Pollack, C., 1987. *Mining Urban Wastes: The Potential for Recycling.* Washington, DC: Worldwatch Institute.

Rathje, William, 1989. Rubbish! *The Atlantic Monthly* 264(6):99–109.

Showers, Victor, 1989. *World Facts and Figures.* New York: John Wiley & Sons.

United Nations, 1988. United Nations Environment Programme. *Industry and Environment* 11(1):1–3.

U.S. Environmental Protection Agency, 1986. Summary of Results for TSDR Facilities in 1985. Study by the Research Triangle Institute, Raleigh, North Carolina.

U.S. Environmental Protection Agency, 1987. *National Air Quality and Emissions Trends Report, 1987.* Office of Air Quality Planning and Standards. Research Triangle Park, NC: U.S. Environmental Protection Agency.

U.S. Environmental Protection Agency, 1989. *The Toxics-Release Inventory: A National Perspective, 1987.* Office of Toxic Substances. Washington, DC: U.S. Government Printing Office.

CHAPTER 5

Chemicalweek, 1989. December 6.

Du Pont, 1989. Testimony by Joseph P. Glas before the Senate Committee on Environment and Public Works. May 19.

Glas, Joseph P., 1989. Protecting the Ozone Layer: A Perspective from Industry. In *Technology and Environment,* ed. J.H. Ausubel and H.E. Sladovich, pp. 137–155. Washington, DC: National Academy Press.

National Toxics Campaign, n.d. *A Consumer's Guide to Protecting the Ozone.* Boston: National Toxics Campaign.

United Nations, 1989. Draft UNEP Integrated Report.

U.S. Public Interest Research Group, 1989. *Du Pont Fiddles While the World Burns—Industry Inaction on Ozone Depletion.* Washington, DC: U.S. Public Interest Research Group.

Wolf, K., and C.W. Myers, 1987. *Hazardous Waste Management by Small Quantity Generators—Chlorinated Solvents in the Dry Cleaning Industry.* Santa Monica, CA: RAND Corporation.

Wray, Tom, 1989. An Ozone Breakdown. *HazmatWorld* 2(1):31.

CHAPTER 6

Andrews, Richard N., and Alvis G. Turner, 1987. Controlling Toxic Chemicals in the Environment. In *Toxic Chemicals, Health, and the Environ-*

ment, ed. L.B. Lave and A.C. Upton, pp. 5–37. Baltimore: Johns Hopkins University Press.

Carson, Rachel, 1962. *Silent Spring.* Boston: Houghton Mifflin Co.

Ciba-Geigy, 1988. *Protection of the Environment at Ciba-Geigy—Policy and Programme for the Company's Plants.*

Colgan, Michael, 1990. "Without Water!" *Muscle and Fitness* 109, 197–198.

Consumers Union, 1987. *Are Home Use Pesticides Safe?* Mt. Vernon, NY: Consumers Union.

Dempster, William, 1990. Quoted in Curt Suplee, Brave Small World. *Washington Post Magazine,* January 21.

Gray, Paul E., 1989. The Paradox of Technological Development. In *Technology and Environment,* ed. J.H. Ausubel and H.E. Sladovich, pp. 192–204. Washington, DC: National Academy of Engineering.

Hess, Charles, E., 1989. Quoted by Will Lepkowski, Debate Builds Over Alternative Agriculture. *Chemical and Engineering News* 67(48):38–39.

Meyer, Marcia, 1989. Pest Control and Regulation: Closing the Gap. *Chemical Times & Trends* 12(4):27–28.

Meyerhoff, Albert H., 1990. Quoted in *New York Times,* February 22.

Moses, Marion, 1988. *A Field Survey of Pesticide-Related Working Condition in the U.S. and Canada.* San Francisco: Pesticide Education and Action Project.

National Research Council, 1987. *Regulating Pesticides in Food: The Delaney Paradox.* Washington, DC: National Research Council.

National Research Council, 1989. *Alternative Agriculture.* Washington, DC: National Research Council.

National Resources Defense Council, 1989. Quoted in Margaret Carlson, Do You Dare to Eat a Peach? *Time,* March 27, pp. 24–38.

National Science Foundation, 1989. A National Science Foundation Strategy for Compliance with Environmental Law in Antarctica. Washington, DC: National Science Foundation.

Postel, Sandra, 1987. *Defusing the Toxics Threat: Controlling Pesticides and Industrial Waste.* Washington, DC: Worldwatch Institute.

Rawls, Thomas H., 1989. Editorial. *Harrowsmith* 4(21):5.

Reetz, Harold F., Jr., 1990. Quoted in Bette Hileman. Alternative Agriculture. *Chemical and Engineering News,* March 5, pp. 26–40.

U.S. Department of Agriculture, *1989 Yearbook of Agriculture—Farm Management—How to Achieve Your Farm Business Goals.* Washington, DC: U.S. Department of Agriculture.

U.S. General Accounting Office 1986a. *Nonagricultural Pesticides.* Washington, DC: U.S. General Accounting Office.

U.S. General Accounting Office, 1986b. *Pesticides: EPA's Formidable Task to Assess and Regulate Their Risks.* Washington, DC: U.S. General Accounting Office.

U.S. Public Research Interest Group, 1989. *A Study of Carcinogenic Pesticides Allowed in 16 Foods and Found in Water: Guess What's Coming to Dinner?* Washington, DC: U.S. Public Research Interest Group.

Virginia Cooperative Extension Service, 1987. *The National Evaluation of Extension's Integrated Pest Management (IPM) Programs.* Virginia Tech and Virginia State Universities. Washington, DC: U.S. Department of Agriculture.

Wall Street Journal, 1990. Return of the Fly, editorial. February 22.

CHAPTER 7

Business Week, 1988. Why the Heat-and-Eat Market Is Really Cooking. June 27.

Casper, Jennifer, 1990. Quoted in Sporting Life Concentrates on Catalogue Retailing, *Washington Post,* January 1.

Consumer Survey Reveals Surprises, 1987. *Packaging* 32(8):47–55.

Darnay, Arsen, and William E. Franklin, 1969. *The Role of Packaging in Solid Waste Management 1966 to 1976.* Written for the Bureau of Solid Waste Management. Rockville, MD: U.S. Department of Health, Education, and Welfare.

Deighton, John, n.d. A White Paper on Packaging. Prepared for the Schechter Group.

Erickson, Eric, 1988. Microwaves Shake Up Frozen Foods. *Packaging,* 33(4):38–41.

Erickson, Greg, 1988. Consumers Get Tough with Packaging. *Packaging,* 33(8):42–51.

Erickson, Greg, 1989. Consumer Opinions Point to Packaging Profits. *Packaging,* 34(8):45–53.

Holmgren, R. Bruce, 1987. Packaging Technology: Suspectors in Packages Improve Microwaving, *Packaging,* 32(9):120–122.

Holmgren, R. Bruce, 1988. CAP Improves Shelf Life . . . But Be Careful. *Packaging,* 33(6):55–57.

James, David, 1975. A Food Industry Perspective. *Proceedings, 1975 Conference on Waste Reduction,* U.S. Environmental Protection Agency, April 2–3. Washington, DC: U.S. Government Printing Office.

Kleiman, Dena, 1989. Fast Food? It Just Isn't Fast Enough Anymore. *The New York Times.* December 6.

Lai, Christopher, C., et al., 1987. Impact of Plastic Packaging on Solid Waste. Presented at Conference on Solid Waste Management and Materials Policy, New York City.

Larson, Melissa, 1988. Microwave Technology Heats Up. *Packaging* 33(8):66–69.

Marriott Corporation, 1989. Quoted in *Washington Post,* Marriott to Sell 800 Restaurants, December 19, p. A1.

McDonald, John J., 1990. Quoted in Latest in Electronic Gizmos Jam Trade Show. *Washington Post,* January 9.

National Science Foundation, 1978. The Application of Technology-Directed Methods to Reduce Solid Waste and Conserve Resources in the Packaging of Non-Fluid Foods, pp. 1–2. Report prepared by Franklin Associates. Washington, DC: National Science Foundation.

Presto! The Convenience Industry: Making Life a Little Simpler, 1987. *Business Week,* April 27.

Quarles, John R., Jr., 1975. The Need for Action Grows. *Proceedings, 1975 Conference on Waste Reduction,* U.S. Environmental Protection Agency, April 2–3. Washington, DC: U.S. Government Printing Office.

Reinemund, Steven S., 1989. Quoted in *Parade Magazine,* November 12, p. 10.

Selke, Susan, 1990. Quoted by Jay Stuller, The Politics of Packaging. *Across the Board* 17(1,2):45.

Simon, Julian, 1981. *How to Start and Operate a Mail-Order Business,* p. 278. New York: McGraw Hill Book Co.

Slocum, Richard D., 1989. New Thresholds in Can Decoration. *Aerosol Age* 34(12):32–35.

Spaulding, Mark, 1987. Eight Hot Trends in Food Packaging. *Packaging* 32(14):40–44.

U.S. Department of Commerce, 1988. *1988 U.S. Industrial Outlook.* Washington, DC: U.S. Department of Commerce.

U.S. Department of Commerce, 1989. *1989 U.S. Industrial Outlook,* pp. 39–15. Washington, DC: U.S. Department of Commerce.

U.S. Office of Technology Assessment 1989. *Facing America's Trash: What Next for Municipal Solid Waste?* Washington, DC: U.S. Government Printing Office.

CHAPTER 8

American Chemical Society, 1989. *Chemical Risk: Personal Decisions.* Washington, DC: American Chemical Society.

Binenstock, Alan, 1989. Disinfectants. *Soap/Cosmetics/Chemical Specialties* 65(11):26.

Caruba, Alan, 1989. Quoted in The Pesticide Industry Needs to Organize and Fight Back. *Chemical Trends & Times* 12(1):28–29.

Consumers Union, 1987. *Are Home Use Pesticides Safe?* Mt. Vernon, NY: Consumers Union.

CSMA's Hiznay Addresses CMCS Annual Meeting, 1989. *Soap/Cosmetics/Chemical Specialties* 65(11):50.

Dawson, Carol, 1988. Safe Products Through Cooperation. *Chemical Trends & Times* 11(1):35–37.

Etter, Robert M., 1988. Reformulating Products and Proper Waste Disposal: Meeting Consumer Interests and Making Ecological Sense. *Chemical Trends & Times* 11(1):32–35.

Gilbert, Pamela, 1987. Testimony by the U.S. Public Interest Research Group before the Subcommittee on Commerce, Consumer Protection, and Competitiveness, U.S. House of Representatives. October 27.

IARC, 1987. *IARC Monographs on the Evaluation of Carcinogenic Risks to Humans, Supplement 7.* Paris: World Health Organization.

Karpatkin, Rhoda H., 1988. Memo to Members. *Consumer Reports* 53(5):283.

Llewellyn, Thomas O., 1988. *Cadmium.* Pittsburgh: U.S. Bureau of Mines.

Lord, John, 1986. *Hazardous Wastes from Homes.* Santa Monica, CA: Enterprise for Education.

National Toxicity Program, 1989. *Fifth Annual Report on Carcinogens, Summary 1989,* p. 58. Washington, DC: U.S. Department of Health and Human Services.

Purin, Gina, et al., 1987. *Alternatives to Landfilling Household Toxics.* Prepared for the Golden Empire Health Planning Center. Sacramento, CA: Local Government Commission.

Reese, Robert G., Jr., 1988. *Mercury.* Pittsburgh: U.S. Bureau of Mines.

Shane, Barbara S., 1989. Human Reproductive Hazards. *Environmental Science and Technology* 23(10):1188–1196.

U.S. Congress, 1989. *Potential Health Hazards of Cosmetic Products.* Hearings before the Subcommittee on Regulations and Business Opportunities, House of Representatives. Washington, DC: U.S. Government Printing Office.

U.S. Environmental Protection Agency, 1989a. Captan: Intent to Cancel Registrations. *54 Federal Register* 54(36):8116.

U.S. Environmental Protection Agency, 1989b. *The Toxics-Release Inventory.* Washington, DC: Office of Toxic Substances.

U.S. Environmental Protection Agency, 1990. *Nonoccupational Pesticide Exposure Study (NOPES).* Research Triangle Park, NC: Atmospheric Research and Exposure Assessment Laboratory.

U.S. General Accounting Office, 1986. *Pesticides: EPA's Formidable Task to Assess and Regulate Their Risks.* Washington, DC: U.S. General Accounting Office.

U.S. General Accounting Office, 1990. *Lawn Care Pesticides: Risks Remain Uncertain While Prohibited Safety Claims Continue.* Washington, DC: U.S. General Accounting Office.

U.S. Office of Technology Assessment, 1989. Technologies for Reducing Dioxin in the Manufacture of Bleached Wood Pulp. Background paper, U.S. Congress. Washington, DC: U.S. Government Printing Office.

Wirka, Jeanne, 1988. *Wrapped in Plastic.* Washington, DC: Environmental Action Foundation.

Woodbury, William D., 1987. *Lead.* Preprint from the Bureau of Mines Yearbook. Pittsburgh: U.S. Bureau of Mines.

CHAPTER 9

Baldwin, Don, 1990. Quoted in Constituencies Are Clashing on Clean Air. *Washington Post,* February 4.

Barnet, Richard J., 1990. Reflections After the Cold War. *The New Yorker* LXV (46):65–77.

Environment Canada, 1990. *Reduction and Reuse: The First 2Rs of Waste Management.* Ottawa: Environment Canada.

Environmental and Energy Study Institute, 1990. *Municipal Solid Waste Management: Issues for the 101st Congress,* Washington, DC.

Netherlands Protection Agency, 1989. *National Environmental Policy Plan—To Choose or to Lose.* Ministry of Housing, Physical Planning, and Environment. The Hague: SDU Publishers.

Resource Recovery Report, 1990. XIV(5):12.

U.S. Environmental Protection Agency, 1990. *Methods to Manage and Control Plastic Wastes.* Washington, DC: U.S. Environmental Protection Agency.

U.S. Office of Technology Assessment, 1986. *Serious Reduction of Hazardous Waste: For Pollution Prevention and Industrial Efficiency.* U.S. Congress. Washington, DC: U.S. Government Printing Office.

U.S. Office of Technology Assessment, 1990. *Making Things Better.* U.S. Congress. Washington, DC: U.S. Government Printing Office.

Tucker, Richard F., 1990. "Holographic" Science to Meet Energy Needs. *Scientific American* 262(3):128.

Varnado, Bruce, 1990. Memorandum on Waste Minimization Workshop Summary. Albuquerque, NM: Environmental and Energy Services Company.

Vig, Norman, and Michael Kraft, 1990. *Environmental Policy in the 1990s—Toward a New Agenda.* Washington, DC: Congressional Quarterly Press.

Wirth, Senator Timothy E., 1990. Hotter and Hotter in the Greenhouse. *Washington Post,* February 20.

Worldwatch Institute, 1990. *State of the World 1990.* New York: W.W. Norton & Co.

INDEX

INDEX

371